AF848290

Multi-Criteria Decision Analysis

Multi-Criteria Decision-Making (MCDM) includes methods and tools for modeling and solving complex problems. MCDM has become popular in the production and service sectors to improve the quality of service, reduce costs, and make people more prosperous. This book illustrates applications through case studies focused on disaster management.

With a presentation of both Multi-Attribute Decision-Making (MADM) and Multi-Objective Decision-Making (MODM) models, this is the first book to merge these methods and tools with disaster management. The book raises awareness for society and decision-makers on how to measure readiness and what necessary preventive measures need to be taken. It offers models and case studies that can be easily adapted to solve complex problems and find solutions in other fields.

Multi-Criteria Decision Analysis: Case Studies in Disaster Management will offer new insights to researchers working in the areas of industrial engineering, systems engineering, healthcare systems, operations research, mathematics, business, computer science, and disaster management, and, hopefully, the book will also stimulate further work in MCDM.

Multi-Criteria Decision Analysis

Case Studies in Disaster Management

Edited by

Muhammet Gul, Melih Yucesan,
and Melike Erdogan

CRC Press
Taylor & Francis Group
Boca Raton London New York

CRC Press is an imprint of the
Taylor & Francis Group, an informa business

First edition published 2023
by CRC Press
6000 Broken Sound Parkway NW, Suite 300, Boca Raton, FL 33487-2742

and by CRC Press
4 Park Square, Milton Park, Abingdon, Oxon, OX14 4RN

CRC Press is an imprint of Taylor & Francis Group, LLC

© 2023 selection and editorial matter, Muhammet Gul, Melih Yucesan, and Melike Erdogan; individual chapters, the contributors

Reasonable efforts have been made to publish reliable data and information, but the author and publisher cannot assume responsibility for the validity of all materials or the consequences of their use. The authors and publishers have attempted to trace the copyright holders of all material reproduced in this publication and apologize to copyright holders if permission to publish in this form has not been obtained. If any copyright material has not been acknowledged please write and let us know so we may rectify in any future reprint.

Except as permitted under U.S. Copyright Law, no part of this book may be reprinted, reproduced, transmitted, or utilized in any form by any electronic, mechanical, or other means, now known or hereafter invented, including photocopying, microfilming, and recording, or in any information storage or retrieval system, without written permission from the publishers.

For permission to photocopy or use material electronically from this work, access www.copyright.com or contact the Copyright Clearance Center, Inc. (CCC), 222 Rosewood Drive, Danvers, MA 01923, 978-750-8400. For works that are not available on CCC please contact mpkbookspermissions@tandf.co.uk

Trademark notice: Product or corporate names may be trademarks or registered trademarks and are used only for identification and explanation without intent to infringe.

ISBN: 978-1-032-08094-9 (hbk)
ISBN: 978-1-032-08095-6 (pbk)
ISBN: 978-1-003-21290-4 (ebk)

DOI: 10.1201/9781003212904

Typeset in Times
by Newgen Publishing UK

*To my dad **Ali**, mom **Fevziye**, wife **Tuğba**, and children,*
***Ahmet Faruk** and **Zeynep Sena** (M. Gul)*
*To my wife **Begüm**, dad **Orhan**, and mom **Mükerrem**, (M. Yucesan)*
*To my sister **Şevval** (M. Erdogan)*

Contents

Preface xi
Editor Biographies xiii
List of Contributors xv

Chapter 1 A Brief Review of the Applications of MCDA in Disaster Management 1

Sumeyye Ozbey, Dilber Baskak, Halit Serdar Saner, Melih Yucesan, Melike Erdogan, and Muhammet Gul

Chapter 2 Multi-Criteria Decision Analysis Models for Emergency and Disaster Management: A Systematic Review 17

Gulhan Sen and Bektas Sari

Chapter 3 Volunteer Assignment Optimization: A Systematic Literature Review with Analysis and Future Directions 31

Umit Ozdemir, Süleyman Mete, and Muhammet Gul

Chapter 4 Flood Risk Management Through Multi-Criteria Decision-Making: A Review 43

Ahmad Ferdowsi, Babak Zolghadr-Asli, Sayed-Farhad Mousavi, and Kourosh Behzadian

Chapter 5 Maritime Supply Chain Resilience Applications in the COVID-19 Pandemic 55

Umur Bucak and Hakan Demirel

Chapter 6 Assessment and Reinforcement of Airport Climate Resilience with a BWM-Enhanced Hoshin Kanri Method 71

Turan Paksoy, Enes Demiralay, and Engin Hasan Çopur

Chapter 7 The Evaluation of Social Vulnerability Factors for Pandemics Based on Fuzzy DEMATEL and the Delphi Method 93

Sukran Seker

Chapter 8 Drought Vulnerability Assessment Based on IVIF AHP and IVIF WASPAS: A Case Study in Turkey 107

Esra Ilbahar

Chapter 9 Evaluation of the Performance of the Health Systems of 12 OECD Countries in Terms of Preparedness for Pandemics 123

Esra Ayhan and Serap Pelin Türkoğlu

Chapter 10 The Use of VIKOR in Location Selection of Field Hospitals: An Application from Çanakkale, Turkey 135

Mehtap Kılıç, Miraç Karakoç, and Cüneyt Çalışkan

Chapter 11 Drone Evaluation and Selection for Training in Disaster Preparedness Phases with Interval MCDA Techniques 155

Zekiye Göktekin and Ahmet Bahadır Şimşek

Chapter 12 Emergency Department Preparedness Assessment for Outbreaks by a Combined Bayesian BWM–RIM Model 169

Huai-Wei Lo, Melike Erdogan, Melih Yucesan, and Muhammet Gul

Chapter 13 Ranking of Districts of Istanbul According to Earthquake Loss Estimates with Entropy and CoCoSo Integration 191

Pelin Toktaş

Chapter 14 Evaluation of Criteria Affecting the Preparedness Levels of Hospitals in Pandemic Periods by the Entropy Weighting Method: A Case Study for the COVID-19 Pandemic 211

Elif Kılıç Delice and Gülin Feryal Can

Chapter 15 Prioritizing Post-Disaster Debris for Recycling 223

Özer Eroglu and Cevriye Gencer

Chapter 16 A GIS-Based Multi-Criteria Decision Analysis Framework for Evaluation of Emergency Assembly Points 235

Ahmet Bahadır Şimşek

Chapter 17 A Novel Interval Valued Neutrosophic AHP-WASPAS Methodology for Emergency Supply Depot Location Selection Problems 251

Ertugrul Ayyildiz and Alev Taskin

Chapter 18 UAV Selection for Post-Disaster Medical Supply Distribution Using a Two-Level Interval-Valued Pythagorean Fuzzy AHP Integrated Interval-Valued Pythagorean Fuzzy VIKOR Methodology, and an Application 267

Pelin Gulum Tas, Bahar Yalcin Kavus, Ertugrul Ayyildiz, and Alev Taskin

Chapter 19 A Simulation-Based Approach for Decision-Making in Earthquake Crisis Management 281

Hassan Ahmadi Choukolaei, Mustafa Jahangoshai Rezaee, Samuel Yousefi, and Morteza Saberi

Chapter 20 Routing of Vehicles to Disaster Relief Tents 303

Engin Pekel

Chapter 21 Disaster Management with Fuzzy Sets and Engineering Economic Analysis 313

Eda Boltürk

Index 325

Preface

The world has come face to face with many natural and man-made disasters in the last century. Dealing with disasters effectively while mitigating risks and damage requires using scientific methods in all phases of the integrated disaster management cycle. Recent trends such as efficient use of resources and combating chronic and epidemic diseases can be effectively controlled with the help of modern and systematic multi-criteria decision-making (MCDM) approaches. How to develop models and methods that can support the decision-making process of decision- and policymakers in disaster management is considered a very important issue by researchers and practitioners. MCDM is the process of developing methods and procedures in which multiple and often conflicting criteria can be included in the planning and decision-making process. It has developed over the years as an important field of operations research and can be easily applied to the basic and sub-activities of disaster management. In an environment where the whole world is faced with disasters such as epidemics and earthquakes, it is obvious that researchers should carefully consider disaster management and bring new and original methods to the MCDM literature.

For this reason, *Multi-Criteria Decision Analysis: Case Studies in Disaster Management* aims to reveal decision-making models that contribute to disaster management. Using multi-attribute decision-making and multi-objective decision-making models under the MCDM approach, important road maps in the fight against all disasters caused by both natural and man-made will emerge. Especially in the ongoing pandemic process, it can be claimed that this book can serve as a reference book in both academic and practical fields, with contributions that will positively impact on daily life.

The book includes chapters with state-of-the-art reviews and case studies regarding mitigation, preparedness, response, and recovery phases: the four phases of integrated disaster management. It raises awareness in this context for society and decision-makers. The findings obtained guide decision-makers, and the book investigates how societies at different cultural and economic levels can affect disasters.

Finally, as the three editors of this book, we are grateful to our families for their constant love, patience, and support, and our reviewers for their objective and constructive comments. Without their outstanding support, we would not have been able to complete this book. With the best wishes that the book will be useful to all concerned.

May 2022
Muhammet Gul
Melih Yucesan
Melike Erdogan

Editor Biographies

Muhammet Gul has been working as Associate Professor at the Department of Transportation and Logistics, Istanbul University, Istanbul, Turkey. He has received his M.Sc. and Ph.D. in Industrial Engineering from Yildiz Technical University. His research interests are in simulation modeling, healthcare system management, occupational safety and risk assessment, multi-criteria decision-making, and fuzzy sets. He is the editor-in-chief of two books released in 2020: *Computational Intelligence and Soft Computing Applications in Healthcare Management Science* (IGI-Global) and *Fine–Kinney-Based Fuzzy Multi-Criteria Occupational Risk Assessment: Approaches, Case Studies and Python Applications* (Springer). He is also the Associate Editor of *Complex & Intelligent Systems* (IF: 4.927), *Mathematical Problems in Engineering* (IF: 1.305), and *Journal of Healthcare Engineering* (IF: 2.686).

Melih Yucesan completed his Ph.D. and has been working as Associate Professor at the Department of Emergency Aid and Disaster Management, Munzur University, Tunceli, Turkey. He received his Ph.D. in Econometrics from Karadeniz Technical University. His research interests are in planning, forecasting, multi-criteria decision-making, and fuzzy sets. His papers appeared in international journals such as *International Journal of Disaster Risk Reduction*, *Applied Soft Computing*, *Energy Policy*, *Soft Computing*, and *International Journal of Healthcare Management*.

Melike Erdogan completed her Ph.D. at the Department of Industrial Engineering, Yıldız Technical University, in 2018 and has been working as Assistant Professor in Düzce University, Turkey, in the Department of Industrial Engineering. Her research interests are decision making, fuzzy logic, and risk analysis. Her papers have appeared in journals such as *Applied Soft Computing*, *Sustainable Cities and Societies*, *Soft Computing*, and *Neural Computing and Applications*.

Contributors

Ertugrul Ayyildiz
Karadeniz Technical University
Turkey

Esra Ayhan
Ankara University
Turkey

Dilber Baskak
Munzur University
Turkey

Kourosh Behzadian
University of West London
UK

Eda Boltürk
Turkey

Umur Bucak
Zonguldak Bulent Ecevit University
Turkey

Cüneyt Çalışkan
University of Health Science
Turkey

Gülin F. Can
University of Başkent
Turkey

Hassan Ahmadi Choukolaei
Urmia University of Technology
Urmia, Iran

Engin Hasan Çopur
Necmettin Erbakan University
Turkey

Elif K. Delice
University of Atatürk
Turkey

Enes Demiralay
Konya Technical University
Turkey

Hakan Demirel
Istanbul Technical University
Turkey

Melike Erdogan
Duzce University
Turkey

Özer Eroğlu
Turkish Land Forces
Turkey

Ahmad Ferdowsi
University of Applied Science and Technology Iran and
Semnan University
Iran

Cevriye Gencer
Gazi University
Turkey

Zekiye Göktekin
University of Gümüşhane
Turkey

Muhammet Gul
Istanbul University
Turkey

Esra Ilbahar
Yildiz Technical University
Turkey

Miraç Karakoç
Investment and Construction Directorate
Turkey

Bahar Yalcin Kavus
Nisantasi University
Turkey

Mehtap Kılıç
Ardahan University
Turkey

Huai-Wei Lo
Chaoyang University of Technology
Taiwan

Süleyman Mete
Gaziantep University
Turkey

Sayad-Farhad Mousavi
Semnan University
Iran

Umit Ozdemir
Munzur University
Turkey

Turan Paksoy
Necmettin Erbakan University
Turkey

Konya, Turkey
Engin Pekel
Hitit University
Turkey

Mustafa Jahangoshai Rezaee
Urmia University of Technology
Iran

Morteza Saberi
University of Technology Sydney
Australia

Bektas Sari
Ege University
Turkey

Sukran Seker
Yildiz Technical University
Turkey

Gulhan Sen
Burdur Mehmet Akif Ersoy University
Turkey

Halit Serdar Saner
Univ of Munzur
Turkey

Ahmet Bahadır Şimşek
University of Gümüşhane
Turkey

Pelin Gulum Tas
Delft University of Technology
The Netherlands

Alev Taskin
Yildiz Technical University
Turkey

Pelin Toktaş
Başkent University
Turkey

Serap Pelin Türkoğlu
Ankara Yıldırım Beyazıt University
Turkey

Sumeyye Ozbey
Munzur University
Turkey

Samuel Yousefi
University of British Columbia, Okanagan Campus
Canada

Melih Yucesan
Munzur University
Turkey

Babak Zolghadr-Asli
The University of Queensland
Australia
and
The University of Exeter
UK

1 A Brief Review of the Applications of MCDA in Disaster Management

Sumeyye Ozbey,[1] Dilber Baskak,[1] Halit Serdar Saner,[2] Melih Yucesan[1], Melike Erdogan,[3] and Muhammet Gul[4]

[1]Department of Emergency Aid and Disaster Management, Munzur University

[2]Department of Engineering Management, University of Munzur

[3]Department of Industrial Engineering, Duzce University

[4]Department of Transportation and Logistics, Istanbul University

CONTENTS

1.1 Introduction ... 1
1.2 MCDA in the Disaster Management Cycle ... 2
1.2.1 Pairwise Comparison-Based Methods ... 4
1.2.2 Outranking-Based Methods ... 4
1.2.3 Distance-Based Methods ... 4
1.2.4 Interaction-Based Methods ... 4
1.2.5 Utility-Based Methods ... 5
1.2.6 Other Methods ... 5
1.3 Prisma Literature Review ... 5
1.4 Results of the Review ... 8
1.5 Conclusion and Future Directions ... 8
References ... 11

1.1 INTRODUCTION

Disaster can be expressed as a crisis situation that exceeds the available recovery capacity and causes widespread damage. For this reason, "disaster management" can be expressed as the management of organizations and responsibilities by using resources in the most effective way, which deals with the mitigation, preparedness, response, and recovery phases of emergencies (Coombs, 2014; Yu et al., 2018). Disasters can be divided into two groups: man-made and natural. Natural disasters can be classified as climatological, biological, meteorological, hydrological, and geophysical hazards. Natural hazards are caused by such as earthquakes, tsunamis, hurricanes, and floods.

DOI: 10.1201/9781003212904-1

Man-made disasters can be grouped as terrorist and technological disasters that do not occur naturally (Guha-Sapir et al., 2015).

Changing climate and environmental conditions in the world cause natural disasters such as floods, landslides, storms, earthquakes, volcanoes, and tsunamis. Therefore, disaster management is a critical issue affecting the whole world, and it highlights the importance of preventive measures against disasters. In particular, attention and interest in disaster management have increased due to the COVID-19 pandemic, a global threat that the entire world has been trying to deal with recently. At this point, it can be seen that there is a serious need to put forward studies in order to take preventive measures for disasters, to be prepared for disasters, and to recover quickly after disasters. Analytical, holistic, and systematic approaches need to be introduced for many decisions to be made at different stages of disaster management.

Multi-criteria decision analysis (MCDA) is one of these approaches and a crucial area in operations research and management science that contributes more to disaster-management operations. It consists of five components: (1) goal, (2) preferences of the decision-maker, (3) alternatives encountered, (4) criteria, and (5) results (Wang et al., 2009).

The literature on the use of MCDA in disaster management has increased, especially in recent years, and the need for a systematic classification of studies using MCDA analysis in the field of disaster management has emerged. For this reason, this chapter presents the studies that integrate disaster management and MCDA approaches over the last 10 years. These studies include both different phases of the disaster and different MCDA methods (such as analytic hierarchy process, VIKOR, technique for order of preference by similarity to ideal solution (TOPSIS), and best–worst method (BWM)) and studies that contain the fuzzy extensions of these methods together are handled.

The study comprises both descriptive and thematic classification in order to better observe the latest status of the relevant literature. In order to make the compilation process more understandable, classifications are made on the basis of databases for keywords.

The paper is organized into the following sections: disaster management lifecycle, research methodology, an overview of the literature, opportunities for future research, and concluding remarks.

1.2 MCDA IN THE DISASTER MANAGEMENT CYCLE

Disaster management aims to reduce potential losses due to hazards and related risks, provide prompt and appropriate assistance to disaster victims, and ensure a rapid and effective recovery. The stages of the disaster-management cycle, which are the creation of disaster plans, a reduction of disaster risks, the response during the disaster, and recovery after the disaster, are carried out together with the government, businesses, and non-governmental organizations (NGOs). These processes are: (1) mitigation, (2) preparedness, (3) response, and (4) recovery. The first stage aims to minimize the effects of the disaster and involves necessities to reduce the risk and increase resilience, while the second stage includes planning details on how to respond such as setting up an early-warning system, preparedness plans, emergency exercises/training, and the pre-position of supplies, and aims to reach a satisfactory level of readiness (Coban et al., 2021; Warfield, 2013). In the third stage, efforts are made to organize timely assistance to victims, distribute relief, and evacuate the affected victims to a safe zone (Rana et al., 2021; Coban et al., 2021). The last stage concerns returning the community to normal and it contains activities to remove debris, rebuild damaged buildings, and repair essential infrastructure (Rana et al., 2021; Coban et al., 2021; Warfield, 2013).

The world has come face to face with many natural and man-made disasters in the last century. Dealing with disasters effectively and reducing risks and damage without experiencing disasters requires using scientific methods in all phases of the disaster-management cycle. The need to develop the scientific ability to deal with these natural and man-made emergencies is evident. How to develop models and methods that can support the decision-making process of decision-makers and policy-makers in disaster management is considered a very important issue by researchers and practitioners. MCDA is the process of developing methods and procedures for the problems in which multiple and often conflicting criteria are included in the planning and decision-making process. It has evolved over the years as an important field of operations research and can be easily applied to the basic and sub-activities of disaster management. MCDA methods are used in situations where it is necessary to select the most suitable one among several alternatives, or even to rank or prioritize them according to different evaluation criteria. It provides a unique convenience to decision-makers and policy-makers at the point of making the most appropriate decision with appropriate solutions. This feature makes MCDA attractive in addressing the different problems that emerge in the activities following all stages of the disaster-management cycle. In an environment where the whole world is faced with disasters such as pandemics and earthquakes, it is obvious that researchers should carefully consider emergency aid and disaster management and bring new and original approaches to the MCDA literature. With the help of modern and systematic MCDA approaches solutions to current trends such as efficient use of resources and combating chronic and pandemic diseases can be offered. The use of these approaches by academics is increasing, to improve processes, reduce costs, and make people more prosperous. For these reasons, this study focuses on reviewing the MCDA methods during different stages of the disaster-management cycle. Before presenting the methodology and results of the literature review, which will be discussed in detail in this chapter, it is useful to take a look at the MCDA methods. These methods, which are classified according to certain features (Yalcin et al., 2022) and are said to be more than 50 in the literature (Wątróbski et al., 2019), are frequently used in all stages of the disaster-management cycle. In the classification, we follow the six-class taxonomy of Yalcin et al. (2022). We handle MCDA methods under six classes as follows: (1) pairwise comparison-based methods, (2) outranking-based methods, (3) distance-based methods, (4) interaction-based methods, (5) utility-based methods, and (6) other methods not investigated under any of the other five classes.

1.2.1 Pairwise Comparison-Based Methods

This class includes methods based on pairwise evaluation of criteria or alternatives. The main methods in this class are analytic hierarchy process (AHP) and analytic network process (ANP) proposed by Saaty (2005), and best–worst method (BWM) proposed by Rezaei (2015). They have been applied to disaster operations from an early-warning system to vulnerability and resilience assessment (Javari, 2021; Mohamed, 2021; Rehman et al., 2021; Nguyen et al., 2019; Alshehri et al., 2015; Karamouz et al., 2015; Sambah and Miura, 2014). Moreover, they have also been applied to location decision-making (Geng, 2021; Junian and Azizifar, 2018; Yang et al., 2015), post-disaster temporary housing (Hosseini et al., 2016), emergency and humanitarian aid logistics performance (Munyaka and Yadavalli, 2021; Li et al., 2021; Celik and Gumus, 2015; Celik et al., 2014), and search-and-rescue problems (Hassanzadeh and Nedovic-Budic, 2016). The difference between AHP and ANP is the absence of interaction in AHP. The innovation brought by BWM is that it offers less pairwise comparison and higher consistency in the experts' judgments.

1.2.2 Outranking-Based Methods

Under this class, the methods at the forefront are Elimination Et Choice Translating Reality (ELECTRE; Roy, 1990) and Preference Ranking Organization Method for Enrichment Evaluations (PROMETHEE; Brans and Vincke, 1985). Both methods follow a procedure that considers the superiority of one alternative over the other. They are applied to disaster-management operations, especially for the mitigation and preparedness stages, as in the pairwise comparison-based methods (Sun et al., 2020; Khalid and Ali, 2020; Nassereddine et al., 2019; Caroleo et al., 2018).

1.2.3 Distance-Based Methods

In this MCDA class, the most appropriate alternative to select, sort, or prioritize is determined based on the shortest distance to the ideal solution and the farthest distance to the negative ideal solution. The most applied methods include the Technique for Order of Preference by Similarity to Ideal Solution (TOPSIS; Hwang and Yoon, 1981) and VlseKriterijumska Optimizacija I Kompromisno Resenje (VIKOR; Opricović, 1998). TOPSIS has been frequently used in disaster management for the reconstruction of transport networks damaged by flooding (Jiang et al., 2021), for developing hospital emergency and disaster management index (Mojtahedi et al., 2021), for disaster risk management and response (Otay and Jaller, 2020), and for seismic vulnerability assessment (Sadrykia et al., 2017). VIKOR has been applied to solve the post-earthquake reconstruction problem in Taiwan (Opricovic and Tzeng, 2002).

1.2.4 Interaction-Based Methods

In this class, criteria in the decision-making hierarchy have effects on each other. The well-known methods of this class are grey relational analysis (Deng, 1982) and

Decision-Making Trial and Evaluation Laboratory (DEMATEL; Gabus and Fontela, 1972). DEMATEL has been used as a crucial decision-making method in disaster management and disaster risk reduction (Zhao et al., 2021; de Moura et al., 2020; Sen et al., 2021; Budak et al., 2020; John et al., 2019; Yang et al., 2018; Yang et al., 2015). Zhao et al. (2021) evaluated critical success factors for COVID-19 emergency management at universities in China. Sen et al. (2021) studied identification of the important resilience factors in housing infrastructure against flood hazards; and Budak et al. (2020) investigated selection of the most appropriate real-time location systems.

1.2.5 Utility-Based Methods

This class contains the MCDA methods that use multi-attribute utility theory. The best-known methods in this class are "Step-wise Weight Assessment Ratio Analysis (SWARA") and "Weighted Aggregated Sum Product Assessment (WASPAS)", and complex proportional assessment (COPRAS; Keršuliene et al., 2010; Zavadskas et al., 2012). These are also applied to some miscellaneous disaster-management problems (Agarwal et al., 2020; Rahman et al., 2020; Sotoudeh-Anvari et al., 2019). Agarwal et al. (2020) evaluated solutions to overcome humanitarian supply chain-management barriers using a hybrid fuzzy SWARA–fuzzy WASPAS approach. On a different track, Rahman et al. (2020) developed models to identify the optimal spatial distribution of emergency evacuation centers to improve flood emergency planning in the Sylhet region of northeastern Bangladesh. Sotoudeh-Anvari et al. (2019) also focused on search-and-rescue operations.

1.2.6 Other Methods

Unlike the methods mentioned in the first five sub-sections, there are some individual methods such as entropy, Ordered Weighted Averaging (OWA), Data Envelopment Analysis (DEA), Multi-Objective Optimization on the Basis of Ratio Analysis (MOORA), QUALItative FLEXible (QUALIFLEX), Weighted Sum Model (WSM), and Stochastic Multicriteria Acceptability Analysis (SMAA) applied in the disaster-management cycle (Zhu et al., 2017; Song et al., 2018; Kou et al., 2014; Moradi et al., 2013).

1.3 PRISMA LITERATURE REVIEW

The literature review was carried out following the PRISMA methodology. The data analysis procedures included in the literature review are explained in Figure 1.1. In this context, we first identified the research directory ("Multi-Criteria Decision Making" OR "Multi-Criteria Decision Analysis" AND "Disaster Management" OR "Emergency Aid" OR "Humanitarian Aid"). Next, we systematically searched for English-language studies published between 2000 and 2021 in seven electronic databases: Web of Science, Scopus, Google Academic, ScienceDirect, Taylor & Francis, Wiley, and Emerald Insight.

Various publications such as journal articles, conference proceedings, reviews, and research articles were included in the literature review. The research consisted of the stages of scanning the literature, examining the titles and abstracts, and reading the full texts. Studies thought to be irrelevant to the subject were excluded from the review. At the end of the evaluation, full-text analysis was carried out on 90 articles.

As a result of the literature review obtained with the PRISMA approach, the most used multi-criteria decision-making (MCDM) methods in 90 studies were determined to be AHP and TOPSIS. In addition, DEMATEL, PROMETHEE, ELECTRE, Entropy, BWM, QUALIFLEX, VIKOR, ANP, OWA, and SWARA emerged as frequently used methods.

When the studies using the geographic information system (GIS)-based AHP method are examined, it was found that it had been used in multiple studies for the analysis of flood risk and the development of early-warning systems (Nguyen et al., 2019; Mohamed, 2021; Rehman et al., 2021; El-Magd et al., 2020; Souissi et al., 2020; Toosi et al., 2019). In another study, the effect of some climatic and non-climatic factors triggering flash floods was evaluated with the same method (Munyaka and Yadavalli, 2021). In another similar study, it was used to create flood hydrographs of rivers (Bayazıt et al., 2021). There are also some studies using other methods together with AHP. To construct a flood vulnerability map, Feloni et al. (2020) used GIS-based AHP and fuzzy-AHP. Bouamrane et al. (2020) additionally used the remote-sensing method.

The AHP method was used to locate and evaluate the effectiveness of earthquake evacuation shelters (Junian and Azizifar, 2018) and to prioritize damaged areas for Urban Search and Rescue after an earthquake (Hassanzadeh and Nedovic-Budic, 2016). It has also been used to evaluate transportation network performance in disaster

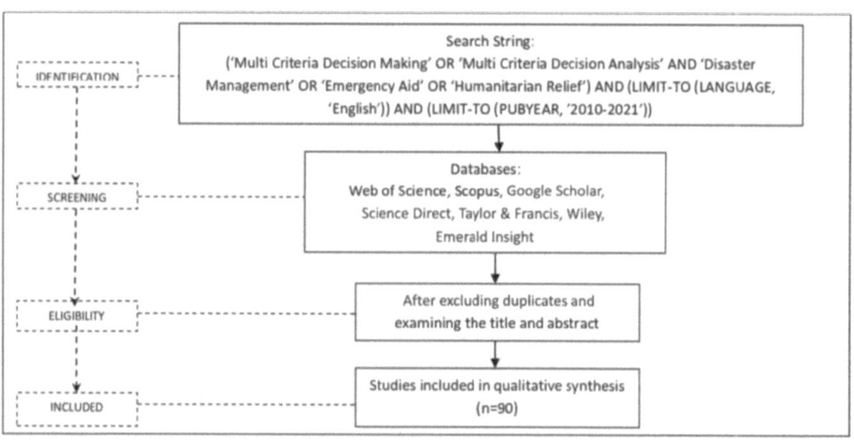

FIGURE 1.1 Literature review with the PRISMA method.

situations (Ghavami, 2019) and to create a drought vulnerability map (Karamouz et al., 2015). In another study, the same method was used to evaluate the applicability of disaster managers to collective decision-making processes (Oh, 2020). The AHP method was also used to evaluate the effectiveness of regional aid operations, to determine the location for temporary housing after the disaster, and to evaluate tsunami vulnerability (Dano, 2022; Hosseini et al., 2016; Sambah and Miura, 2014).

Other studies have used alternative methods together with AHP. Ortiz-Barrios et al. (2017) used AHP, DEMATEL, and TOPSIS to evaluate the disaster preparedness of a hospital emergency department. In Liu et al. (2018), AHP and entropy methods were used to assess water resource scarcity. Wang et al. (2018) evaluated uninterrupted communication in disasters using AHP–TOPSIS methods. Saksrisathaporn et al. (2016) used the same AHP–TOPSIS combination to select the most suitable supplier, warehouse, and vehicle in disasters. In another study, the same methods were applied to identify the security vulnerability of buildings and to evaluate the effectiveness of humanitarian aid operations (Harirchian et al., 2020; Celik and Gumus, 2015).

Considering the studies that evaluated the AHP method with the PROMETHEE method, it was found that wastewater treatment plants have been adopted to assess flood resistance and to improve the preparedness, response capacity, and performance of NGOs (Karamouz et al., 2016; Celik and Gumus, 2016).

The TOPSIS method has been developed and used to evaluate the emergency preparedness of hospitals (Mojtahedi et al., 2021), and to rebuild transportation networks destroyed by flooding (Jiang et al., 2021). In other studies, the fuzzy–TOPSIS method has been applied to measure flood-risk sensitivity (Jun et al., 2013), and to create a flood hazard map (Kim et al., 2019). The fuzzy–TOPSIS method has been used together with the fuzzy–AHP method to evaluate post-earthquake shelters (Trivedi and Singh, 2017). The Delphi-based TOPSIS method has been used to determine the locations of COVID-19 isolation hospitals (Aydin and Seker, 2021). TOPSIS and COPRAS have been combined to increase the effectiveness of search-and-rescue activities in earthquake disasters (Sotoudeh-Anvari et al., 2019). AHP and F-TOPSIS (Roh et al., 2015) have also been adopted in the selection of warehouse locations for humanitarian aid organizations.

The DEMATEL method has been used in evaluating the critical success factors of COVID-19 emergency management (Zhao et al., 2021), in evaluating logistics activities (De Moura et al., 2020), and in determining resilience factors of housing infrastructure against flood hazards (Sen et al., 2021). In addition, DEMATEL and ANP have been used together in the field of humanitarian aid in disasters (John et al., 2019) and in a study of a flood disaster information system for volunteers (Yang et al., 2018). The PROMETHEE method has been utilized to assess emergency response systems (Nassereddine et al., 2019) and the vulnerability of flood-affected sectors (Khalid and Ali, 2020). The PROMETHEE-II method has been applied to design a humanitarian aid logistics network (Rezaei-Malek et al., 2016).

In another study, ELECTRE was implemented in evaluating supply chain management in disasters (Roy et al., 2021). In addition, a decision-support system for the crisis (Caroleo et al., 2018) was developed with the ELECTRE III method. When other studies were examined, WSM, TOPSIS, and ELECTRE (Sun et al., 2020) have been used together to create a flood-risk analysis model. In post-disaster temporary shelter location selection, TOPSIS, ELECTRE, SAW, and AHP (Omidvar et al., 2013) have been used together with GIS. Li et al. (2021) used the BWM method to present a post-flood hybrid emergency decision-making approach; and Song et al. (2019) adopted the rough QUALIFLEX method to address the sustainable shelter site selection problem.

1.4 RESULTS OF THE REVIEW

MCDM methods allow for the systematic solution of problems that arise in disasters and emergencies. In this context, different problems for each stage of disaster management have been evaluated with similar methods and similar problems with different methods in the literature. In the literature review, 90 studies were examined. When these studies were examined in terms of the stages they focus on in the disaster-management cycle in Figure 1.2, they were used in the steps of mitigation, preparedness, response, and recovery, respectively. It was determined that these studies, in which MCDM methods are used, focused more on the risk-management stage in order to prevent disaster risks before they occur. However, another important point here is that the number of studies that focused on risk management was similar to the number of studies carried out to prevent the existing risks and to reduce the losses that may occur during the crisis-management phase. From this point of view, although most of the studies focused on prevention before the risk occurs, it is a positive development in terms of disaster management that preparatory studies are not neglected in case the risk occurs.

The distribution of the studies in the literature review by year is given in Figure 1.3 (a). The three papers that were out of the standard deviation were not included in the distribution table according to year. When the table is evaluated, it can be seen that the use of MCDM methods in solving the problems that arise in the disaster-management phase has increased in recent years. When the distribution of studies according to regions is examined in Figure 1.3 (b), the three countries with the most studies are Iran, China, and Turkey. When the studies conducted according to the type of disaster they focus on in Figure 1.3 (c) are examined, it can be determined that the number of studies on multiple hazards is higher. The other three most focused on disaster types are floods, earthquakes, and COVID-19. In terms of journal distribution, it is shown in Figure 1.3 (d) that most studies are published in the *International Journal of Disaster Risk Reduction*. When Figure 1.3 (e) is analyzed, it can be seen that 45% of the studies focused on mitigation and 22% focused on the preparation phase of disaster management. In other words, it can be said that the majority of the studies presented in the literature are risk-management focused.

1.5 CONCLUSION AND FUTURE DIRECTIONS

There are many natural and man-made disasters in the world such as floods, landslides, storms, earthquakes, volcanoes, and tsunamis. With effective disaster management, preventive measures can be taken against disasters, and loss of life and property can be reduced. As a result of the recent COVID-19 pandemic, the importance of

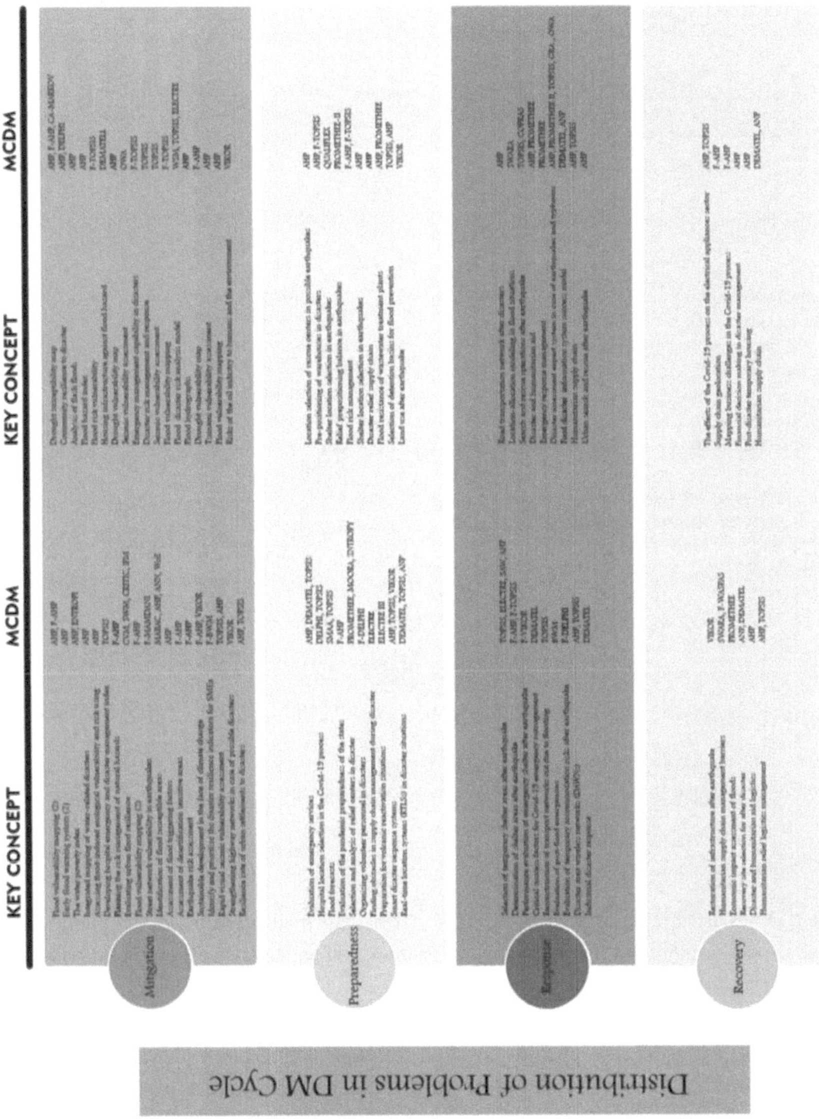

FIGURE 1.2 An outline of the reviewed studies.

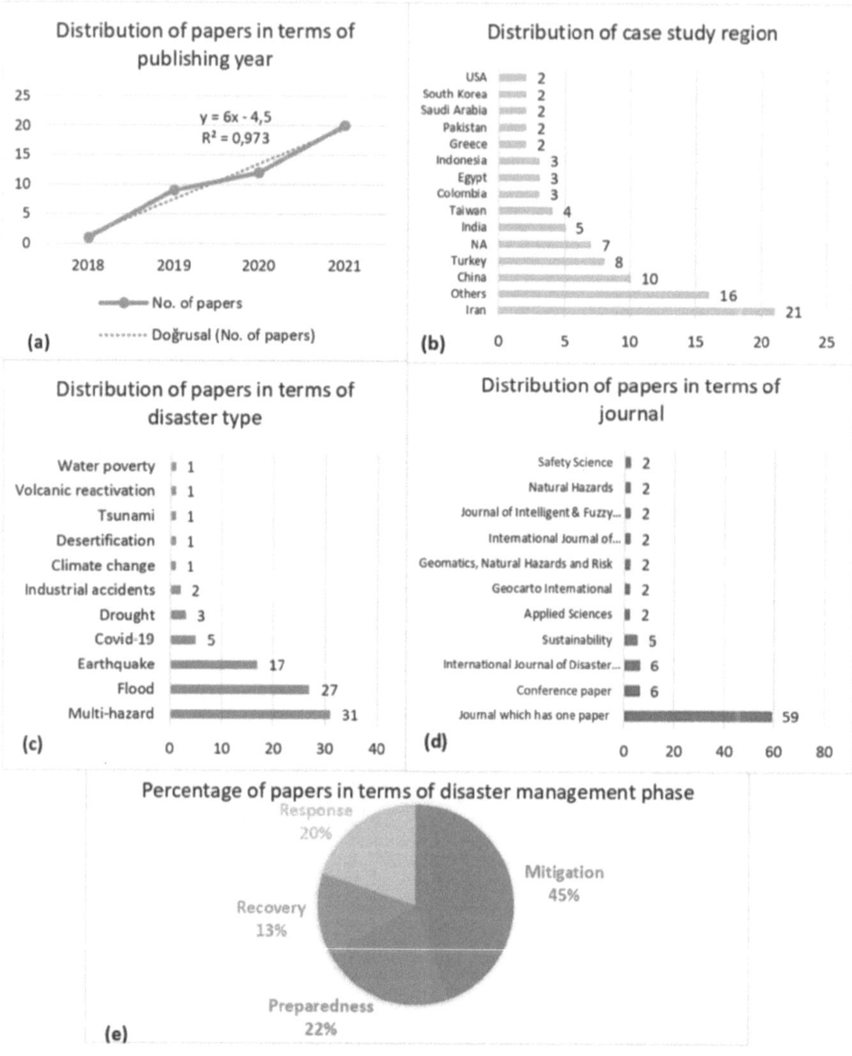

FIGURE 1.3 Distribution of problems in disaster management in terms of (a) publishing year, (b) case study region, (c) disaster type, (d) journal, and (e) disaster management phase.

disaster management has also been better understood. It is essential to apply analytical methods in order to take and implement effective decisions in disaster management. At this point, one of the most popular approaches encountered in the literature is MCDA. Studies on the use of MCDA approaches in disaster management have been particularly popular in recent years, and, therefore, the need for a systematic classification of studies using MCDA analysis in the area of disaster management has emerged. In this chapter, the studies put forward by applying MCDA approaches in the field of disaster management and its phases have been examined and classified from a systematic point of view. In this regard, a PRISMA approach was used for the methodological literature search. First, MCDA studies were examined according to the classifications accepted in the literature and important studies in the field of disaster management were summarized. Then, the studies were investigated in terms of the year, journal, region where the case study was executed, and disaster-management phases. As a result of all these examinations, the use of MCDA approaches in the field of disaster management has been revealed in detail and a literature review was conducted for researchers who want to work in this field, where they can easily see the gaps and innovations.

For future studies, an extended literature search could be done for each of the phases of disaster management separately, or a review including multi-objective decision-making techniques could be carried out.

REFERENCES

Agarwal, V., S. Sharma, L. Gupta, et al. 2020. COVID-19 and psychological disaster preparedness—an unmet need. Disaster Medicine and Public Health Preparedness, 14(3): 387–390.

Alshehri, S. A., Y. Rezgui, and H. Li. 2015. Delphi-based consensus study into a framework of community resilience to disaster. Natural Hazards 75(3): 2221–2245.

Aydin, N. and S. Seker. 2021. Determining the location of isolation hospitals for COVID-19 via Delphi-based MCDM method. International Journal of Intelligent Systems 36(6): 3011–3034.

Bayazıt, Y, C. Koç, and R. Bakış, R. 2021. Urbanization impacts on flash urban floods in Bodrum Province, Turkey. Hydrological Sciences Journal 66(1): 118–133.

Bouamrane, A, O. Derdous, N. Dahri, et al. 2020. A comparison of the analytical hierarchy process and the fuzzy logic approach for flood susceptibility mapping in a semi-arid ungauged basin (Biskra basin: Algeria). International Journal of River Basin Management 20: 203–213.

Brans, J. P. and P. Vincke. 1985. Note—A Preference Ranking Organisation Method: (The PROMETHEE Method for Multiple Criteria Decision-Making). Management Science 31(6): 647–656.

Budak, A, İ. Kaya, A. Karaşan, and M. Erdogan. 2020. Real-time location systems selection by using a fuzzy MCDM approach: an application in humanitarian relief logistics. Applied Soft Computing 92, 106322.

Caroleo, B., E. Palumbo, M. Osella, et al. 2018. A knowledge-based multi-criteria decision support system encompassing cascading effects for disaster management. International Journal of Information Technology & Decision Making 17, 1469–1498.

Celik, E. and A. T. Gumus. 2015. An assessment approach for non-governmental organizations in humanitarian relief logistics and an application in Turkey. Technological and Economic Development of Economy 24(1): 1–26.

Celik, E. and A. T. Gumus. 2016. An outranking approach based on interval type-2 fuzzy sets to evaluate preparedness and response ability of non-governmental humanitarian relief organizations. Computers & Industrial Engineering 101: 21–34.

Celik, E., A. T. Gumus, and M. Alegoz. 2014. A trapezoidal type-2 fuzzy MCDM method to identify and evaluate critical success factors for humanitarian relief logistics management. Journal of Intelligent & Fuzzy Systems 27(6): 2847–2855.

Coban, B., M. P. Scaparra, and J. R. O'Hanley. 2021. Use of OR in earthquake operations management: a review of the literature and roadmap for future research. International Journal of Disaster Risk Reduction 65: 102539.

Coombs, W. T. 2014. Ongoing crisis communication: Planning, managing, and responding. Thousand Oaks, CA: Sage Publications.

Dano, U. L. 2022. An AHP-based assessment of flood triggering factors to enhance resiliency in Dammam, Saudi Arabia. GeoJournal 87: 1945–1960.

De Moura, E. H., T. B. R. e Cruz, and D. M. De Genaro Chiroli. 2020. A framework proposal to integrate humanitarian logistics practices, disaster management and disaster mutual assistance: a Brazilian case. Safety Science 132: 104965.

Deng, Z. 1982. Fuzzy pseudo-metric spaces. Journal of Mathematical Analysis and Applications 86(1): 74–95.

El-Magd, S. A. A, R. A. Amer, and A. Embaby. 2020. Multi-criteria decision-making for the analysis of flash floods: a case study of Awlad Toq-Sherq, Southeast Sohag, Egypt. Journal of African Earth Sciences 162(1): 103709.

Feloni, E., I. Mousadis, and E. Baltas. 2020. Flood vulnerability assessment using a GIS-based multi-criteria approach—the case of Attica region. Journal of Flood Risk Management 13, e12563.

Gabus, A. and E. Fontela. 1972. World problems, an invitation to further thought within the framework of DEMATEL. Battelle Geneva Research Center, Geneva, Switzerland.

Geng, J., H. Hou, and S. Geng. 2021. Optimization of warehouse location and supplies allocation for emergency rescue under joint government–enterprise cooperation considering disaster victims' distress perception. Sustainability 13(19): 10560.

Ghavami, S. M. 2019. Multi-criteria spatial decision support system for identifying strategic roads in disaster situations. International Journal of Critical Infrastructure Protection 24: 23–36.

Guha-Sapir, D., R. Below, and P. Hoyois. 2015. EM-DAT: international disaster database. Annual Disaster Statistical Review 27: 57–58.

Harirchian, E., K. Jadhav, K. Mohammad, et al. 2020. A comparative study of MCDM methods integrated with rapid visual seismic vulnerability assessment of existing RC structures. Applied Sciences 10: 6411.

Hassanzadeh, R. and Z. Nedovic-Budic. 2016. Where to go first: prioritization of damaged areas for allocation of Urban Search and Rescue (USAR) operations (PI-USAR model). Geomatics, Natural Hazards and Risk 7(4): 1337–1366.

Hosseini, S. A., A. de la Fuente, and O. Pons. 2016. Multicriteria decision-making method for sustainable site location of post-disaster temporary housing in urban areas. Journal of Construction Engineering and Management 142(9): 04016036.

Hwang, C. L. and K. Yoon. 1981. Methods for multiple attribute decision making. In Multiple attribute decision making (pp. 58–191). Springer, Berlin, Heidelberg.

Javari, M. 2021. Rainfall random variability and its effects on flood risk management in Iran. Modeling Earth Systems and Environment 8: 1109–1133.

Jiang, G. J., H. X. Chen, H. H. Sun, et al. 2021. An improved multi-criteria emergency decision-making method in environmental disasters. Soft Computing 25: 10351–10379.

John, L., A. Gurumurthy, G. Soni, and V. Jain. 2019. Modelling the inter-relationship between factors affecting coordination in a humanitarian supply chain: a case of Chennai flood relief. Annals of Operations Research 283(1): 1227–1258.

Jun, K. S, E. S. Chung, Y. G. Kim, and Y. Kim. 2013. A fuzzy multi-criteria approach to flood risk vulnerability in South Korea by considering climate change impacts. Expert Systems with Applications 40(4): 1003–1013.

Junian, J. and V. Azizifar. 2018. The evaluation of temporary shelter areas locations using geographic information system and analytic hierarchy process. Civil Engineering Journal 4(7): 1678–1688.

Karamouz, M., E. Rasoulnia, Z. Zahmatkesh, et al. 2016. Uncertainty-based flood resiliency evaluation of wastewater treatment plants. Journal of Hydroinformatics 18: 990–1006.

Karamouz, M, A. Zeynolabedin, and M. A. Olyaei. 2015. Mapping regional drought vulnerability: a case study. International Archives of the Photogrammetry, Remote Sensing & Spatial Information Sciences 40. https://doi.org/10.5194/isprsarchives-XL-1-W5-369-2015

Keršuliene, V, E. K. Zavadskas, and Z. Turskis. 2010. Selection of rational dispute resolution method by applying new step-wise weight assessment ratio analysis (SWARA). Journal of Business Economics and Management 11(2): 243–258.

Khalid, M. A. and Y. Ali. 2020. Economic impact assessment of natural disaster with multi-criteria decision making for interdependent infrastructures. Environment, Development and Sustainability 22(8): 7287–7311.

Kim, T. H., B. Kim, and K.Y. Han. 2019. Application of fuzzy TOPSIS to flood hazard mapping for levee failure. Water 11(3): 592.

Kou, G. and W. Wu. 2014. Multi-criteria decision analysis for emergency medical service assessment. Annals of Operations Research 223(1): 239–254.

Li, H., J. Y. Guo, M. Yazdi, et al. 2021. Supportive emergency decision-making model towards sustainable development with fuzzy expert system. Neural Computing and Applications 33(22): 15619–15637.

Liu, W, M. Zhao, W. Hu, and Y. Cai. 2018. Spatial-temporal variations of water poverty in rural China considered through the KDE and ESDA models. Natural Resources Forum 42(4): 254–268.

Mohamed, S. A. 2021. Development of a GIS-based alert system to mitigate flash flood impacts in Asyut governorate, Egypt. Natural Hazards 108(3): 2739–2763.

Mojtahedi, M., R. Y. Sunindijo, F. Lestari, et al. 2021. Developing hospital emergency and disaster management index using TOPSIS method. Sustainability 13(9): 5213.

Moradi, M, M. R. Delavar, and B. Moshiri. 2013. Sensitivity analysis of ordered weighted averaging operator in earthquake vulnerability assessment. In: Proceedings of SMPR 2013 Conference (pp. 5–8).

Munyaka, J. C. B. and V.S. S. Yadavalli. 2021. Decision support framework for facility location and demand planning for humanitarian logistics. International Journal of System Assurance Engineering and Management 12(1): 9–28.

Nassereddine, M, A. Azar, A. Rajabzadeh, and A. Afsar. 2019. Decision making application in collaborative emergency response: A new PROMETHEE preference function. International Journal of Disaster Risk Reduction 38: 101221.

Nguyen, T. N, C. X. Pham, H. Q. Nguyen, and T. N. B. Dang. 2019. Establishing an early warning system for flash floods in Hoang Su Phi District, Ha Giang Province, Vietnam. Singapore Journal of Tropical Geography 40(2): 312–333.

Oh, N. 2020. Collective decision-making for developing emergency management capabilities. International Journal of Emergency Services. https://doi.org/10.1108/IJES-06-2019-0025

Omidvar, B, M. Baradaran-Shoraka, and M. Nojavan. 2013. Temporary site selection and decision-making methods: a case study of Tehran, Iran. Disasters 37(3): 536–553.

Opricović, S. 1998. VIKOR method. Multicriteria optimization of civil engineering systems (pp. 142–175). University of Belgrade, Faculty of Civil Engineering, Belgrade.

Opricovic, S. and G.H. Tzeng. 2002. Multicriteria planning of post-earthquake sustainable reconstruction. Computer-Aided Civil and Infrastructure Engineering 17(3): 211–220.

Ortiz-Barrios, M. A., B. A. Aleman-Romero, J. Rebolledo-Rudas, et al. 2017. The analytic decision-making preference model to evaluate the disaster readiness in emergency departments: the ADT model. Journal of Multi-Criteria Decision Analysis 24(5–6): 204–226.

Otay, I. and M. Jaller. 2020. Multi-expert disaster risk management & response capabilities assessment using interval-valued intuitionistic fuzzy sets. Journal of Intelligent & Fuzzy Systems 38(1): 835–852.

Rahman, M. A., N. Zaman, A. T. Asyhari, F. Al-Turjman, M. Z. A. Bhuiyan, and M. F. Zolkipli. 2020. Data-driven dynamic clustering framework for mitigating the adverse economic impact of Covid-19 lockdown practices. Sustainable Cities and Society 62: 102372.

Rana, I. A., M. Asim, A. B. Aslam, and A. Jamshed. 2021. Disaster management cycle and its application for flood risk reduction in urban areas of Pakistan. Urban Climate 38: 100893.

Rehman, S., M. S. U. Hasan, A. K. Rai, R. Avtar, and H. Sajjad. 2021. Assessing flood-induced ecological vulnerability and risk using GIS-based in situ measurements in Bhagirathi sub-basin, India. Arabian Journal of Geosciences 14(15): 1–17.

Rezaei, J. 2015. Best-worst multi-criteria decision-making method. Omega 53: 49–57.

Rezaei-Malek, M., R. Tavakkoli-Moghaddam, N. Cheikhrouhou, and A. Taheri-Moghaddam. 2016. An approximation approach to a trade-off among efficiency, efficacy, and balance for relief pre-positioning in disaster management, Transportation Research Part E: Logistics and Transportation Review 93: 485–509.

Roh, S., S. Pettit, I. Harris, and A. Beresford. 20 15. The pre-positioning of warehouses at regional and local levels for a humanitarian relief organisation, International Journal of Production Economics 170(Part-B): 616–628.

Roy, B. 1990. The outranking approach and the foundations of ELECTRE methods. In Readings in multiple criteria decision aid (pp. 155–183). Springer, Berlin, Heidelberg.

Roy, S., A. Paul, A. Paul, S. Kashyap, and A. Jana. 2021. Ranking barriers of supply chain management by MCDM method during disaster management: a case study of India. International Journal of System Dynamics Applications 10: 1–16.

Saaty, T. L. 2005. Making and validating complex decisions with the AHP/ANP. Journal of Systems Science and Systems Engineering 14(1): 1–36.

Sadrykia, M., M. R. Delavar, and M. Z are. 2017. A GIS-based fuzzy decision making model for seismic vulnerability assessment in areas with incomplete data. ISPRS International Journal of Geo-Information 6(4): 119.

Saksrisathaporn, K., A. Bouras, N. Reeveerakul, and A. Charles. 2016. Application of a decision model by using an integration of AHP and TOPSIS approaches within humanitarian operation life cycle. International Journal of Information Technology & Decision Making 15(04): 887–918.

Sambah, A. B. and F. Miura. 2014. Remote sensing and spatial multi-criteria analysis for tsunami vulnerability assessment. Disaster Prevention and Management 23(3): 271–295.

Sen, M. K., S. Dutta, and G. Kabir. 2021. Evaluation of interaction between housing infrastructure resilience factors against flood hazard based on rough DEMATEL approach. International Journal of Disaster Resilience in the Built Environment 12(5): 555–575.

Song, W., Z. Chen, A. Liu, Q. Zhu, W. Zhao, S.B. Tsai, and H. Lu. 2018. A study on green supplier selection in dynamic environment. Sustainability 10: 1226.

Song, S., H. Zhou, and W. Song. 2019. Sustainable shelter-site selection under uncertainty: a rough QUALIFLEX method. Computers & Industrial Engineering 128: 371–386.

Sotoudeh-Anvari, A., S. J. Sadjadi, S. M. Hadji Molana, and S. Sadi-Nezhad. 2019. A stochastic multi-objective model based on the classical optimal search model for searching for the people who are lost in response stage of earthquake. Scientia Iranica 26(3): 1842–1864.

Souissi, D., L. Zouhri, S. Hammami, M. H. Msaddek, A. Zghibi, and M. Dlala. 2020. GIS-based MCDM–AHP modeling for flood susceptibility mapping of arid areas, southeastern Tunisia. Geocarto International 35(9): 991–1017.

Sun, R., Z. Gong, G. Gao, and A. A. Shah. 2020. Comparative analysis of Multi-Criteria Decision-Making methods for flood disaster risk in the Yangtze River Delta. International Journal of Disaster Risk Reduction 51: 101768.

Toosi, S. A., G. H. Calbimonte, H. Nouri, and S. Alaghmand. 2019. River basin-scale flood hazard assessment using a modified multi-criteria decision analysis approach: a case study. Journal of Hydrology 574: 660–671.

Trivedi, A. and A. Singh. 2017. Prioritizing emergency shelter areas using hybrid multi-criteria decision approach: a case study. Journal of Multi-Criteria Decision Analysis 24(3–4): 133–145.

Wang, J. J., Y. Y. Jing, C. F. Zhang, and J. H. Zhao. 2009. Review on multi-criteria decision analysis aid in sustainable energy decision-making. Renewable and Sustainable Energy Reviews 13(9): 2263–2278.

Wang, X., D. Li, X. Zhang, and Y. Cao. 2018. MCDM-ECP: multi criteria decision making method for emergency communication protocol in disaster area wireless network. Applied Sciences 8(7): 1165.

Warfield, C. 2013. The disaster management cycle. www.gdrc.org/uem/disasters/1-dm_cycle.html

Wątróbski, J., J. Jankowski, P. Ziemba, A. Karczmarczyk, and M. Zioło. 2019. Generalised framework for multi-criteria method selection. Omega 86: 107–124.

Yalcin, A. S., H. S. Kilic, and D. Delen. 2022. The use of multi-criteria decision-making methods in business analytics: a comprehensive literature review. Technological Forecasting and Social Change 174: 121193.

Yang, C. L., M. C. Shieh, C. Y. Huang, and C. P. Tung. 2018. A derivation of factors influencing the successful integration of corporate volunteers into public flood disaster inquiry and notification systems. Sustainability 10(6): 1973.

Yang, M., F. Khan, and P. Amyotte. 2015. Operational risk assessment: a case of the Bhopal disaster. Process Safety and Environmental Protection 97: 70–79.

Yu, M., C. Yang, and Y. Li. 2018. Big data in natural disaster management: a review. Geosciences 8(5): 165.

Zavadskas, E. K., Z. Turskis, J. Antucheviciene, and A. Zakarevicius. 2012. Optimization of weighted aggregated sum product assessment. Elektronika ir Elektrotechnika 122(6): 3–6.

Zhao, J., P. Song, S. M. Mubarak, and S. M. Taresh. 2021. Critical success factors for epidemic emergency management in colleges and universities during COVID-19: a study based on DEMATEL method. Safety Science 145: 105498.

Zhu, F., P. A. Zhong, Y. N. Wu, Y. Sun, J. Chen, and B. Jia. 2017. SMAA-based stochastic multi-criteria decision making for reservoir flood control operation. Stochastic Environmental Research and Risk Assessment 31(6): 1485–1497.

2 Multi-Criteria Decision Analysis Models for Emergency and Disaster Management

A Systematic Review

Gulhan Sen[1] *and Bektas Sari*[2]

[1]Burdur Mehmet Akif Ersoy University, Department of Emergency Aid and Disaster Management, Burdur, Turkey

[2]Ege University, Atatürk Vocational School Of Health Services, Izmir, Turkey

CONTENTS

2.1 Introduction 17
2.2 Literature Review 19
2.3 Methodology 20
2.4 Findings 21
2.5 Conclusion 25
References 27

2.1 INTRODUCTION

Nowadays, we live in a world where nature-based or man-made disasters and emergencies that adversely affect people and society have become frequent. The important features of these events are that they may cause material and social losses for people or the environment. Regarding its terminology, disaster is a phenomenon resulting in great destruction and damage (Abdullah et al., 2021). Disasters, which are unforeseen and generally occur suddenly, are predicaments that exceed local capacity, require national or international assistance, and cause enormous loss of life and property (CRED, 2019). The United Nations defines disasters as events that occur because of the interaction of hazardous events with the exposure, vulnerability, and capacity of societies, disrupting the functioning of society and causing physical, economic, social, and environmental losses and impacts (UN, 2016). Besides disaster, emergency is an unexpected event that endangers life or property, requiring an immediate response using routine community resources and procedures (Drabek, 1996).

DOI: 10.1201/9781003212904-2

In many cases, it can be seen that emergency and disaster are used interchangeably. However, the main difference between the two is whether the capacity is exceeded and local facilities sufficient. In emergencies, it is possible to use routine procedures and intervene with local facilities. The damages caused by disasters in the physical, social, economic, ecological, and political environments are more than in emergencies. If emergencies are not responded to promptly and effectively, they can turn into disasters (Şen, 2017). In addition, it is seen that the COVID-19 pandemic in 2020 brought the world to a near standstill in many areas from economy to tourism to culture to sports (Koçak and Sarı, 2021).

In 2020, over 350 natural disasters were reported by the Emergency Database (EM-DAT). These devastating events negatively affected almost 100 million people and cost over US$170 billion (CRED, 2021).

Table 2.1 and Table 2.2, prepared with the data obtained from CRED, list the disasters that caused the most loss of life and affected the most people physically, socially, and economically in 2020 (CRED, 2021). In addition, the countries most affected by heatwaves are in Europe, and the countries most affected by flooding are in Asia.

According to the Table 2.2, disasters, including floods, droughts, and hurricanes, affected the greatest number of people in 2020. These disasters can be associated with meteorology and climate. It also appears that almost all the affected people live in the Asia region. Regarding the occurrence of these disasters around the world, disaster-management processes are required to concentrate on the pre-, during, and

TABLE 2.1
Top 10 Causes of Mortality by Disaster in 2020

Country	Disaster	Total Death Toll	Country	Disaster	Total Death Toll
UK	Heatwave	2,556	Pakistan	Flood	410
France	Heatwave	1,924	Netherlands	Heatwave	400
India	Flood	1,922	Kenya	Flood	285
Belgium	Heatwave	1,460	China	Flood	280
Nepal	Flood	448	Bangladesh	Flood	257

TABLE 2.2
Top 10 Populations Affected by Disasters in 2020

Country	Disaster	Total Affected (millions)	Country	Disaster	Total Affected (millions)
India	Cyclone Amphan	18	Honduras	Hurricane Eta	4.6
China	Flood	10	China	Flood	4.2
Mali	Drought	6.8	Niger	Drought	3.7
Bangladesh	Flood	5.4	Philippines	Typhoon Rolly	3.4
Philippines	Typhoon Ulysses	4.9	Burkina Faso	Drought	2.9

post-disaster periods. Disaster management is the process of organization, planning, and implementation of studies for the prevention and mitigation of disasters, rapid and effective response, and returning to normal life as soon as possible. Although there is a great deal of overlap between emergency management and disaster management, it differs whether the event is an emergency or a disaster (UNDRR, 2022). In a more exact statement, disaster management has its own disciplinary phases, including mitigation, preparedness, response, and recovery. These phases are central for managing disasters and all have their essential elements, including planning and decision-making.

The decision-making process is the prominent part of each phase of disaster management. Decision-makers (DMs) need to make effective decisions at every stage of emergency and disaster management. At this stage, making decisions based on the experience and knowledge of the DM may cause irreversible results. Therefore, in emergency and disaster management, where there are many criteria, DMs should make the best decision not only based on their experience and intuition, but also by using analytical and quantitative tools. Therefore, it is important to use multi-criteria decision analysis (MCDA) models.

Systematic review studies in the literature did not address MCDA models for both emergency and disaster management. In addition, many studies are narrowed down by restricting the type of disaster or method. Therefore, this chapter aims to systematically review the use of MCDA techniques in emergency and disaster management. Thus, a general framework has been presented on the frequency, place, and importance of the MCDA methods used effectively in emergency and disaster management. For this reason, the findings are more comprehensive and instructive, and fill the gap in the literature.

2.2 LITERATURE REVIEW

Decision-making procedure is an area that includes different steps. These steps are recognizing the complication; building the priority; assessing the other possibilities; and deciding the best alternatives (Kleindorfer et al., 1993; Simon, 1977). From a broad perspective, three different analysis methods could be applied to resolve decision-making problems. First, an illustrative method is covered by the problems that DMs solve; second, a prescriptive survey considers the models that DMs should use to ameliorate their decisions; and, last, normative analysis concentrates on the issues that DMs ought to refer to (Bell et al., 1988; Kleindorfer et al., 1993). When DMs evaluate many dissimilar criteria and problems, such as the weights of criteria, preference dependence, and conflicts among criteria, which seem to complicate the problems and need to be overcome by more sophisticated methods (Tzeng and Huang, 2011).

MCDA or multi-criteria decision-making (MCDM) is regarded as one of the most important scientific methods used by many experts (Alinezhad and Khalili, 2019). When a DM considers more than one attribute, the discussion of MCDM/MCDA is proposed, which allocates a large part of everyday decisions in organizations and human societies to the DM (Alinezhad and Khalili, 2019). To deal with MCDM/MCDA difficulties, the first approach ought to be to figure out how many related standards prevail in the issue and how to deal with the issues. The next step is to accumulate the suitable data or figures in which the predisposition of DMs can be accurately indicated and contemplated (Tzeng and Huang, 2011).

Emergencies and disasters are complex events that require different institutions and stakeholders to come together and work on an issue accordingly. Furthermore, managers and DMs ought to take coordinated action in case of disasters. Emergencies often demand high-quality decision-making results, because a wrong decision may cause incalculable losses. Therefore, in such cases, the views of all DMs should be fully considered (Xu et al., 2015). In terms of disaster management and disaster risk management, the majority of researchers have used probability statistics, fuzzy mathematics, and grey system theory as approaches to MCDM/MCDA (Gong and Forrest, 2014). A few other scholars have used different MCDM/MCDA methods to study and analyze uncertainty (Gong et al., 2020). Furthermore, in the disaster field, they have been used to determine disadvantages and future disaster risks to support decision-making (Sun, Gao, et al., 2020).

In disaster management and other decision-making processes, many distinct MCDM/MCDA methods have been used widely, including analytic hierarchy process (AHP), analytic network process (ANP), weighted sum model (WSM), the Technique for Order of Preference by Similarity to Ideal Solution (TOPSIS), Elimination Et Choice Translating Reality (ELECTRE), VlseKriterijumska Optimizacija I Kompromisno Resenje (VIKOR), Decision-Making Trial and Evaluation Laboratory (DEMATEL), and Preference Ranking Organization Method for Enrichment Evaluations (PROMETHEE). In particular, with the advantages of solving multi-attribute decision-making problems, the MCDM/MCDA methods are being increasingly used in the disaster field (Sun, Gong, et al., 2020). Also, it can be seen from the literature that MCDM/MCDA methods are used to analyze risks, evaluate vulnerability, determine locations, map a geographic area, and model best decision-making in the field.

2.3 METHODOLOGY

This study aimed to systematically review the use of MCDM/MCDA techniques in emergency and disaster management. The study was carried out within the scope of the Preferred Reporting Items for Systematic Reviews and Meta-Analysis (PRISMA) protocol. The eight keywords provided in the Table 2.3 were used.

The keywords were chosen to extend a comprehensive inquiry related to MCDM, MCDA, emergency management, and disaster management. The Web of Science (WoS) database was questioned as "topic (title, abstract, author keywords, and Keywords Plus)" without time limitation using eight compounds of keywords pertinent to the topic. Only research articles written in English were included; book chapters, papers, theses, non-English articles, and review articles were excluded.

The steps and explanations for selecting the articles according to the exclusion criteria are provided in Table 2.4. As a result of the keywords scanned in the first stage, a total of 1034 documents were obtained. After eliminating these documents according to the exclusion criteria, 798 articles remained. In total, 455 duplicate articles were excluded, leaving 343 remaining. Then, the authors evaluated the abstracts of the

TABLE 2.3
Keywords and Keyword Codes Used for Online Article Search

Keyword (KW) Code	Keyword
KW1	"Multi-criteria decision making" and "disaster"
KW 2	"Multi-criteria decision making" and "emergency"
KW3	"Multi-criteria decision analysis" and "disaster"
KW4	"Multi-criteria decision analysis" and "emergency"
KW5	"Multi-criteria decision making" and "disaster management"
KW6	"Multi-criteria decision making" and "emergency management"
KW7	"Multi-criteria decision analysis" and "disaster management"
KW8	"Multi-criteria decision analysis" and "emergency management"

TABLE 2.4
Phases of Excluded and Included Articles

	Explanation	Remaining articles (n)	Keywords (KW)	
Step 1	Total articles in the database as a result of 8-keyword search	1,034	KW 1: 226	KW 5: 116
			KW 2: 165	KW 6: 85
			KW 3: 181	KW 7: 97
			KW 4: 110	KW 8: 55
Step 2	Book chapter, thesis, paper, etc. Documents, inaccessible articles, and non-English articles excluded	798	KW 1: 177	KW 5: 91
			KW 2: 133	KW 6: 65
			KW 3: 134	KW 7: 70
			KW 4: 85	KW 8: 43
Step 3	Duplicate articles (n = 455) removed	343		
Step 4	Non-relevant articles and review articles excluded (n = 144)	199		

remaining articles. The authors then decided that 144 articles were out of the scope of the study, leaving 199 pertinent articles for review.

Metadata of the 199 articles included in the study were obtained and organized. The metadata comprised the authors' names, publication title, year of publication, the journal of publication, MCDM/MCDA techniques mentioned, emergency and disaster management stages, findings, and results. A comprehensive analysis of this metadata was managed using both quantitative and qualitative approaches.

2.4 FINDINGS

In this systematic review, MCDM/MCDA techniques in emergency management and disaster management were queried using eight keyword combinations in the WoS database. All articles up to August 2021 were scanned without any date limitation.

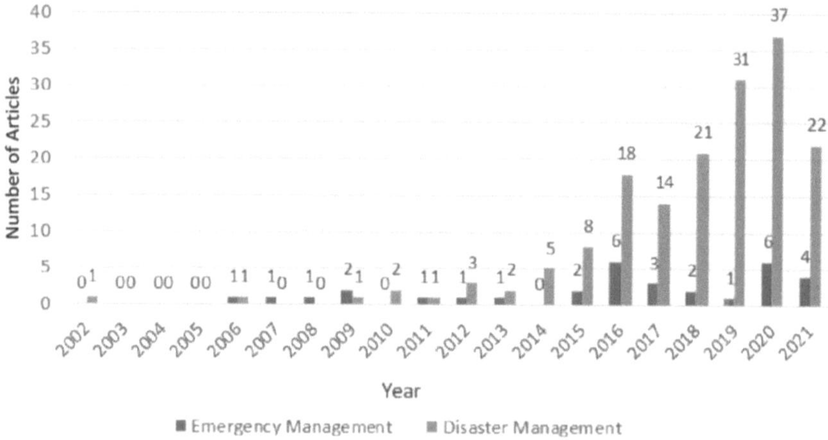

FIGURE 2.1 Distribution of emergency and disaster management studies using multi-criteria decision-making (MCDM)/multi-criteria decision analysis (MCDA) techniques by year.

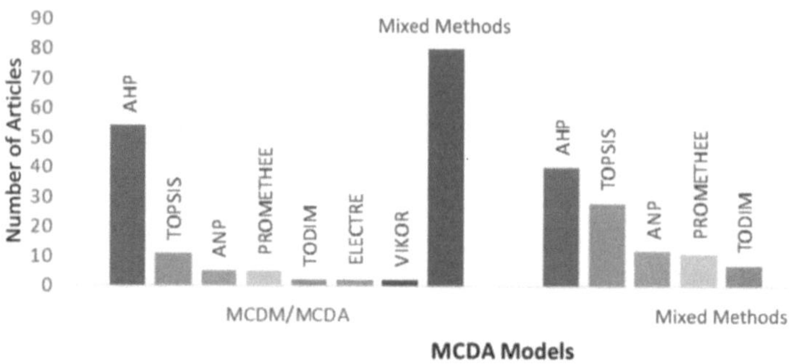

FIGURE 2.2 Distribution of multi-criteria decision-making (MCDM)/multi-criteria decision analysis (MCDA) techniques included in selected articles.

According to the findings, MCDM/MCDA techniques are used in 16.1% of emergency management situations, while their rate of use in disaster management is 83.9%. The distribution of the studies according to years is provided in Figure 2.1. As of 2016, the use of MCDM/MCDA techniques in emergency and disaster management has increased significantly. It has continued to increase in the period from 2016 to the present. In recent years, we can say that MCDM/MCDA techniques have been significant in disaster management. We also anticipate an increase for 2021; however, we will not be able to see it until the end of 2022, as the study covered work done up to August 2021.

The distribution of MCDM/MCDA methods in the 199 articles included in the study is shown in Figure 2.2. Mixed methods have been generally preferred in disaster and emergency management. The preferred methods alone are mostly AHP and TOPSIS. Among the mixed methods, combinations with AHP (50%) and TOPSIS

TABLE 2.5
Distribution of Multi-Criteria Decision-Making (MCDM)/Multi-Criteria Decision Analysis (MCDA) Approaches Based on Emergency Management Phases

	Emergency Management Phases			
MCDM/MCDA technique	Mitigation and Preparedness	Response	Recovery	Total
AHP	4	–	1	5
TOPSIS	1	2	1	4
Mixed	6	6	1	13
Others	5	4	1	10
Total	**16**	**12**	**4**	**32**

(35%) are most common. Most of these combinations are composed of both AHP and TOPSIS (22.5%) methods. Combinations with ANP (15%), DEMATEL (13.8%), and VIKOR (8.8%) were the next most common mixed methods. The ANP method is most often combined with other methods rather than being used alone. However, the PROMETHEE method is most often used as a single method.

Table 2.5 shows the MCDM/MCDA techniques used according to the stages of emergency management. Mixed methods are preferred in emergency management. Combinations of mostly AHP and TOPSIS are also used. In addition, MCDM/MCDA techniques in emergency management are the most often used for preparedness and mitigation, followed by the response phase. The purpose of using MCDA in emergency management is generally to develop a new model and approach. According to this, Busini et al. (2011) developed a model to analyze industry accident risk, Hoscan and Cetinyokus (2021) developed a model to determine emergency assembly areas in industrial accidents with the AHP method, and Nyimbili and Erden. (2020) developed a model to determine the suitability of urban emergency facilities with entropy AHP. Xiong et al. (2020) developed a three-stage decision support approach in case of an emergency at sea with combinations of the TOPSIS method.

Table 2.6 shows the MCDM/MCDA techniques used according to the stages of disaster management. MCDM/MCDA techniques are most often wielded in studies covering the pre-disaster period (67.6%). Again, mixed methods from MCDA are at the center of disaster studies. In addition, the most preferred single MCDM/MCDA method in disaster studies is AHP.

Figure 2.3 shows the distribution of mixed MCDM/MCDA methods based on disaster management. Accordingly, combinations (24.6%) of AHP and TOPSIS are the most often used among the mixed methods. Another frequently used combination is ANP plus DEMATEL. Combinations with any of the AHP and TOPSIS methods are also frequently favored in disaster-management studies. Accordingly, while mixed methods using the AHP method are used in 52.1%, mixed methods using the TOPSIS method are used in 30%. The point here is that the TOPSIS method is more often used in multiple combinations and not on its own. In disaster management studies,

TABLE 2.6
Distribution of Multi-Criteria Decision-Making (MCDM)/Multi-Criteria Decision Analysis (MCDA) Techniques Based on Disaster-Management Phases

	Disaster Management Phases			
MCDM/MCDA technique	Mitigation and Preparedness	Response	Recovery	Total
AHP	36	6	5	47
TOPSIS	2	1	2	5
ANP	3	1	–	4
DEMATEL	1	1	–	2
PROMETHEE	3	1	–	4
VIKOR	1	1	–	2
Entropy	2	–	–	2
ELECTRE	2	–	–	2
Mixed	46	11	12	69
Others	18	4	8	30
Total	**114**	**26**	**27**	**167**

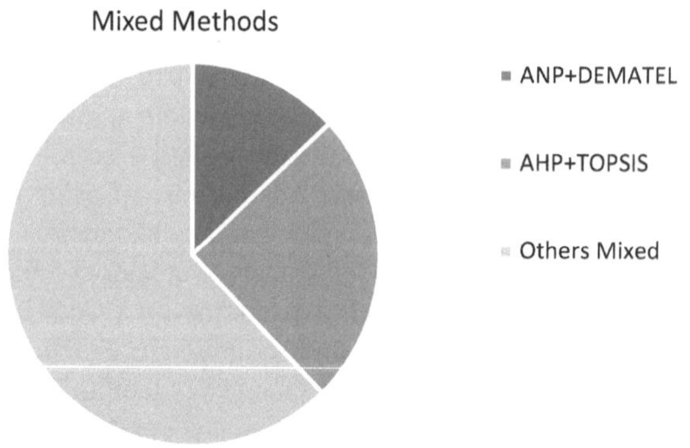

FIGURE 2.3 Distribution of mixed methods based on disaster management.

while TOPSIS is used on its own in 3%, its use in mixed methods increases to 30%. Likewise, ANP, PROMETHEE, and DEMATEL are favored more in mixed methods rather on their own in disaster management.

MCDM/MCDA methods are mostly used in the pre-disaster phase. Studies of this phase typically aim to analyze risk, create risk and sensitivity maps, and identify vulnerability. Ali et al. (2021) used fuzzy DEMATEL and ANP to determine landslide-susceptible areas, while Feloni et al. (2020) determined flood-prone areas using AHP and fuzzy AHP. Gosh and Mistri (2021) evaluated the vulnerability of beaches against environmental hazards with the AHP method, while Ekmekcioglu et al. (2021) created flood vulnerability, hazard, and risk maps of Istanbul with the hybrid fuzzy AHP–TOPSIS model. MCDM/MCDA has also been used frequently in disaster logistics (Munyaka and Yadavalli, 2021; Praneetpholkrang and Kanjanawattana, 2021; Kim et al., 2019; Saksrisathaporn et al., 2016; Connelly et al., 2016).

Table 2.7 shows the distribution of MCDM methods according to disaster type: natural disasters (34.7%), floods (31.13%), earthquakes (15%), and landslides (7.7%). Mixed methods are mostly preferred in studies dealing with all disasters. These studies have dealt with disasters in general, and not any particular type of disaster. AHP and mixed methods are mostly used in flood studies.

In addition, Figure 2.4 shows which disaster types and disaster management phases the MCDA technique is used for. Accordingly, 60% of earthquake-related studies, 86.5% of flood-related studies, and 48.2% of studies that did not specify any disaster type and addressed all disasters used MCDA for preparedness and mitigation studies.

Another frequently preferred method in the field of disaster management is geographic information system (GIS)-based MCDA. Disaster-management studies in recent years have stressed the importance of GIS-based integration in MCDA (14.3%). As seen in Figure 2.5, the use of this technique, which is especially used in risk analysis and in creating sensitivity maps, has been increasing since 2016.

GIS-based studies are mostly preferred in flood research, often together with the AHP method. These studies generally focus on identifying flood hazard zones, determining vulnerability, creating flood-risk and susceptibility maps, and performing flood-risk analysis (Hussain et al., 2021; Komolafe et al., 2020; Msabi and Makonyo, 2021; Morea and Samanta, 2020).

2.5 CONCLUSION

This study systematically examined the place of MCDM/MCDA techniques in emergency and disaster management within a broad framework. Accordingly, eight keywords were searched in the WoS database, and 199 articles were included in the study. The articles mainly covered disaster management.

As far as the findings go, MCDM/MCDA techniques in emergency and disaster management studies started to increase especially in 2016. In particular, pre-disaster risk-reduction studies started to increase after 2016. In addition, it has been observed that mixed methods are preferred rather than single-method MCDM/MCDA techniques in this field. AHP and TOPSIS are mostly opted for in combination with other methods. Among the single methods, AHP, TOPSIS, ANP, PROMETHEE, TODIM, ELECTRE, and VIKOR are used most often. Therefore, AHP and TOPSIS are most often used in emergency and disaster management. GIS-based integration of MCDA is gaining importance, especially as of 2014. In particular, this method has been used frequently in flood disasters and in combination with AHP to conduct risk analysis and create sensitivity maps.

MCDM/MCDA methods have mostly been preferred in mitigation and preparedness studies of emergency and disaster management. These studies mainly aim to carry out risk analyses, create risk and sensitivity maps, and identify vulnerability. In addition, MCDM in disaster management is common in studies that deal with all disasters, mostly floods.

TABLE 2.7
Distribution of Multi-Criteria Decision-Making (MCDM)/Multi-Criteria Decision Analysis (MCDA) Techniques Based on Disaster Types

MCDM/MCDA	Disaster Type												
Technique	Earthquake	Flood	Landslide	Fire	Hurricane	Tsunami	Pandemic	Climate Change	War	Migration	Volcanic Eruption	Natural Disasters	Total
AHP	5	22	6	2	2	1	1	1	1	–	–	6	47
TOPSIS	–	–	–	–	–	–	–	–	–	–	–	5	5
PROMETHEE	–	2	–	–	–	–	–	–	–	–	–	2	4
ANP	–	–	–	–	1	–	–	–	–	–	–	3	4
Mixed	8	22	5	3	2	1	1	–	–	1	–	26	69
Others	12	6	2	–	–	1	–	–	–	–	1	16	37
Total	**25**	**52**	**13**	**5**	**5**	**3**	**2**	**1**	**1**	**1**	**1**	**58**	**167**

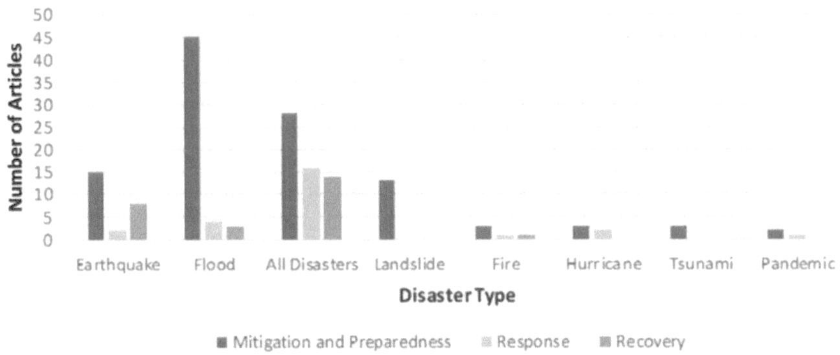

FIGURE 2.4 Distribution of disaster types based on disaster phases.

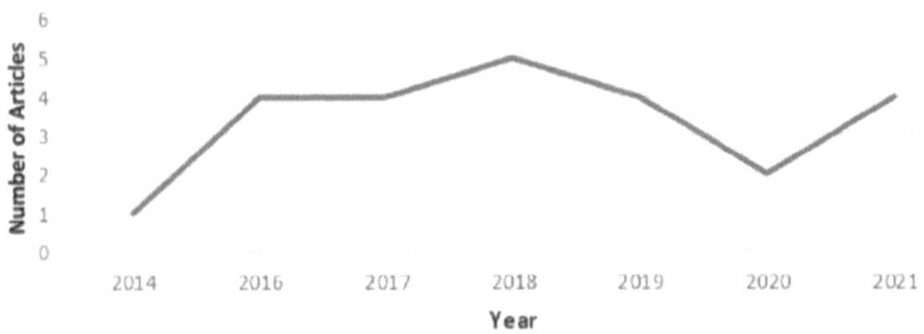

FIGURE 2.5 Distribution of GIS-based multi-criteria decision-making (MCDM)/multi-criteria decision analysis (MCDA) techniques by year.

As a result, the importance of MCDM/MCDA methods in creating risk-reduction strategies and building disaster-resilient cities and societies has been recognized more in recent years. In particular, the creation of different models using these methods draws attention. At this point, MCDA/MCDM in disaster management should be integrated into public policies, DMs should master these methods, and decision-making tools should be enriched with new models. In addition, for future research, we suggest that MCDA/MCDM should be used more frequently for risk management and resilience studies in man-made disasters.

As the limitation of the study, only the WoS database was queried for "topic" and research articles in English. However, a it is a current issue, the number of studies is constantly increasing. The study considered articles up to August 2021.

REFERENCES

Abdullah, M. F., Siraj, S., & Hodgett, R. E. 2021. An Overview of Multi-Criteria Decision Analysis (MCDA) Application in Managing Water-Related Disaster Events: Analyzing 20 Years of Literature for Flood and Drought Events. Water (Switzerland), 13(10): 13101358.

Ali, S.A., Parvin, F., Vojteková, J., Costache, R., Linh, N.T.T., Pham, Q. B., … and Ghorbani, M. A. 2021. GIS-based landslide susceptibility modeling: A comparison between fuzzy multi-criteria and machine learning algorithms. Geoscience Frontiers 12(2):857–876.

Alinezhad, A. and Khalili, J. 2019. *New methods and applications in multiple attribute decision making (MADM)* (Vol. 277). Cham: Springer.

Bell, D. E., Raiffa, H., and Tversky A. 1988. *Decision making: descriptive, normative, and prescriptive interactions.* Cambridge: Cambridge University Press.

Busini, V., Marzo, E., Callioni, A., and Rota, R. 2011. Definition of a short-cut methodology for assessing earthquake-related Na-Tech risk. *Journal of Hazardous Materials* 192(1):329–339.

Connelly, E. B., Lambert, J. H., and Thekdi, S. A. 2016. Robust investments in humanitarian logistics and supply chains for disaster resilience and sustainable communities. *Natural Hazards Review* 17(1):04015017.

CRED. 2019. Centre for Research on the Epidemiology of Disasters: natural disasters 2019. www.cred.be/publications?field_publication_type_tid=19.

CRED. 2021. Disaster year in review 2020 global trends and perspectives. https://cred.be/sites/default/files/CredCrunch62.pdf.

Drabek, T. 1996. The social dimensions of disaster. (FEMA Emergency Management Higher Education Project College Course Instructor Guide). http://training.fema.gov/EMIWeb/edu/completeCourses.htm.

Ekmekcioglu, Ö., Koc, K., and Özger, M. 2021. Stakeholder perceptions in flood risk assessment: A hybrid fuzzy AHP-TOPSIS approach for Istanbul, Turkey. *International Journal of Disaster Risk Reduction* 60:102327.

Feloni, E., Mousadis, I. and Baltas, E. 2020. Flood vulnerability assessment using a GIS-based multi-criteria approach—The case of Attica region. *Journal of Flood Risk Management* 13:e12563.

Ghosh, S. and Mistri, B. 2021. Assessing coastal vulnerability to environmental hazards of Indian Sundarban delta using multi-criteria decision-making approaches. *Ocean & Coastal Management* 209:105641.

Gong, Z. and Forrest, J. Y. L. 2014. Special issue on meteorological disaster risk analysis and assessment: on basis of grey systems theory. *Natural Hazards* 71, 995–1000.

Gong, Z., Wang, Y., Wei, G., Li, L., and Guo, W. 2020. Cascading disasters risk modeling based on linear uncertainty distributions. *International Journal of Disaster Risk Reduction* 43:101385.

Hoscan, O. and Cetinyokus, S. 2021. Determination of emergency assembly point for industrial accidents with AHP analysis. *Journal of Loss Prevention in the Process Industries* 69:104386.

Hussain, M., Tayyab, M., Zhang, J., Shah, A. A., Ullah, K., Mehmood, U., and Al-Shaibah, B. 2021. GIS-based multi-criteria approach for flood vulnerability assessment and mapping in district Shangla: Khyber Pakhtunkhwa, Pakistan. *Sustainability* 13(6):3126.

Kim, S., Ramkumar, M., and Subramanian, N. 2019. Logistics service provider selection for disaster preparation: a socio-technical systems perspective. *Annals of Operations Research* 283(1):1259–1282.

Kleindorfer, P. R., Kunreuther, H. C., and Schoemaker, P. J. H. 1993. *Decision sciences: an integrative perspective* (Vol. 91, Issue 2). Cambridge: Cambridge University Press.

Koçak, H. and Sarı, B. 2021. Türkiye'de covi̇d-19 ile mücadele süreci̇ne afet yöneti̇mi̇ açısından bi̇r yaklaşım. *Resilience* 5(1):37–49.

Komolafe, A. A., Awe, B. S., Olorunfemi, I. E., and Oguntunde, P. G. 2020. Modelling flood-prone area and vulnerability using integration of multi-criteria analysis and HAND model in the Ogun River Basin, Nigeria. *Hydrological Sciences Journal* 65(10):1766–1783.

Morea, H. and Samanta, S. 2020. Multi-criteria decision approach to identify flood vulnerability zones using geospatial technology in the Kemp-Welch Catchment, Central Province, Papua New Guinea. *Applied Geomatics* 12(4):427–440.

Msabi, M. M. and Makonyo, M. 2021. Flood susceptibility mapping using GIS and multi-criteria decision analysis: a case of Dodoma region, central Tanzania. *Remote Sensing Applications: Society and Environment* 21:100445.

Munyaka, J. C. B. and Yadavalli, V. S. S. 2021. Decision support framework for facility location and demand planning for humanitarian logistics. *International Journal of System Assurance Engineering and Management* 12(1):9–28.

Nyimbili, P. H. and Erden, T. 2020. A hybrid approach integrating entropy-AHP and GIS for suitability assessment of urban emergency facilities. *ISPRS International Journal of Geo-Information* 9(7):419.

Praneetpholkrang, P. and Kanjanawattana, S. 2021. A multi-objective optimization model for shelter location-allocation in response to humanitarian relief logistics. *The Asian Journal of Shipping and Logistics* 37(2):149–156.

Saksrisathaporn, K., Bouras, A., Reeveerakul, N. and Charles, A. 2016. Application of a decision model by using an integration of AHP and TOPSIS approaches within humanitarian operation life cycle. *International Journal of Information Technology & Decision Making* 15(4):887–918.

Şen, G. 2017. *Afet triajında etik karar verme-izmir ili örneği*. İzmir: Kitapana.

Simon, H.A. 1977. The logic of heuristic decision making. In: *Models of discovery* (pp. 154–175). Springer, Dordrecht.

Sun, R., Gao, G., Gong, Z., and Wu, J. 2020. A review of risk analysis methods for natural disasters. *Natural Hazards* 100: 571–593.

Tzeng, G.-H. and Huang, J.-J. 2011. Multiple attribute decision making: methods and applications. Boca Raton, FL: CRC Press.

UN. 2016. Report of the open-ended intergovernmental expert working group on indicators and terminology relating to disaster risk reduction (Vol. 21184, Issue December). www.preventionweb.net/files/50683_oiewgreportenglish.pdf.

UNDRR. 2022. Disaster management. www.undrr.org/terminology/disaster-management.

Xiong, W., Van Gelder, P. H. A. J. M., and Yang, K. 2020. A decision support method for design and operationalization of search and rescue in maritime emergency. *Ocean Engineering* 207:107399.

Xu, X. H., Du, Z. J., and Chen, X. H. 2015. Consensus model for multi-criteria large-group emergency decision making considering non-cooperative behaviors and minority opinions. *Decision Support Systems* 79:150–160.

3 Volunteer Assignment Optimization

A Systematic Literature Review with Analysis and Future Directions

Umit Ozdemir,[1] *Süleyman Mete,*[2] *and Muhammet Gul*[3]
[1]Munzur University
[2]Gaziantep University
[3]Istanbul University

CONTENTS

3.1 Introduction 31
3.2 Review Methodology 32
  3.2.1 Key Definitions of Volunteer Management 32
  3.2.2 Methodology 33
3.3 Results 34
3.4 Future Research Directions 40
3.5 Conclusion 40
References 41

3.1 INTRODUCTION

In the last few decades, the world has been affected by different types of man-made and natural disasters. There were 905 natural disasters in Turkey in 2020. When we examine the recorded number of earthquakes larger than 5 in magnitude from 1900 to 2020, there have been 1796 recorded in Turkey (Afad, 2020). The risk index identifies countries at risk from humanitarian crises and disasters and the capacity of the national response. It is based on three parts: hazards and exposure; vulnerability; and lack of coping capacity. In 2020, the risk index of Turkey was 5, and it is increasing (Inform, 2020). This means that the number of disasters in Turkey will increase in coming years and affect increasing numbers of people. During and after any disaster, many people are needed to help with many tasks, such as research, rescue, or the distribution of vital items. Volunteers are a very important part of helping the response and recovery, and they can provide invaluable assistance to official professionals.

DOI: 10.1201/9781003212904-3

In humanitarian organizations, official professionals are generally supported by volunteers. According to the Turkish Red Crescent, almost 2000 registered volunteers and many spontaneous volunteers attended to help rescue victims in the Elazig earthquake in 2020. This human resource provides an important source both financially and socially; however, the management of volunteers is different to that of professionals. Volunteers are generally unpaid workers and their numbers are not predictable, especially spontaneous volunteers. Moreover, this resource can be used ineffectively, which leads to congestion and results in some victims being unable to get help. To tackle this problem, different volunteer organizations have to manage what type of work to allocate volunteers in response to the resource needs of the organization. This leads to the requirement of effective disaster management.

In recent years, the scientific community has witnessed volunteerism and disaster management. A review paper was published by Altay and Green (2006) on disaster management, which included articles up to 2004. Similarly, a review paper was published by Galindo and Batta (2013), which covered articles from the years 2005 to 2010. Both review papers showed that, despite the significant growth in disaster management, there are areas that should be expanded in future research. Neither of the review papers included the problem of volunteer assignment, location allocation, or scheduling, including the planning of volunteers' tasks and their preferences.

This study aimed to examine research on volunteer assignment problems in the literature, to explain which methods are used, to examine the methods according to their aims and criteria, and to reveal the pros and cons of each study. To achieve this, we began by searching keywords to retrieve articles written on volunteer assignments in disaster management and present it as a framework. The framework has been designed for the problem of volunteer management. According to each paper, the methodology is explained, objectives are given with sub-criteria, limitations are discussed, and suggestions for future studies are made.

The rest of the chapter is structured as follows: first, the definition of volunteerism and the review methodology is given with the search technique and sub-criteria; second, the related literature is examined in detail, and all the papers are discussed according to problem type, aims, methodology, and limitations; and, finally, the results of the study and research opportunities for volunteer assignments are provided.

3.2 REVIEW METHODOLOGY

3.2.1 Key Definitions of Volunteer Management

Some definitions of volunteering and spontaneous volunteering are provided in the literature. One definition of volunteering is that the "Volunteer is an individual who offers him/herself a service without an expectation of monetary compensation" (Shin and Kleiner, 2003) while another definition is "Volunteerism involves long-term, planned, prosocial behaviors that benefit strangers, and usually occur in an organizational setting" (Penner, 2002).

The definition of spontaneous volunteerism is "the spontaneous volunteers as civilians who impulsively contribute to the disaster response without previous affiliations to recognized volunteer organizations (e.g., the Red Cross) or other typical first responders, such as firefighters" (Lowe and Fothergill, 2003). Similarly, the four types of volunteering in disaster conditions were defined by Shaskolsky (1967) as follows:

(i) Anticipated individual volunteers fulfil the general expectations of society on an individual basis, such as a doctor who comes to the aid of victims.
(ii) Anticipated organization volunteers are regularly associated with an organization, such as a volunteer fire brigade.
(iii) Spontaneous individual volunteers assist individuals, usually in the early stages of a disaster, for example in search and rescue activities.
(iv) Spontaneous organization volunteers place themselves at the service of an organization only once an emergency or disaster has occurred.

3.2.2 Methodology

In this study, a systematic quantitative literature review was implemented, which related to the subject, problem type, and research technique. Papers and conference proceedings were searched, examined, and categorized according to keywords. Then, they were sequenced for the implemented solution methodology under year order, illustrated a pivot table, and evaluated by the topic questions. Article reviews find the answers to the following questions:

(1) Who has published the paper?
(2) Where and when was it published?
(3) Which method was implemented?
(4) What was the aim of the papers?
(5) Which sub-criteria was used in their model?
(6) What was the limitation of the papers?

Original conference papers, articles, and book chapters written in English or Turkish, on the topic of volunteer or spontaneous volunteer-management problems in humanitarian organizations were added to the research. A keyword search of Web of Science, Scopus, and Google Scholar was performed. The year of the publication was limited from 2004 to now. The keywords used included the following terms: "volunteering," "spontaneous volunteer," "location," "allocation," "humanitarian," "aid," "participation," "organization," "recruitment," "disaster", "labor", and "assignment." Many published papers are cited in reference list, so we double-checked to ensure the originality of authors, journal, and DOI numbers.

Each paper was evaluated according to the problem type, objective, methodology, criteria, and limitations and then recorded in a Microsoft Excel spreadsheet. In this study, 14 articles from 27 papers and conference papers on volunteer assignments were examinded in detail (some were added to the other articles in Figure 3.1, because they did not directly examine the volunteer-assignment problem but contributed to our study). Volunteer assignment, location, and allocation of volunteers, and spontaneous volunteer problems were deeply reviewed and then each article was listed in order of year, from oldest to newest.

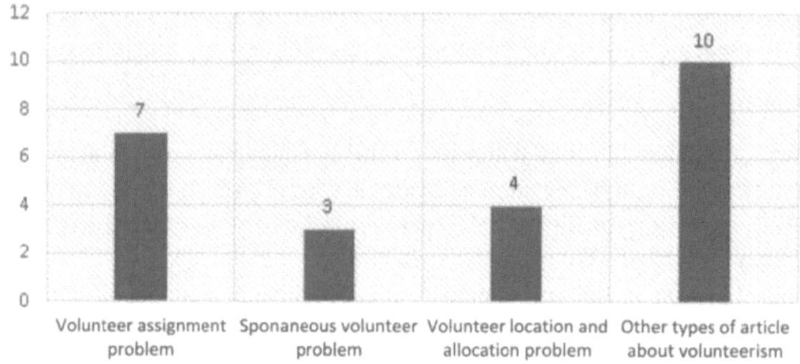

FIGURE 3.1 The number of papers chosen from the literature.

3.3 RESULTS

After a deep literature review, a total of 27 papers and conference papers on volunteer management from 2004 to 2020 were added to our research. Analysis of the articles included the following steps: (1) the problem discussed in the paper; (2) the purpose of the article; (3) what criteria were used; (4) the solution methodology; and (5) the limitations of the article. To achieve this end goal, we created Table 3.1, which lists each component of the articles. Before beginning to examine each paper's aims, criteria, and limitations according to the type of volunteer problem (voluntary assignment; spontaneous volunteer; and place and allocation problems), an explanation of each method is provided.

In these articles, the integer programming method was used most often. Integer programming, where one or more of the decision variables must be an integer, is generally used for the minimization or maximization of volunteer-assignment problems. Three papers applied this method. It was used for minimization of surplus volunteers at the Edmond Folk Festival (Gordon and Erkut, 2004). Their model was a spreadsheet-based decision support tool that scheduled volunteers under their preferences and time blocks with adjusted shift times, but their model was quite simple. It did not consider the per diem or shortages of volunteers. Falasca and Zober (2012) published a paper on volunteer assignment problems and applied the bi-criteria integer programming model to volunteer schedules. Their model targeted the minimization of scheduling costs, undesired assignments according to the volunteers' preferences (working alone or in a group), labor costs, and multiple locations without per diem of volunteers. Garcia et al. (2018) studied the volunteer resource allocation problem by using integer programming for maximization of the total allocation benefit and minimization of the total shortage and deviation costs. In their model, volunteer resource allocation problems were identified under six key headings, which were decided on by a combination of literature reviews and interviews of the volunteer managers. Their framework under these headings approached the dynamic resource reallocation problem, which aims to minimize the undesirable impacts of changes, and to supply the desired and changing objectives. Using model criteria, they examined the duration of volunteering time, consumable resources, and shortage penalties for resource and

TABLE 3.1
Review of Previous Studies

Article	Year	Problem Type	Objective function	Criteria	Solution methodology	Limitation
Improving volunteer scheduling for the Edmond Folk Festival	2004	Volunteer assignment	Minimization of surplus volunteers	Time period, working preferences	Integer programming	Not considering the per diem and volunteer shortages
Optimization of volunteer labor assignments	2006	Volunteer assignment	Minimization of the task shortages, undesired assignments, under- or over-utilization of volunteers	Volunteers and preferences task demand	Goal programming	Not consider the labor costs
Winning volunteer scenarios: the soul of a new machine	2007	Volunteer assignment	Generating multiple winning volunteer scenarios	Different types of volunteers with their assets and availabilities	Volunteer scenario machine	Not giving the best volunteer matching or volunteer preferences
An optimization model for volunteer assignment in humanitarian organizations	2012	Volunteer assignment	Minimization of the scheduling cost, undesired assignments according to volunteer preferences	Task demand, period, schedule costs, location, undesired assignment	Bi-criteria integer programming	Not considering the per diem of volunteers
OR/MS applications in Mt. Merapi disaster management	2012	Volunteer assignment	To minimize penalty costs incurred by deviating on the number of volunteers, shifts and tasks	The number of volunteers and their skills, tasks, and time block	Goal programming	Not considering the non-utilization of volunteer labor and their preferences
Building a new global search algorithm for managing Chinese volunteer relief organizations	2013	Spontaneous volunteer assignment	Minimize the travel time and cost of rescue work	Number of volunteers, distance from volunteer to the affected area	Global search algorithm	Not considering volunteer conditions such as fatigue, injury, or availability, and their preferences

(Continued)

TABLE 3.1 (Continued)
Review of Previous Studies

Article	Year	Problem Type	Objective function	Criteria	Solution methodology	Limitation
Reorganizing an existing volunteer fire station network in Germany	2014	Volunteer allocation problem	Minimize the number of required volunteer fire stations	Number of volunteers, travel time, each node must be assigned one fire station	Mathematical modeling	Not considering volunteer schedule, per diem, and uncertainty of rescue demand
Improving volunteer productivity and retention during humanitarian relief efforts	2014	Volunteer assignment	Minimization of unmet task demands, volunteer assignment costs, and mismatching	Number of volunteers, period, the skill levels of volunteers, training period	Volunteer management model	Not considering volunteer availability, and arrival and departure time of volunteers
Allocation of volunteers in non-governmental organizations aided by non-supervised learning	2016	Volunteer allocation problem	Decide the best clustering algorithm	Different number of clusters and types	K-means and K-medoid	Not considering the large group of volunteers and not mentioning other clustering methods
The optimal assignment of spontaneous volunteers	2017	Spontaneous volunteer assignment	Maximization of volunteer assignment	Volunteer arrival and service time, volunteer skill, different task queue	Markov decision process	Only considering two types of queue and one type of volunteer, without preferences
Dynamic resource allocation and coordination for high-load crisis volunteer management	2018	Volunteer allocation problem	Maximize the total allocation benefit and minimize the total shortage and deviation costs	Period, consumable resource, shortage penalty for resource, transportation cost	Integer programming	Integer programming not suitable for a complex problem, did not address how volunteer and resource optimization should best be integrated

Disaster Relief system designed for spontaneous volunteers	2018	Spontaneous volunteer assignment	Minimize travel time between volunteers and victims	Period, distance, volunteer number	General assignment problem	Only used with mobile applications and the internet
Volunteer management in charity storehouses: experience, congestion and operational performance	2019	Volunteer assignment	Decide the best matching of volunteers under different conditions	Different type of volunteers (experienced or inexperienced),different conditions (steady or congested)	Multi-method research	Only considered allowing or impeding congestion, not contributions to the organization in financial or time donations

transportation costs. The limitations of the study are that integer programming is not suitable for a complex problem, and how volunteer and resource optimization should best be integrated was not addressed.

Goal programming is an extension of mathematical programming that enables the user to develop models that "satisfice". It is often an attempt by the user to extend linear programming models to include more realistic multiple objectives and constraints, defined by O'Leary et al. (1981). Two papers were modeled using this programming method. Sampson (2006) presented the differences between traditional labor assignment and volunteer labor assignment, with four suppositions, in a table, in order to discuss the two types of assignment problem. Goal programming was then applied to their model on volunteer assignment problems for the minimization of the task shortages, undesired assignments, and under- or over-utilization of volunteers. The number of volunteers with preferences and task demand was included in the model criteria, but labor costs were not considered. Aman et al. (2012) studied both the distribution of logistics and humanitarian aid workforce assignment problems. This paper's assignment model included five different types of teams, and each team required different skills. They aimed to minimize penalty costs incurred by deviating on the number of volunteers, shifts, and tasks in their objective function by using the sub-criteria (number of volunteers and their skills, tasks, and time block). Their study did not consider the non-utilization of volunteer labor and their preferences.

Degel et al. (2014) applied mathematical modeling when deciding to minimize the number of the required volunteer fire stations. They aimed to reduce the effect of decreasing the volunteer number and available budget of the organization. In their model, the relocation or closing of a current station was examined and the effect of both situations checked. The study did not consider the volunteer schedule, per diem, and the uncertainty of rescue demand.

K-means aims to find a partition that minimizes the squared distance between the empirical mean of a cluster and the points in the cluster, considering all the clusters (Celebi et al., 2013). K-medoid has a main flow, the same as in K-means, However, in K-medoid, clusters are represented using an object whose average dissimilarity to all the objects in the cluster is minimal, instead of the mean (Kaufman and Rousseeuw, 1987)

The K-means and K-medoid methods were used by Bezerra et al. (2016) in a paper on volunteer allocation problems to decide the best clustering algorithm. Their study included the created recommender systems techniques and examined two types of clustering under two methods of measurement. The effectiveness of the two clusters was then evaluated by using the different volunteer groups, but they did not consider the large group of volunteers and their skills, and did not mention another type of clustering method.

Littman (2001) defined the Markov decision process, which is a decision-making model in discrete, stochastic, and sequential environments. The essence of the model is that a decision-maker, or agent, inhabits an environment, which changes state randomly in response to action choices made by the decision-maker. Mayorga et al. (2017) published a paper studying the maximization of spontaneous volunteer assignments. They used the multiple queening system with different adjusted demand and stochastic arrival times of the spontaneous volunteers. In their framework, they selected the volunteers from one of the several queues to supply the maximum benefit to the organization at the reception area. The continuous-time Markov decision process was used to achieve a close optimal and practical alternative, but they only considered two types of the queue and one type of volunteer, without preferences.

The general assignment problem was studied by Özel and Sezen (2018) to minimize the travel time from spontaneous volunteers to sufferers, and the waiting time of the assigned sufferers, by using a mobile phone application between volunteers and sufferers. Moreover, they created a web application for the center of disaster coordination. However, their model is only used with mobile applications and the internet.

Urrera et al. (2019) studied multi-method research to decide the best matching of volunteers under different conditions. They observed two types of volunteers arriving at different times at the storehouse: experienced (Exp) and inexperienced (Inexp). These volunteers work in a team under different conditions, of which there were three types: a mixed team (Exp-Inexp), and two types of non-mixed teams ((Exp-Exp) and (Inexp-Inexp)). They designed 32 (4×2×4) experimental treatments according to volunteer experience with storehouse congestion policies (allowing/impeding congestion) under four conditions (steady, high supply, high demand, high supply and high demand). Their results showed that a mixed-pairing (Exp-Inexp) is better, on average, than a non-mixed pairing in steady conditions. However, the non-mixed pairing is better than mixed-pairing under conditions of a high supply of volunteers and high demand for orders, with congestion allowed. The limitation of their study was that they only considered allowing or impeding congestion, and not financial or time contributions to the organization.

Conversely, some papers developed a new model for their studies on volunteer-management problems. Meijs and Brudney (2007) created a slot machine to generate multiple winning volunteer scenarios, which were defined as three types: the assets of potential volunteers; the availability of volunteers; and a particular volunteer assignment. Their model aimed to optimized winning volunteer scenarios, for example the potential and availability of volunteers to be matched with a particular volunteer assignment by the organization. The different types of volunteers, with their assets and availabilities, were added to the criteria, but the limitation of this study was that it did not give the best volunteer matching and the volunteer preferences were not included in their model. Liu and Wang (2013) published a paper on spontaneous volunteer scheduling problems with a target of minimizing travel time and the cost of rescue work. Their model organized the spontaneous volunteers to enable them to participate immediately and effectively when any type of disaster occurred. In addition, they suggested a new global search algorithm that helped to find the best path between volunteers and disaster areas and decided on the most appropriate number of volunteers in humanitarian organizations, but they did not consider volunteer conditions such as fatigue, injury, or availability, or their preferences. Lassiter et al. (2014) aimed to achieve minimization of unmet task demands, volunteer assignment costs, and mismatching of volunteer tasks according to skill. They created a model called the volunteer management model (VMM), with four real-life humanitarian disaster scenarios with short- and long-term task demands, and discussed the solutions. Their article included several useful features, including volunteer skill level and training, scenario-based costing, and additional task demand generated by an unmet task demand from prior periods. However, the availability of volunteers, and the arrival and departure times of the volunteers were not included in their VMM.

3.4 FUTURE RESEARCH DIRECTIONS

In this study, we have tried to highlight the pros and cons of the examined studies related to the solution methodology and types of volunteer management. Some studies were based on a programming method, especially mathematical modeling, and developed and tested by these models. Others created original models, and we applied them as case studies in simple examples to test their consistency.

The articles were focused on a single type of problem, such as the assignment or location allocation; there was no consideration of the problem of volunteer training and location allocation together. Moreover, two different models were not applied in the same studies, which would give better results. Similarly, whether the models used in the assignment of volunteers are applicable in a disaster situation has never been examined, or a limited number of applications have been made. Some studies should be applied comprehensively. According to Altay and Green (2006), there are four types of activities in disaster management: mitigation; preparedness; response; and recovery. Many papers focused on the recovery phase of the disaster. The other phases can be developed by volunteer assignment or location problems with suitable modeling to target the objectives under the sub-criteria.

3.5 CONCLUSION

Volunteer optimization in humanitarian organizations has been studied with regard to the assignment, training, management, and location of volunteers. This chapter suggests that volunteers need to be managed sufficiently under a crisis because emergency services and non-profit organizations have limited professional and budgets to realize effective responses. This chapter has found that the research on volunteer optimization has increased in recent decades, because researchers are witnessing the importance of volunteer management and how it open up new research opportunities. More research is needed to evaluate the various research objectives in the field of volunteer optimization and to provide an adequate initial framework. Volunteer management can be reanalyzed for different disaster areas such as mitigation or response. In addition, the assignment, location or allocation, and training of volunteers can be studied and applied to many different types of natural or man-made disasters. Finally, the adequacy of the studies can be checked by using different optimization models under real-life scenarios. This systematic literature review analysis may help provide researchers with an opportunity to expand the area of volunteerism, a better understanding of volunteer management, and a more effective application of future research.

REFERENCES

Afad. "Disaster statistics for Turkey in 2020". www.afad.gov.tr

Altay, N. and W.G. Green. 2006. "OR/MS research in disaster operations management". European Journal of Operational Research, 175: 475–493.

Aman, A., T. Bakhtiar, F. Hanum, and P. Supriyo. 2012. "OR/MS applications in Mt. Merapi disaster management". Journal of Mathematics and Statistics, 8: 264–273.

Bezerra, M., B. Aráujo, and V. Macari. 2016. "Allocation of volunteers in non-governmental organizations aided by non-supervised learning". 5th Brazilian Conference on Intelligent Systems (BRACIS), 223–228.

Degel, D., L. Wiesche, S. Rachuba, and B. Werners. 2014. "Reorganizing an existing volunteer fire station network in Germany." Socio-Economic Planning Sciences, 48: 149e157.

Emre Celebi, M., H.A. Kingravi, and P.A. Vela. 2013. "A comparative study of efficient initialization methods for the k-means clustering algorithm". Expert Systems with Applications, 40: 200–210.

Falasca, M. and C. Zober. 2012. "An optimization model for volunteer assignment in humanitarian organizations". Socio-Economic Planning Sciences 46: 250e260.

Galindo, G. and Batta, R. 2013. "Review of recent developments in OR/MS research in disaster operations management", European Journal of Operational Research, 230: 201–211.

Garcia C., G. Rabadi, and F. Handy. 2018. "Dynamic resource allocation and coordination for high-load crisis volunteer management", Journal of Humanitarian Logistics and Supply Chain Management, 8: 533–556.

Gordon, L. and E. Erkut. 2004."Improving volunteer scheduling for the Edmonton Folk Festival." Interfaces, 34: 367–376.

INFORM. 2020. www.inform-index.org

Kaufman, L. and P. Rousseeuw. 1987. "Clustering by means of medoids." In Proceedings of the Statistical Data Analysis Based on the L1 Norm Conference, Neuchatel, Switzerland, 31.

Lassiter, K., A. Alwahishie, and K. Taaffe. 2014. "Improving volunteer productivity and retention during humanitarian relief efforts." International Journal of Supply Chain Management, 3: 1–10.

Littman M. L. 2001. International Encyclopedia of the Social and Behavioral Sciences, pp. 9240–9242. New York: Elsevier.

Liu J. and S. Wang. 2013. "Building a new global search algorithm for managing the Chinese volunteer relief organisations." International Journal of Emergency Management 9: 187–204.

Lowe, S. and A. Fothergill. 2003. "A need to help: emergent volunteer behaviour after September 11th." In Beyond September 11th: An account of post-disaster research. pp. 293–314. Boulder, CO: University of Colorado.

Mayorga M., E. Lodree, and J.Wolczynski. 2017. "The optimal assignment of spontaneous volunteers." The Journal of the Operational Research Society, 68: 1106–1116.

Meijs M. and L. Brudney. 2007. "Winning volunteer scenarios: the soul of a new machine." The International Journal of Volunteer Administration, 24: 68–79.

O'Leary D.E. and P. Marwick. 1981. "A Goal Programming Model for The Cask Management Problem." American Institute for Decision Sciences 10th Annual Meeting Western Regional Conference, Hawaii.

Ozel, B. and B. Sezen. 2018. "Disaster Relief system designed for spontaneous volunteers." Natural Hazards and Disaster Management 2018: 528–536.

Penner, L.A. 2002. "Dispositional and organizational influences on sustained volunteerism: an interactionist perspective." The Journal of Social Issues, 58: 447.

Sampson, S.E. 2006. "Optimization of volunteer labor assignments." Journal of Operations Management, 24: 363–377.

Shaskolsky, L. 1967. "Volunteerism in disaster situations." Disaster Research Center, Ohio State University.

Shin, S. and B. H. Kleiner. 2003. "How to manage unpaid volunteers in organisations." Management Research News, 26: 63.

Urrera, G., P. Martinez, and M. Besiou. 2019. "Volunteer management in charity storehouses: experience, congestion and operational performance." Production and Operation Management, 28: 2653–2671.

4 Flood Risk Management Through Multi-Criteria Decision-Making

A Review

Ahmad Ferdowsi,[1] *Babak Zolghadr-Asli,*[2] *Sayed-Farhad Mousavi,*[3] *and Kourosh Behzadian*[4]

[1]University of Applied Science and Technology, Tehran, Iran; and Department of Water Engineering and Hydraulic Structures, Faculty of Civil Engineering, Semnan University, Semnan

[2]Centre for Water in the Minerals Industry, Sustainable Minerals Institute, The University of Queensland, Brisbane Australia; and The Centre for Water Systems (CWS), The University of Exeter, Exeter, UK

[3]Department of Water Engineering and Hydraulic Structures, Faculty of Civil Engineering, Semnan University, Semnan, Iran

[4]School of Computing and Engineering, University of West London, UK

CONTENTS

4.1 Introduction 43
4.2 Floods: Hazard and Risk Management 45
4.3 MCDM: Framework and Methods 46
4.4 MCDM Applications in Flood Risk 48
4.5 A Deeper Look at the Landscape of MADM Methods for Flood Management 49
4.6 Concluding Remarks 51
References 52

4.1 INTRODUCTION

Based on the Emergency Events Database report, some 315 natural disasters occurred worldwide in 2018, affecting around 70 million people, and leading to 11,804 fatalities and $131.7 billion in financial losses (EM-DAT, 2019). The intensity (and frequency) of natural disasters, including floods and droughts, has increased, and

DOI: 10.1201/9781003212904-4

climate change and anthropogenic activities have exacerbated these problems. As per Figure 4.1, the number of flood events between 2010 and 2020 has increased significantly (Abdullah et al., 2021).

With respect to the number of affected people, floods are the first-ranked natural disasters of the twenty-first century (EM-DAT, 2019). In most cases, floods are considered to be natural disasters created by torrential rainfall. There are numerous reasons for floods that can be classified according to type. In a broad category, floods can be classified as dam-break floods, rising ground water floods, snowmelt floods, tsunamis, flash floods, tidal floods, urban sewer floods, fluvial (riverine) flood, and pluvial flood (Schanze, 2006). Flow velocity, water depth, and temporal and spatial dynamics are important characteristics of floods. Due to the specific features of floods that affect residential areas, the number of the people who suffer from floods can be different from a region to another one. Floods may also cause water-quality and environmental problems.

Traditionally, floods have been controlled by massive structures, which are usually associated with expensive construction and maintenance costs. Flood-risk management (FRM) has attracted a lot of attention due to problems caused by floods, including loss of life and extensive damage (Chitsaz and Banihabib, 2015). FRM is the ability to manage the risk of an existing flood to decrease the flood risk (Plate, 2002). Water-related problems entail addressing a raft of inherent conflicting objectives that should be evaluated simultaneously in decision-making processes (Ferdowsi et al., 2021a, b).

FRM decision-makers and stakeholders may have various options for flood mitigation and control. The available options may consist of structural and non-structural actions, which may not be easily prioritized. This complexity can be streamlined by using multi-criteria decision-making (MCDM) approaches that can be helpful tools for decision-makers and emergency services in dealing with multi-criteria tasks and prioritizing the alternatives. MCDM techniques have been widely employed in mitigating the risk of natural hazards (El-Magd et al., 2020). MCDM

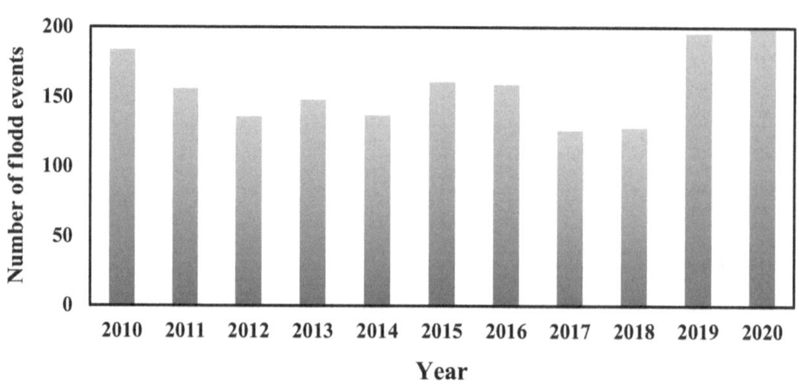

FIGURE 4.1 The number of flood events between 2010 and 2020.

methods contribute to solving a large range of water-related problems (Zolghadr-Asli et al., 2021a).

This chapter reviews the relevant literature and gives an in-depth assessment of MCDM applications in FRM. The current chapter reviews published works from January 2000 to December 2020. We used a combination of keywords ranging from flood management, multi-attribute decision-making (MADM), natural hazards, and water resources to topical MADM methods to identify papers in relevant journals. We then manually scanned each paper to determine the relevancy of the manuscript to the current study. While articles that were not deemed relevant were filtered out, the reference sections of the selected papers were also reviewed manually to identify any other potentially relevant studies. Without doubt, many MADM-based studies have been published as conference proceeding and reports. The current review covered the majority of mainstream publishers, including, but not limited to, Elsevier, the International Water Association, the Multidisciplinary Digital Publishing Institute, Springer, the American Society of Civil Engineers, Taylor & Francis, and Wiley & Sons. What is crucial to note here is that this study aimed to shed light on the general trend and provocative ideas found during this review. As such, we made a deliberate attempt to shed light on a few handpicked articles to make these ideas abundantly clear.

The chapter is presented in three main sections. The first discusses flood risk-management-related issues. It is followed by a presentation of the importance of decision-making and the role of MCDM methods. Some of the recently proposed and popular MCDM models are also introduced briefly. In the last section, the literature is critically reviewed to show how MCDM methods have contributed to the mitigation of flood-related hazards, and we present the most appropriate alternatives for FRM.

4.2 FLOODS: HAZARDS AND RISK MANAGEMENT

Flood is known as one of the severe natural disasters. The consequences of these natural events, as a part of hydrological cycle, are numerous. For instance, they may cause fatalities; forcibly displace people; damage cities, historic buildings and sites, and infrastructure; harm the environment; and so on. The formation of floods depends on several factors. A rising water level caused by climate change, for example, is a reason for coastal flooding. Human activities can exacerbate the flooding in many cases. For example, the capability of a basin's natural retention can be affected by human activities (Vojtek and Vojteková, 2016); in the perspective of urban flooding, impermeable pavements may increase flooding risks.

Schanze (2006) stated that the old expression of 'flood protection' is not an appropriate paradigm because it is unsustainable and unachievable to withstand flood events all the time due to high costs and the inherent uncertainties of flood events. Dams, drainage systems, and levees are some traditional engineering techniques for the mitigation and control of floods, but they can increase the risk of damage to the environment (Sörensen et al., 2016). Therefore, FRM as a suitable alternative to flood protection has been proposed in some studies (Plate, 1999, Hooijer et al., 2004), which focused on a broad spectrum of topics from flood hazards prediction to social issues (Schanze, 2006). Considering the complexity and variety of problems in flooding, systematization and integration are needed in the planning of FRM. From another perspective, Tan et al. (2021) divided the studies related to FRM into six different groups, including flood forecasting, streamflow prediction, rainfall runoff simulation, analysis of flood frequency, water level prediction, and flood risk assessment. FRM involves three main parts: 1) analysis of risk; 2) assessment of risk; and 3) risk-reducing measures (Hall et al., 2003).

Generally, the alternatives to FRM can be divided into structural and non-structural measures (Chitsaz and Banihabib, 2015). Physical and socio-economic conditions should also be considered in FRM alternatives (Yazdandoost and Bozorgy, 2008). Structural measures mainly refer to physical activities such as the construction of dams, floodwalls, levees, rainwater-harvesting systems, flood-control reservoirs, and runoff collection systems (Ferdowsi et al., 2019, 2020a, 2020b; Abdullah et al., 2021).

4.3 MCDM: FRAMEWORK AND METHODS

Decision-making for real-world engineering problems can be a very complex process because it may include many conflicting objectives and options (Ferdowsi et al., 2021a, 2021b). If the decision-maker wants to choose from the available, discrete, and feasible options, the problem entails the MCDM process. Various approaches to obtaining a logical and optimal decision-making process have been recommended so far. MCDM presents practical applications through numerous methods for solving such problems (Bozorg-Haddad et al., 2021) by assuming that there is no unique solution that can be optimal with respect to all performance criteria. Some of the well-known and most-often used MCDM techniques for engineering problems are briefly reviewed below.

SAW. SAW (simple additive weighting) is among the most well-known MCDM methods. It works based on a weighted score that is calculated for an alternative (Ustinovichius et al., 2007). Other terms and approaches, such as weighted sum method, weighted linear combination, and weighted linear average, refer to the same method (de Brito and Evers, 2016; Bozorg-Haddad et al., 2021).

VIKOR. VIKOR (*VlseKriterijumska Optimizacija I Kompromisno Resenje*) was proposed based on the closeness of a solution to the ideal point (Opricovic, 1998).

AHP. AHP (analytic hierarchy process) follows a pairwise comparison matrix to compare alternatives with respect to each criterion or sub-criterion separately (Saaty, 1977; Saaty, 1980) and is a MCDM technique that takes both the weights and optimization of the objective function. AHP divides the problem into different layers; each is dealt with as a single-objective problem and finally converts the objectives to weighting-based methods such as SAW (Bozorg-Haddad et al., 2021).

ANP. ANP (analytic network process) proposed by Saaty (2004) is an enhanced version of the AHP method. In comparison, the AHP method benefits from a hierarchy structure, but the ANP method uses a network structure.

ELECTRE. ELECTRE (*Elimination Et Choix Traduisant la Realite*—in English, "elimination and choice expressing reality") is a member of the MCDM family based on the ranking of alternatives. The members of this family group include ELECTRE I, II, III, and IV (Benayoun et al., 1966; Roy and Bertier, 1971; Roy, 1978; Roy and Hugonnard, 1982).

TOPSIS. TOPSIS (Technique for Order Preferences by Similarity to an Ideal Solution) can be considered as a popular MCDM approach that defines the superior solution that has the least distance from the ideal one, and the farthest length from the inferior one. In fact, TOPSIS lets a decision-maker avoid selection of the most expensive solution.

BWM. BWM (best–worst method) is a method proposed by Rezaei (2015) according to the pairwise comparison of alternatives (best or worst).

PROMETHEE. PROMETHEE (Preference Ranking Organization METHods for Enrichment Evaluation) is a member of the MCDM family of methods classified under outranking-oriented MCDM approaches (Brans, 1982; Brans et al., 1986).

Table 4.1 summarizes the MCDM methods discussed above.

TABLE 4.1
Some MCDM Methods in Order of Year of Publication

Year	Method	Research
1738	Utility function	Bernoulli (1738)
1954	WSM	Churchman and Ackoff (1954)
1966	ELECTERE I	Benayoun et al. (1966)
1971	ELECTERE II	Roy and Bertier (1971)
1977	AHP	Saaty (1977)
1978	ELECTERE III	Roy (1978)
1981	TOPSIS	Hwang and Yoon (1981)
1982	ELECTERE IV	Roy and Hugonnard (1982)
1982	PROMETHEE I	Brans (1982)
1985	PROMETHEE II	Vincke and Brans (1985)
1986	PROMETHEE III	Brans et al. (1986)
1987	PROMETHEE IV	Mladineo et al. (1987)
1989	Grey Relational Analysis (GRA)	Deng (1989)
1996	ANP	Saaty (1996)
1998	VIKOR	Opricovic (1998)
2001	Superiority and Inferiority Ranking (SIR)	Xu (2001)
2008	Pairwise Rankings of All Possible Alternatives (PAPRIKA)	Hansen and Ombler (2008)
2015	BWM	Rezaei (2015)
2019	Combined Compromise Solution (CoCoSo)	Yazdani et al. (2019)

4.4 MCDM APPLICATIONS IN FLOOD RISK MANAGEMENT

MCDM is a practical term that refers to a series of mathematical frameworks, which allows decision-makers to investigate the potential options by taking into account conflicting criteria (Zolghadr-Asli et al., 2021a). MCDM methods have been employed to solve a vast range of conflicting problems. FRM, as a complex issue that may cause environmental and financial problems, needs careful analysis. MCDM methods have been used to solve FRM problems many times. It has been proven time and again in such studies that these methods, though sometimes vastly differing from one another in theory and fundamental assumptions, have great potential when it comes to handling water resource planning and management in general, and flood management specifically (Zolghadr-Asli et al., 2021a; Abdullah et al., 2021). In order to better understand the nuances and status of these methods with regard to flood management, a comprehensive review has been conducted to identify and, in turn, analyze the most relevant studies in this field. In order to identify these studies, a combination of keywords ranging from flood management, multi-attribute decision-making, natural hazards, water resources planning, and management were used. To ensure that the selected papers were relevant to this study, the authors then manually scanned each given paper and handpicked the articles for the current study. In order to expand the search net, the reference sections of the selected papers were also reviewed manually to identify any other potentially relevant study. As a disclaimer, one should bear in mind that while this is, indeed, a comprehensive review, it might not be an exhaustive one per se. It should be noted, however, that this study aimed to shed light on the general trend and thought-provoking ideas found during this review. As such, we made a deliberate attempt to shed light on a few handpicked articles, to make these ideas abundantly clear. In the next section, some of the applications of MCDM methods in FRM are discussed.

Using MCDM models, Margeta and Knezic (2002) evaluated some alternatives for managing floods in a karstic zone. A karstic field was chosen for the study as floods occur in most months of the year. Other actions in the region, like constructing a tunnel, were done for flood mitigation in 1985, but they were not completely practical and did not eliminate the flood problem. Hence, for sustainable flood management, six different alternatives were considered in this study, using two MCDM methods: PROMETHEE and AHP. They reported that the integral MCDM approach could be helpful and practical in obtaining a compromise for such a complex karstic water system.

Ghanbarpour et al. (2013) developed a flood-risk mitigation methodology based on a TOPSIS-based MCDM framework that was based on geographic information system (GIS) and using the Hydrologic Engineering Center's River Analysis System (HEC-RAS) as a river model. The proposed approach was applied to the Zaremroud River in Tajan Basin, Iran. Construction of a levee (on the right side of the river) was selected as the best scenario for the study area, considering three criteria: 1) flood risk, based on the inundation of area and the level of flood; 2) environmental impact (the flow velocity and stream power); and 3) economic issues (construction costs).

Zagonari and Rossi (2013) developed a decision-support system approach, using the TOPSIS method to rank the alternatives for flood-mitigation strategies or recovery measures. The results of this study revealed that, from a mitigation perspective, constructing sand dunes and the development of evacuation programs were the best options. Sand dune construction, evacuation programs, and insurance schemes were proposed as the best options for recovery.

Chitsaz and Banihabib (2015) compared eight MCDM methods (i.e., SAW, Compromise Programming, VIKOR, TOPSIS, modified TOPSIS, ELECTRE I, ELECTRE III, and AHP) to select the best measures for FRM in the case of Gorganrood River—a susceptible river with a high risk of flooding. The case study was chosen as more than 60 floods happened in a decade, with more than 300 fatalities and almost US$115 million in damage. The study concluded that ELECTRE III provided the most suitable option for flood management.

Malekian and Azarnivand (2016) investigated flood risk, focusing on the prioritization of watershed hydrological units of Shemshak located in Tehran, Iran. This study was performed using a combination approach based on Shannon's entropy method and VIKOR. The proposed framework was able to identify the most hazardous zone of the watershed. It was recommended that this methodology should be applied to the watershed of interest prior to the implementation of flood-control measures (structural and non-structural ones) for reducing flood risks.

Kafle and Shakya (2018) studied the prioritization of some alternatives for flood risk and sediment management of a real-world case in Koshi River, Nepal. The study considered five alternatives for the FRM, eight alternatives for the sediment control and seven assessment criteria covering environmental, economic, political, and social issues. Three MCDM methods—ELECTRE, TOPSIS, and SAW—were used to analyze the alternatives for FRM. Finally, the most and the least prioritized options were using the old Koshi channel and raising embankments, respectively.

Akay and Baduna Koçyigit (2020) used five MCDM methods (AHP, ANP, TOPSIS, VIKOR, and ELECTRE I) and several traditional approaches for watershed prioritization, focusing on the potential of flash floods in Akcay's sub-basins (in Turkey). They concluded that an efficient management strategy should be based on water resources and soil-conservation policies.

Ghaleno et al. (2020) performed a comprehensive study on 13 sub-basins and 11 evaluation indices for prioritizing flood hazards and sedimentation in a river basin by using four MCDM methods: TOPSIS, SAW, ELECTRE, and VIKOR. Since the most vulnerable sub-basins (based on the results of the MCDM analysis) also had the highest erosion, poor soil, and gradient (based on field data), it was concluded that the results of this study were reliable enough.

Sun et al. (2020) evaluated flood risk by using three MCDM methods—ELECTRE, TOPSIS, and WSM—in four Chinese cities (Shanghai, Jiangsu, Zhejiang, and Anhui). The study identified the city with the highest flood risk and the one with the least flood risk. The study also concluded that agricultural factors, population density, drainage density, and the number of medical and health institutions were the most effective parameters (in the Yangtze River Delta area).

4.5 A DEEPER LOOK AT THE LANDSCAPE OF MADM METHODS FOR FLOOD MANAGEMENT

MADM methods have been repeatedly proven to be effective and proficient ways to tackle the intricate decision-making challenges that are often associated with the realm of water resource planning and management (Enayati et al., 2021; Zolghadr-Asli et al., 2021c). The flexible nature of these methods, in general, makes them an ideal vessel to capture the multi-dimensional nature of these types of problems (Zolghadr-Asli et al., 2021a, 2021b). That said, in order to obtain reasonable and reliable solutions out of these methods, one must always use the proper method for the problem at hand. This is due to the fact that, while these methods are grouped under the umbrella term MADM, they have different assumptions and perceptions with regard to the general paradigm of decision-making (Bozorg-Haddad et al., 2021). Ignoring these simple yet crucial facts may lead to unreliable and unrealistic solutions that would do little good in practical cases.

As for the application of these methods for the case of FRM, based on the findings of the current review, three crucial factors need to be addressed at this juncture. These are arguably the three main pillars that could determine the degree whether the set framework is, indeed, a suitable alternative to handle the said problem or not.

The first point is the subjective versus objective MADM methods. Based on the findings of this review and similar research that has been conducted in this field (e.g., de Brito and Evers, 2016; Abdullah et al., 2021; Zolghadr-Asli et al., 2021a), subjective-based methods such as AHP or ANP seem to be gaining popularity as researchers tend to lean more toward this branch of MADM, which is based on drastically different principles, than toward objective-based MADM methods such as TOPSIS or VIKOR. While these methods are sound in principle and are in and of themselves full of merits, one should be cautious when interpreting the solutions obtained from these frameworks. The idea is that such methods provide an environment that helps decision-makers to implement and, in turn, utilize the subjective assessments of stakeholders, experts, and analysts. While one cannot put a price on the years of experience and knowledge that goes into such assessments, decision-makers must always ensure that these are unbiased assessments in line with the reality of the case, which, more often than not, is easier said than done. As such, while it is absolutely beneficial to keep implementing these opinions in the decision-making process, it is better to have a mechanism in place to check the legitimacy of such options, whether by coupling these methods with objective-based MDAM methods or incorporating multiple views in the process, which brings us to the second point.

Given the multi-dimensional, multi-stakeholder nature of these problems, it is best to implement group decision-making paradigms to handle FRM problems. Of course, more and more researchers seem to be gravitating toward this paradigm of decision-making, as such frameworks enable decision-makers to truly capture the multi-stakeholder nature of the problem. Such solutions would be much more applicable as they can reflect the demands and requirements of all involved parties. As such, it is much more likely that such solutions would be implemented as the stakeholders would all feel heard and understood through the process of decision-making. What is important to note here is that this participation should not be limited to selecting or weighing the alternatives, which is often the case in such situations. Rather, such stakeholders should be able to define their own criteria and even alternatives. However, in all fairness, this seems a bit more unparticle in real-world cases as often there are too many involved parties with drastically different viewpoints.

Last, but certainly not least, is the issue of addressing the uncertainty associated with different aspects of the decision-making process. Uncertainty is an inseparable aspect of most practical cases where a decision must be made in a realistic environment. This should be the case when it comes to any decision-making paradigm for FRM. Unfortunately, and rather strangely, this is not the case. In fact, studies suggest that this seems to be the case for most MADM problems in the field of water resources planning and management (Zolghadr-Asli et al., 2021). Understanding these intricacies, and, in turn, implementing frameworks such as sensitivity analysis, Monte Carlo simulation, or fuzzy logic alongside MADM methods would ensure that the emerging solutions are more in line with the reality of the case.

4.6 CONCLUDING REMARKS

Natural hazards endanger many people's lives and properties worldwide. Flood is one of the most frequent natural hazards and has increased on an unprecedented scale due mainly to climate change and human activities. FRM aims to analyze, assess, and mitigate the risk of flooding. The measures normally includes structural and non-structural actions and the assessment criteria are taken from sustainability framework, which includes environmental, social, political, technical, and economic aspects. To manage this issue, various alternatives proposed by experts can be analyzed by MCDM methods to provide a vast range of viable and logical frameworks for decision-makers and help them to prioritize the alternatives and select the optimal scenario for FRM. The literature review showed that using and comparing different MCDM methods, including the widely used and recently proposed ones, could lead to the most appropriate alternatives being found. Most articles involved real-world case studies focused on multiple alternatives and showed the performance of these tools in the real world. The studies showed that using MCDM methods can be a useful decision-making tool along with other methods for simulation, assessment, and visualization, such as hydraulic and hydrological tools, and GIS. That said, in order to produce reliable and robust results, one should always address three pillars when it comes to applying MADM methods to FRM. First, if one opts for subjective-based MADM methods such as AHP, which seems to be trending in the literature, it is crucial to ensure the integrity of the subjective assessments that are utilized in these paradigms. Such opinions must be unbiased and consistent with the reality of the case. Having an objective-based MADM method on the side for comparison purposes could be a good idea for keeping things in check. Secondly, it could be more beneficial to implement group decision-making paradigms as such frameworks enable decision-makers to incorporate the different viewpoints of all the stakeholders and involved parties to the decision-making process. Lastly, one should always attempt to understand and address the uncertainty that surrounds most real-world problems. Implementing frameworks such as sensitivity analysis, Monte Carlo simulation, or fuzzy logic alongside the MADM methods could be a good practical way to ensure that the emerging solutions are more in line with the reality of the case.

REFERENCES

Abdullah, M. F., Siraj, S., and Hodgett, R. E. 2021. An overview of multi-criteria decision analysis (MCDA) application in managing water-related disaster events: analyzing 20 years of literature for flood and drought events. Water 13(10):1358.

Akay, H. and Baduna Koçyiğit, M. 2020. Flash flood potential prioritization of sub-basins in an ungauged basin in Turkey using traditional multi-criteria decision-making methods. Soft Computing 24(18):14251–14263.

Benayoun, R., Roy, B., and Sussman, B. 1966. ELECTRE: une méthode pour guider le choix en présence de points de vue multiples. Note de travail 49, SEMA-METRA International, Direction Scientifique, Paris, France.

Bernoulli, D. 1738. Specimen theoriae novae de mensura sortis. Comentarii Academiae Scientiarum Imperiales Petropolitanae 1738(5):175–192.

Bozorg-Haddad, O., Zolghadr-Asli, B., and Loaiciga, H. A. 2021. A Handbook on Multi-attribute Decision-making Methods. John Wiley & Sons, New York.

Brans, J. P. 1982. L'ingénierie de la decision. Elaboration d'instruments d'aide a la decision: Methode PROMETHEE. In: L'aide a la Decision: Nature, Instruments et Perspectives D'avenir (eds. R. Nadeau and M. Landry), pp. 183–214. Presses de Universite Laval, Québec, Canada.

Brans, J. P., Vincke, P., and Mareschal, B. 1986. How to select and how to rank projects: the PROMETHEE method. European Journal of Operational Research 24(2):228–238.

Chitsaz, N. and Banihabib, M. E. 2015. Comparison of different multi criteria decision-making models in prioritizing flood management alternatives. Water Resources Management 29(8):2503–2525.

Churchman, C. W. and Ackoff, R. L. 1954. An approximate measure of value. Journal of the Operations Research Society of America 2(2):172–187.

de Brito, M. M. and Evers, M. 2016. Multi-criteria decision-making for flood risk management: a survey of the current state of the art. Natural Hazards and Earth System Sciences 16(4):1019–1033.

Deng, J. 1989. Introduction to grey system theory. The Journal of Grey System 1(1):1–24.

El-Magd, S. A. A., Amer, R. A., and Embaby, A. 2020. Multi-criteria decision-making for the analysis of flash floods: a case study of Awlad Toq-Sherq, Southeast Sohag, Egypt. Journal of African Earth Sciences 162:103709.

Enayati, M., Bozorg-Haddad, O., Fallah-Mehdipour, E., Zolghadr-Asli, B., and Chu, X. (2021). A robust multiple-objective decision-making paradigm based on the water–energy–food security nexus under changing climate uncertainties. Scientific Reports 11:1–14.

EM-DAT. 2019. Natural Disasters 2018. The OFDA/CRED International Disaster Database. Universite Catholique de Louvain, Brussels.

Ferdowsi, A., Farzin, S., Mousavi, S. F., and Karami, H. 2019. Hybrid bat & particle swarm algorithm for optimization of labyrinth spillway based on half & quarter round crest shapes. Flow Measurement and Instrumentation 6:209–217.

Ferdowsi, A., Hoseini, S. M., Farzin, S., Faramarzpour, M., and Mousavi, S. F. 2020a. Shape optimization of gravity dams using a nature-inspired approach. Journal of Soft Computing in Civil Engineering 4(3):65–78.

Ferdowsi, A., Mousavi, S. F., Farzin, S., and Karami, H. 2020b. Optimization of dam's spillway design under climate change conditions. Journal of Hydroinformatics 22(4):916–936.

Ferdowsi, A., Singh, V. P., Ehteram, M., and Mirjalili, S. 2021a. Multi-objective optimization approaches for design, planning, and management of water resource systems. In: Essential Tools for Water Resources Analysis, Planning, and Management (ed. O. Bozorg-Haddad), pp. 275–303, Springer, Singapore.

Ferdowsi, A., Valikhan-Anaraki, M., Mousavi, S. F., Farzin, S., and Mirjalili, S. 2021b. Developing a model for multi-objective optimization of open channels and labyrinth weirs: theory and application in Isfahan Irrigation Networks. Flow Measurement and Instrumentation 80: 101971.

Ghaleno, M. R. D., Meshram, S. G., and Alvandi, E. 2020. Pragmatic approach for prioritization of flood and sedimentation hazard potential of watersheds. Soft Computing 24(20):15701–15714.

Ghanbarpour, M. R., Salimi, S., and Hipel, K. W. 2013. A comparative evaluation of flood mitigation alternatives using GIS-based river hydraulics modelling and multicriteria decision analysis. Journal of Flood Risk Management 6(4):319–331.

Hall, J. W., Meadowcroft, I. C., Sayers, P. B., and Bramley, M. E. 2003. Integrated flood risk management in England and Wales. Natural Hazard Review, ASCE 4(3):126–135.

Hansen, P. and Ombler, F. 2008. A new method for scoring additive multi-attribute value models using pairwise rankings of alternatives. Journal of Multi-Criteria Decision Analysis 15(3–4):87–107.

Hooijer, A., Klijn, F., Pedroli, B., and Van Os, A. 2004. Towards sustainable flood risk management in the Rhine and Meuse River basins: synopsis of the findings of IRMA-SPONGE. River Research and Applications 20:343–357.

Hwang, C. L. and Yoon, K. 1981. Methods for multiple attribute decision making. In: Multiple Attribute Decision making: Lecture Notes in Economics and Mathematical Systems (eds. C. L. Hwang and K. Yoon), pp. 58–191, Springer, Heidelberg.

Kafle, M. R. and Shakya, N. M. 2018. Multi-criteria decision-making approach for flood risk and sediment management in Koshi Alluvial Fan, Nepal. Journal of Water Resource and Protection 10:596–619.

Malekian, A. and Azarnivand, A. 2016. Application of integrated Shannon's entropy and VIKOR techniques in prioritization of flood risk in the Shemshak watershed, Iran. Water Resources Management 30(1):409–425.

Margeta, J. and Knezic, S. 2002. Selection of the flood management solution of karstic field: Vrgorsko polje case study. Water International 27(3):431–441.

Mladineo, N., Margeta, J., Brans, J. P., and Mareschal, B. 1987. Multicriteria ranking of alternative locations for small scale hydro plants. European Journal of Operational Research 31(2):215–222.

Opricovic, S. 1998. Multicriteria optimization of civil engineering systems. Ph.D. thesis. Faculty of Civil Engineering, Belgrade, Serbia.

Plate, E. J. 1999. Flood risk management: a strategy to cope with floods (eds. A. Bronstert, A. Ghazi, J. Hladny, Z. W. Kundzewicz, and L. Menzel). Proceedings of the European Meeting on the Oder Flood 1997. Ribamod concerted action, European Communities, Office for Official Publications of the European Communities, Luxemburg, pp. 115–128.

Plate, E. J. 2002. Flood risk and flood management. Journal of Hydrology 267(1–2):2–11.

Rezaei, J. 2015. Best–worst multi-criteria decision-making method. Omega 53:49–57.

Roy, B. 1978. ELECTRE III: un algorithme de classement fonde sur une representation floue des preferences en presence de criteres multiples. Cahiers du Centre d'Etudes de Recherche Operationnelle 20(1):3–24.

Roy, B. and Bertier, P. 1971. La methode ELECTRE II. In: Note de travail 142, SEMA-METRA. Direction Scientifique Metra International, Paris.

Roy, B. and Hugonnard, J. C. 1982. Ranking of suburban line extension projects on the Paris metro system by a multicriteria method. Transportation Research Part A: General 16 (4):301–312.

Saaty, T. L. 1977. A scaling method for priorities in hierarchical structures. Journal of Mathematical Psychology 15(3):234–281.

Saaty, T. L. 1980. The Analytic Hierarchy Process. McGraw-Hill Education, New York.

Saaty, T. L. 1996. Decision Making with Dependence and Feedback: The Analytic Network Process. RWS Publications, Pittsburgh, PA.

Saaty, T. L. 2004. Fundamentals of the analytic network process – dependence and feedback in decision-making with a single network. Journal of Systems Science and Systems Engineering 13(2):129–157.

Schanze, J. 2006. Flood risk management—a basic framework. In: Flood Risk Management: Hazards, Vulnerability and Mitigation Measures, pp. 1–20, Springer, Dordrecht.

Sörensen, J., Persson, A., Sternudd, C., Aspegren, H., Nilsson, J., Nordström, J., … and Mobini, S. 2016. Re-thinking urban flood management- Time for a regime shift. Water 8(8):332.

Sun, R., Gong, Z., Gao, G., and Shah, A. A. 2020. Comparative analysis of multi-criteria decision-making methods for flood disaster risk in the Yangtze River Delta. International Journal of Disaster Risk Reduction 51:101768.

Tan, L., Guo, J., Mohanarajah, S., and Zhou, K. 2021. Can we detect trends in natural disaster management with artificial intelligence? A review of modeling practices. Natural Hazards 107:2389–2417.

Ustinovichius, L., Zavadkas, E. K., and Podvezko, V. 2007. Application of a quantitative multiple criteria decision making (MCDM-1) approach to the analysis of investments in construction. Control and Cybernetics 36(1):251.

Vincke, J. P. and Brans, P. 1985. A preference ranking organization method: the PROMETHEE method for MCDM. Management Science 31(6):647–656.

Vojtek, M. and Vojteková, J. 2016. Flood hazard and flood risk assessment at the local spatial scale: a case study. Geomatics, Natural Hazards and Risk 7(6):1973–1992.

Xu, X. 2001. The SIR method: a superiority and inferiority ranking method for multiple criteria decision making. European Journal of Operational Research 131(3):587–602.

Yazdandoost, F., and Bozorgy, B. 2008. Flood risk management strategies using multi-criteria analysis. Proceedings of the Institution of Civil Engineers-Water Management 161(5):261–266.

Yazdani, M., Zarate, P., Zavadskas, E. K., and Turskis, Z. 2019. A Combined Compromise Solution (CoCoSo) method for multi-criteria decision-making problems. Management Decision 57(9):2501–2519.

Zagonari, F. and Rossi, C. 2013. A heterogeneous multi-criteria multi-expert decision-support system for scoring combinations of flood mitigation and recovery options. Environmental Modelling & Software 49:152–165.

Zolghadr-Asli, B., Bozorg-Haddad, O., Enayati, M., and Chu, X. 2021a. A review of 20-year applications of multi-attribute decision-making in environmental and water resources planning and management. Environment, Development and Sustainability 23:14379–14404.

Zolghadr-Asli, B., Bozorg-Haddad, O., and van Cauwenbergh, N. 2021b. Multi-attribute decision-making: a view of the world of decision-making. In Essential Tools for Water Resources Analysis, Planning, and Management, pp. 305–322, Springer, Singapore.

Zolghadr-Asli, B., Bozorg-Haddad, O., Enayati, M., and Goharian, E. 2021c. Developing a robust multi-attribute decision-making framework to evaluate performance of water system design and planning under climate change. Water Resources Management 35(1):279–298.

5 Maritime Supply Chain Resilience Applications in the COVID-19 Pandemic

Umur Bucak[1] *and Hakan Demirel*[2]
[1]Zonguldak Bulent Ecevit University
[2]Istanbul Technical University

CONTENTS

5.1 Risk Management in the Supply Chain 55
5.2 Supply Chain Resilience 56
5.3 Maritime Supply Chain Resilience 57
5.4 COVID-19 Strategies of Maritime Transport Companies 59
5.5 MCDM Applications in Supply Chain Disaster Management 64
5.6 Conclusion 65
References 66

5.1 RISK MANAGEMENT IN THE SUPPLY CHAIN

Risk has been defined by the British Standards Institution (BSI) as a "combination of probability or frequency of occurrence of a defined hazard and magnitude of the occurrence" (Gerber and von Solms, 2005). Risk concept has two components: potential losses and the likelihood of those losses. Thus, risk emerges as a result of an unpredicted event; in other words, unpredicted events lead to the existence of risks (Manuj and Mentzer, 2008: 196). Risk definitions most often refer to unexpected or unpredicted events that disrupt the system or organization. The handling of supply chain risks relies on descriptions of past disruptive events in the literature. Tummala and Schoenherr (2011: 474) conceptualized supply chain risks as unpredicted events that affect operations and result in failure to achieve the desired performance level in terms of responsiveness, service quality, cost, and so on. Supply chain risks are multidimensional and its sources are multifarious. Spekman and Davis (2004) revealed the dimensions of the sources of supply chain risks as the inbound supply; flow of information; financial flow; security of the information system; cooperation with stakeholders; and corporate social responsibility (Sodhi et al. 2012: 4). Supply chains have a large number of stakeholders that each of them carries on a great variety of

DOI: 10.1201/9781003212904-5

businesses. Accordingly, supply chain risks may arise from a wide variety of events. Risks throughout the supply chain can be associated with unacceptable products delivered from upstream and also environmental impacts such as natural disasters, or terrorist/pirate attacks (Tummala and Schoenherr, 2011: 474). In other words, supply chain risks may emerge from internal operations carried out by stakeholders through the chain and from uncontrollable external events that happen rarely but which have serious impacts. Supply chain risks may be categorized under two main topics: operational risks and disruption risks. Operational risks mainly stem from ordinary operations that cause failures in delivery, such as lead time and demand fluctuations. Disruption risks stem from extraordinary events that rarely happen but that have serious impacts (El Baz and Ruel, 2021). Moreover, the categorization may be extended to involve risks that stem from supply, demand, process, intellectual property, organizational behaviour, and political/social events (Tang and Tomlin, 2008; Sodhi et al., 2012: 4). As the stakeholders in the supply chain are connected to each other by a chain, the operations carried out are also interconnected. If an unpredictable or uncontrollable risk occurs here, the ripple effect will show itself more quickly and effectively than in other structures. For this reason, it should respond as quickly as possible to disruption that arises from the occurrence of the risks in supply chains. This may be possible with well-designed risk management.

5.2 SUPPLY CHAIN RESILIENCE

Supply chains are one of the most significant elements to constitute a network between stakeholders in international trade, cargo, or public transport. It is expected that the supply chain will provide appropriate goods and services just in time at the right position to the customers (Carvalho et al., 2012: 355). At the same time, supply chains face various challenges such as short product lifecycles, instability in demand, and changes in customer expectations; however, adaptation strategies make supply chains more complex, instable, and unpredictable (Kamalahmadi and Parast, 2016: 116). Apart from threats arising from the nature of the supply chain, supply chains are inherently at risk due to various unpredictable disruptions (Soni et al., 2014: 11). In the literature, globalization and outsourcing are seen as main contributors to disruptions in supply chains (Börjeson and Boström, 2018). Disruptions in the scope of the supply chain can be defined as situations that occur after or in the process of unexpected events, and affect the whole or some elements of the supply chain negatively. Therefore, to better understand the impacts of these disruptions on supply chains, disruptions should be defined in detail. Although Wong et al. (2020) demonstrated that supply chain resilience capability has a positive effect on such performance aspects as competitiveness, financial performance, customer service, risk performance, and market performance, according to de Sa et al. (2019), current supply chain resilience strategies for sale-based firms have been capable of maintaining operations when disruptive events occur but are inadequate in addressing future uncertainties. This may be due to the fact that the supply chain is too sensitive to dynamic processes in today's world. Undoubtedly, the maritime supply chain, which represents 90% of global trade, feels this high sensitivity the most.

5.3 MARITIME SUPPLY CHAIN RESILIENCE

Resilience strategies in supply chains should be built at the strategic level, but shipping lines engage with supply chain partners at the operational level rather than at tactical or strategic levels (Lam and Bai, 2016: 22). This situation makes ship operators, especially those involved in container shipping, more vulnerable against supply chain risks. Recently, the increasing interdependency between maritime actors, intense competition, and special requests of the shippers have made maritime supply chains more vulnerable to potential disruptions throughout the supply chain (Lam and Bai, 2016: 16). Vessel operators charter approximately half of their vessel fleet, and thus they cannot control the information technology (IT) systems of these chartered vessels. These circumstances cause them to be at risk of cyber-attacks, which can generate serious operational disruptions. On the side of cargo and its representatives, cargoes may pass through several countries, be moved through several different systems, and can also be controlled by various organizations such as ports, customs offices, trucking companies, banks, shared-service centers, and industry information portals. These organizations neither provide any IT infrastructure nor mutually agreed cyber security standard (Jensen, 2015: 36). A great variety of risks may be encountered in maritime supply chains. Therefore, each risk should be defined and its impact investigated.

When defining risks in the maritime supply chain, risk sources and risk factors should be separated. Risk sources particular to maritime transportation can be identified as undesirable weather conditions, failures in IT or telecommunication, loss of skilled labor, civil unrest, industrial controversy, fire, a cyber-attack, and so on. These sources may cause risk occurrence and disrupt the whole chain. However, risk factors are more specific and directly associated with the chain, and trigger risk occurrence. Many risk factors exist and threaten maritime supply chains. The most prominent factors may be ordered as port worker strikes, information system breakdown, ice conditions in the ocean, fire (Vilko and Hallikas, 2012), port congestion, equipment or vehicle breakdown, uncleanness, empty container unavailability, customs clearance (Gurning and Cahoon, 2011), terrorist attacks or piracy, and dissimulation on cargo information by shippers (Chang et al., 2014). The literature, in recent years, has ignored the pandemic factor, possibly because an epidemic affecting the whole world has not emerged for a long time. Lam and Bai (2016) categorized maritime supply chain risks as internal, external, and supply chain induced, and concluded that internal risks which arise from the organization are the most important risks in comparison to external and supply chain-induced risks. Additionally, after the 9/11 terrorist attack in New York, importance was attached to physical security in the maritime sector, especially for port sector, with the International Ship and Port Facility Security (ISPS) Code (IMO, 2015; Jensen, 2015: 35). This attack greatly influenced the relevant literature to produce security-oriented studies since the beginning of the 2000s. However, it should not be forgotten that the maritime sector has faced many commercial risks, especially after the 2008 financial crisis. The volume of global trade interruptions has increased since 2012, and a vast number of delayed shipments were reported by 26% of shippers (World Economic Forum, 2013; Yang and Hsu, 2018: 221).

Some of the key ports located in USA were shut down due to a labor strike in 2002. This shut down created financial disruption of US$1 or 2 billion per day, on average (Cohen, 2002; Jensen, 2015: 35). When the 2008 snow disaster occurred in China, ports located in Yangtze River Delta, which is one of the most significant hubs for the global supply chain, were shut down (Dui et al., 2021). After a 9.0 magnitude earthquake in March 2011 hit Japan, several tsunamis occurred. These tsunamis caused disruptions to and shut downs of Japanese ports, and it was concluded that Japanese ports handled 7% less cargo than the previous year (Chen et al., 2017: 282). The ports of Los Angeles, Long Beach, and Oakland on the US west coast were shut down for a while due to a stoppage of labor after negotiations between the International Longshore and Warehouse Union and the Pacific Maritime Association failed. This resulted in a loss of approximately US$2.5 billion (Chen et al., 2017: 283). The Asian coal market was badly affected by disruption to Indonesian ports due to the hurricane disaster in Australia in 2017 (Dui et al., 2021).

Nanyang Technological University and Cambridge University released a report in 2019, which revealed that if 15 particular ports located in five Asian countries become disabled because of a cyber-attack, economic losses to global trade may reach US$100 billion (Dui et al., 2021). The maritime supply chain has a great impact on the world trade economy and risks that can occur through the chain may lead to failures in global logistics services. When the maritime supply chain is disrupted, companies operate in a less optimal way and thus sailing distance and lead times become longer and round-trip frequency is reduced (Zavitsas et al., 2018: 118). For instance, in the first months of 2021, the Suez Canal—one of the most important bottlenecks of global shipping—was blocked for 6 days and 7 hours as a result of grounding of the mega-vessel *Ever Given*. Studies showed that every day the ship blocked this critical waterway, it prevented the flow of $US9.6 billion in world trade. Another study conducted by German insurer Allianz showed that this blockage caused annual global trade growth to be reduced by 0.2 to 0.4 percentage points (Russon, 2021). The increasing representation of maritime transport in global trade and its high dependence on the links of optimized maritime supply chains demonstrates the importance of resilience strategies to weaken vulnerabilities.

From a security perspective there are a number of initiatives for maritime companies, including Authorized Economic Operator (AEO); Customs-trade Partnership Against Terrorism (C-TPAT); the 24-Hour Advance Cargo Manifest Declaration Rule (24-h rule); Container Security Initiative (CSI; Yang and Hsu, 2018: 222); the International Ship and Port Facility Security (ISPS) code; the International Convention for the Safety of Life at Sea (SOLAS); and the revised Seafarers' Identity Documents (SID) Convention of 2003 (Zavitsas et al., 2018: 116). These initiatives are regulatory-level resilience strategies. Firms should adapt these regulations, but they can also develop their own resilience strategies. In their study, Yang and Hsu (2018) found that the security-management practices of maritime firms have a significant relationship with the firms' resilience capabilities. Mansouri et al. (2009) offered some resilience strategies for maritime transportation system, such as a track-and-trace service for containers, the construction of extra facilities, or holding additional equipment in ports (as a redundancy strategy). However, the cost-effectiveness of these strategies should be identified as resilience strategies for maritime transport systems are costly. In the study of Lam and Bai (2016) it was revealed that resilience strategies such as contingency planning, monitoring and maintenance, and supply chain relationship management rise to prominence in recovery and in better sustaining operations in the maritime supply chain. Recent developments in IT (i.e., increasing digitalization) have removed physical borders, especially for global trade. This new borderless era has encouraged Asian countries to cooperate with each other in terms of economic and trade relations. Therefore, the resilience strategies offered above can be coordinated, controlled, or tracked with the help of state-of-the-art technologies related to data exchange. Finally, Schröder-Hinrichs et al. (2016) focused on maritime resilience while sailing in the scope of safety concept. They pointed out the human factor, in addition to the necessity for technological developments in ship design. Although digital processes have entered our lives more rapidly, especially with the COVID-19 pandemic, the human factor is still very important in business processes. Therefore, when a disruption occurs in the supply chain, a skilled workforce is very important in executing resilience strategies. But what if you can't use the people in the organizational design for a while, or if the people in your team where you carry out your resilience strategy have to constantly change? Or what if your partners along the supply chain can't find anyone to carry cargo and yet the demand for your operations is higher than it has ever been in recent years?

5.4 COVID-19 STRATEGIES OF MARITIME TRANSPORT COMPANIES

The maritime industry has already faced challenges in recent years such as trade arrangements counter to the US–China trade war, green shipping and green port applications against International Maritime Organization (IMO) restrictions on sulfur usage, and shipping fleet deployment. COVID-19 added some challenges on top of these (Yazır et al., 2020: 253). The maritime industry faced many challenges in the COVID-19 pandemic process such as the management of disease and close contacts on board, the use of individual protective equipment on board, interaction with personnel on the shore side, crew changes, testing, accessing medical treatment overseas, and protecting mental health (Stannard, 2020). The COVID-19 pandemic has been one of the toughest challenges since World War II as people were faced with losing jobs and income (World Bank, 2021; Narasimha et al., 2021: 192). COVID-19 lockdowns caused a decrease in global gross domestic product (GDP) of 4% and the loss of more than US$5 trillion worth of output in the first 6 months of the pandemic (Maliszewska et al., 2020; Thunstrom et al., 2020; Kuhn et al., 2021: 195). The global economy and trade had never experienced that kind of disruption before and therefore nobody knew what should be done. Stock markets reacted quickly to news related to the COVID-19 pandemic, which included both negative and positive events (Kamal et al., 2021). The maritime industry has also been directly affected by market reactions due to its role in global trade. Some of the main problems in the maritime industry due to the pandemic have been as follows: border restrictions and port closures; decrease in demand for freight and, accordingly, bankruptcy risk; conflict in laytime calculation; and disagreements between ship and cargo owners on lead times and resourcing, and so on. (Kumar and Jolly, 2021: 82). While the measures taken disrupted the supply–demand equilibrium, the supply–demand disequilibrium made global trade, and therefore the maritime sector, volatile. Global and especially local transportation volume were reduced due to COVID-19, causing a decrease in oil consumption and energy utilization (Shehzad et al., 2020; Cengiz and Turan, 2021: 46). In April 2020, the oil price had dropped below zero for the first time as a result of delivery contract expiries and limited storage capacity (Millefiori et al., 2021). One of the main reasons why fuel prices fell at this level was that factories around the world were closed or were operating at half capacity. This situation also reduced the demand for raw materials and directly affected the dry bulk shipping market. The Baltic Dry Index fell to 275 points in May 2020, the lowest level in the previous four years and decreased by 89.1% in comparison to September 2019 (Kamal et al., 2021: 1).

In the early days of the COVID-19 outbreak, sharp decreases were observed in the container shipping and cruise ship industries (Öztürk and Turan, 2020). As a result of a questionnaire conducted by an economist in World Ocean Initiative webinars, the maritime sectors most impacted by COVID-19 were ranked as follows: tourism (70.7%); fisheries (10.4%); offshore oil and gas (7.2%); and shipping (6.2%). Millefiori et al. (2021) revealed that the mobility of all types of vessels fell due to pandemic in the first six months of 2020. In particular, the mobility volumes of container vessels, of liquid bulk carriers, and of cruise ships were measured to be quite lower in comparison to forecasted data based on accumulated data. The COVID-19 pandemic caused great congestion in Chinese ports especially; however, according to Alphaliner, 33 sailings in January 2020 were cancelled in Asia/Northern Europe and 17 sailings more were added in February 2020 (Koyuncu et al., 2021). Therefore, the COVID-19 pandemic has not only affected the shipping supply chain, but has also deeply affected hinterland logistics, and in this way, it restricted maritime transport. The congestion status of ports around the world in July 2021 is shown by the number of ships waiting in anchorage. In this period 116 ports reported that they were disrupted. Vendors did not have the ability to reach goods because of labor-saving applications in ports and warehouses as a result of lockdown measures and restricting movement strategies to restrain the spread of the disease (Menhat et al., 2021). However, the main reason for this congestion is not related to shipping activities or port restrictions, labor shortages, and so on. Although the US president at the time called for ports to work 24/7, port congestion has continued up to the time of writing. The main reason for this is not related to the working principles of ports. Truck driver shortages (especially in the USA as a result of the low vaccination rate (57.8%) and in England as a result of Brexit), labor shortages in distribution centers and factories, and the energy crisis in China contributed more to port congestions. Therefore, strategies should be developed to cover the entire supply chain to overcome these problems.

The picture is not so pessimistic. In order to prevent the spread of the virus, measures that can be applied in all sectors have been taken worldwide. These measures were primarily social, not medical, and had implications for the global economy. Verschuur et al. (2021) tried to measure the effects of non-pharmacologic interventions (NPIs) on global trade. They found that NPIs such as closing schools, shutting workplaces, restricting public gatherings, and shutting down public transport have negative impact on global trade. However, positive effects of NPIs such as stay-at-home requirements, internal movement restrictions, and international travel restrictions on global trade were detected. With COVID-19 restrictions, global demand was reshaped. The demand for some products was increased, for instance electronic equipment, mobile phones, exercise equipment, home-improvement and gardening materials, and medical protective equipment such as surgical masks, gowns, disinfectants, and so on, which are mostly produced by China (Cullinane and Haralambides, 2021: 370). As a result of changing purchasing habits, demand tended to stabilize, especially in the second half of 2020. This situation emerged as an opportunity for maritime transport companies to be managed correctly with the right strategies, within certain precaution-based limitations. Some implementations such as teleworking, flexible working hours, investment postponement, and increasing marketing efforts came to the forefront as resilience strategies for maritime companies against the COVID-19 pandemic (Cengiz and Turan, 2021: 56). In the first months of the pandemic, a decline in the number of port calls was seen as a result of blank sailing applications, which can be described as the cancelation of whole liner services in particular routes or the bypassing of particular ports en route by liner shipping companies. Blank sailings comprised 20% of total sailings (Humphreys, 2020). Eventually, this blank sailing strategy not only provided an important opportunity to optimize routes, but also offered important opportunities to maritime transport companies in terms of demand balancing. Liner (container) shipping companies adjusted supplies in accordance with demand in the second half of 2020 with the help of blank sailings, that is the withdrawal of shipping capacity (Cullinane and Haralambides, 2021: 371). Liner shipping companies implemented additional measures against disruptive impacts of the COVID-19 pandemic on demand such as a slow-steaming strategy and reorganizing routes, preferring the Cape of Good Hope rather than the Suez Canal, for example. In May 2020, transit ship numbers through the Suez Canal declined by 32% (Cullinane and Haralambides, 2021: 371). In this way, an alternative solution was found against the supply–demand imbalance and operating costs were significantly reduced. Tatenhove (2021) proposed transformation strategies such as shortening maritime supply chains in ways that there is less dependence on "just in time" and lean inventory models, and increasing investment in digitalization and automation. In fact, in this "new normal", reliable access applications came to the fore rather than timely access. These strategies and the mobilization in demand suddenly cleared the atmosphere. It is accepted that the COVID-19 pandemic played a major role in the 4.1% decrease in world maritime transport volume in 2020. However, it was forecasted that maritime transport would recover and acquire a 4.8% increase in 2021 (UNCTAD, 2020: 3). This sudden recovery in maritime transport required ports to develop strategies to organize their business processes more efficiently, with fewer people. Mankowska et al. (2021) revealed the response strategies of selected Polish port terminal operators against COVID-19 disruptions. These strategies were grouped under two main headings: proactive and adaptive. In this study, response strategies of terminal operators, such as changing the supply chain and searching for a new one, were handled as proactive strategy. Strategies such as changing storage time, which facilitated adjusting the services on offer, and changing flow directions were evaluated under the heading of adaptive strategies. Shipyards, which intensively served ship operators in a hurry to equip scrubbers due to IMO 2020 restrictions, also known as a 'sulfur cap', just before the pandemic, started to receive new ship orders with the sudden recovery after the second quarter of 2020. With this intensity and the restrictions imposed by COVID-19, shipyards had to do their jobs more efficiently. Praharsi et al. (2021) proposed technology utilization applications such as ship-launching rails, automatic wood-cutting machines, tracker machines, wood-drying machines, cranes, and wood-bending machines to mitigate COVID-19 impacts on the shipbuilding industry.

Like many things that have changed with COVID-19, the purchasing habits of people and the way companies do business have also changed. Working with few people due to various restrictions and quarantine processes has led companies to be more productive. Physical constraints and mobility impossibilities also transformed the way companies do business. The COVID-19 pandemic forced the international business community to quickly arrange online meetings and implement digital transactions. The pandemic brought such digital processes to the maritime industry as ship-to-shore electronic information exchange; electronic bills of lading; general expansion of distributed ledger technology; and so on. However, these digitization applications also brought about threats, including cyber-attacks (Piñeiro et al., 2021: 129–130). New technology utilization increased in the maritime industry in reaction to COVID-19 restrictions to provide resilience, including track-and-trace services, e-document applications, and block chain technologies (Yam, 2021; Kuhn et al., 2021: 196). However, the maritime industry should increase its resilience against the increased cyber risks concomitant with accelerating remote transactions as a result of technological integration. The resilience strategies of maritime companies were thus transferred to digital processes. In addition, in this process, the necessity of some structural solutions in terms of providing resilience in physical work has been revealed. The COVID-19 pandemic re-emphasized risk management and resilience applications in ports as one of the most important approaches to port management. Port authorities were faced with disruptions that necessitated providing organizational resilience, leanness, and agility. Accordingly, port authorities should progress their resilience and adaptive planning, and enhance their adaptive capacity, while challenging economic shocks, market trends, and climate change restrictions (Notteboom and Haralambides, 2020: 344). As a result of COVID-19 outbreaks, ports were taught that they should enhance cooperation and coordination with stakeholders through the supply chain, accelerate their digitization transformation, facilitate trade and operations, and utilize and manage their capacities effectively (Piñeiro et al., 2021: 129–132).

Global container throughput was fell in 2020 for the first time since 1970 by 8.47%; freight and charter rates were decreased by 4.76%; and nearly 6% of the fleet was laid up as a result of the US–China trade war and COVID-19. However, global container throughput was forecasted to rise from –8.47% in 2020 to +7.44% in 2021. The main factors of this rapid recovery include stock-building by the industrial sectors and the rapid economic recovery of the countries which drive global economies (Yazır et al., 2020: 260). In 2020, high profit rates were only seen in liner shipping companies. The world leader in terms of fleet capacity among the liner shipping companies, Maersk Line, reported record profits in Q3 and Q4 of 2020 and Q1 of 2021 (Baker, 2021; Cullinane and Haralambides, 2021: 370). The total profitability of companies in the liner shipping market is expected to exceed US$100 billion for 2021.

The primary reason of this profitable circumstance is that shippers should pay freight rates ranging between 5 and 10 times more than pre-COVID terms (the Baltic Dry Index also reached 5650 points in October 2021). Notwithstanding, shippers had to wait for weeks to secure bookings on ships or to find containers to ship their orders from Asia (Attinasi et al., 2021; Cullinane and Haralambides, 2021: 370). This congestion in the maritime supply chain also reduced the profitability of the companies. In particular, the container shipping market has been turned into an auction room in this congested environment due to the transportation requirements of the products contained in the market. Nevertheless, it should be emphasized that it is not possible to talk about profitability based only on revenues. Figure 5.1 shows us that liner shipping companies have profitability beyond its revenues.

During the COVID-19 pandemic period, liner shipping companies proved that they had learned a lot from the 2008 global financial crisis. In fact, freight charges, which have been gradually reducing since the 2008 financial crisis, with various scale-economy strategies such as horizontal and vertical integrations and operational mergers (shipping alliances), were put on the attack in this period. Various restrictions during this period were considered as an opportunity by liner shipping companies. For example, due to port restrictions, blank sailing practices were used and operational costs were reduced. To comply with quarantine practices and travel restrictions, digital transformation, work-from-home (teleworking) strategies have reduced costs. In fact, thanks to these applications, senior company employees who can travel for a day for a one-hour meeting have been able to manage their time more effectively by connecting to meetings remotely. Thus, more time could be spent on acquiring customers and generating competitive strategies for liner shipping companies. The problem of finding empty containers emerged from the imbalance in supply and demand; the problem of port congestion was managed very well by the companies and the problems were turned into opportunities. Especially with the rapid increase in production and vaccination rate, the expectation of transition to

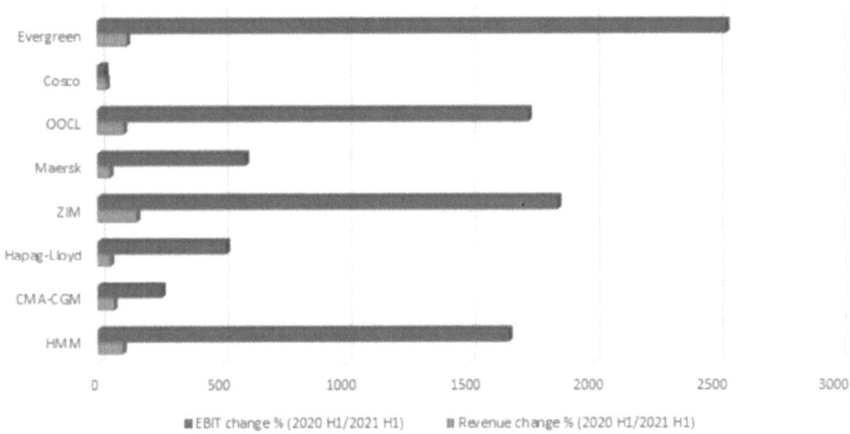

FIGURE 5.1 Operating profits and revenues of liner shipping companies in H1 of 2021.

the new normal has increased the rush to grow products for Christmas shopping. In fact, some carriers' prior container applications during this period turned the market into an open bidding one. On the one hand, the rush of early product supply for Christmas shopping; on the other hand, long-term contracts with some shippers brought about a lack of empty capacity on the container ships. This lack of capacity caused the world's leading manufacturers and retailers such as Coca Cola, IKEA, and Home Depot to buy ships to carry their own cargo (Rahman, 2021). Therefore, many supply chain disruptions experienced during this period, such as the energy crisis in China, the problem of finding containers, the shortage of capacity on ships, the lack of truck drivers, and port congestion, were used as bargaining power for shipping companies.

5.5 MCDM APPLICATIONS IN SUPPLY CHAIN DISASTER MANAGEMENT

Several studies have been published related to supply chain disaster management. However, the use of multi-criteria decision-making (MCDM) methods in this field has intensified in recent years. Yan et al. (2017) investigated inland waterway congestions in the Yangtze River with the help of a hybrid cost–benefit ratio and the fuzzy TOPSIS (Technique for Order of Preference by Similarity to Ideal Solution) method. Al Yami et al. (2017) evaluated risk control options in container terminals and tried to determine appropriate safety strategies with a hybrid fuzzy AHP (analytic hierarchy process)–TOPSIS method. Ding and Chou (2012) employed a MCDM model to identify investment risks in location selection for container terminals. Khorram and Khalegh (2020) investigated the case of Shahid Rajaee Port dredging projects, which caused damage, and tried to develop a risk-analysis method with a FSA (formal safety assessment)-based approach that integrated FST (fuzzy set theory), GOWPA (generalized ordered weighted proportional averaging), and DEMATEL (decision-making trial and evaluation laboratory) techniques. Shen et al. (2021) developed a system based on the selection of hidden risks of target ships during port state controls with the help of a hybrid fuzzy importance performance analysis (IPA)–TOPSIS method. Kharisma and Ardi (2020) evaluated the supply risks of generic medicines by employing DEMATEL-based ANP (analytic network process). Rodbundith and Sopadang (2021) evaluated factors that affected the performance of air cargo terminals during COVID-19 by the best–worst method. Especially in global trade, maritime transportation has a unique role in the supply chain and when risks occur in shipping, sea supply chains become paralyzed. Therefore, some studies have focused on risks that may occur in maritime transportation and ships. Ding (2009) tried to determine the appropriate strategic partner for risk management for container shipping companies by using a fuzzy MCDM algorithm. Başhan and Demirel (2018, 2019) evaluated operational faults in marine diesel generators and marine boilers under sail, respectively. Başhan et al. (2020) assessed risks in ship engine rooms.

Grida et al. (2020) determined the effects of prevention strategies against the spread of the COVID-19 virus on aspects of the supply chain with the help of a neutrosophic TOPSIS method. They revealed that the demand aspect of the supply chain was more prominently affected by prevention strategies than supply and logistics aspects. Jamshidiantehrani et al. (2020) described the factors related to the supply chain agility of pharmaceutical companies and prioritized them with the help of MCDM methods. Paul et al. (2021) proposed that studies related to the sustainability of supply chains may wish to focus on the COVID-19 period. Additionally, they suggested that these studies could employ various MCDM methods to determine the impact levels of a pandemic and to evaluate different aspects of these impacts on sustainable supply chain practices. Sharma et al. (2020) investigated agriculture supply chain risks during the COVID-19 pandemic by using fuzzy linguistic quantifier order weighted aggregation (FLQ-OWA). They revealed that strategies such as the adoption of industry 4.0 technologies, supply chain collaboration, and shared responsibility, will provide resilience in the sustainable agriculture supply chain.

5.6 CONCLUSION

Wieland and Durach (2021) defined the COVID-19 pandemic as a 'black swan' event for supply chains. The pandemic was a source of disruption that the globalizing world had experienced for the first time. For this reason, it has had unpredictable results for countries, decision-making bodies, and companies. In fact, the world was not ready for such a rapid recovery, even though it experienced similar deteriorations during the 2008 global economic crisis. While various measures taken to prevent the spread of the virus had economic consequences for companies, they also changed their business habits as a result. Since the third quarter of 2020, in particular, the rapid recovery in China and the accordingly high volumes of production at the expense of depleting all stocks made China almost the only production center of the world. For this reason, the one-way flow, especially in the supply of semi-finished products and end products, disrupted the supply–demand balance and created the empty container crisis. This axis dislocation in trade has caused maritime transport to reshape the supply chain. In this process, strategies such as blank sailing and directing containers to China came to the fore. Then, the shortage of truck drivers, which started due to quarantine processes and sharpened by Brexit at the beginning of 2021, turned important hub ports into container graveyards. All these developments pushed shippers, who want to deliver their cargoes to the consignees, to find spaces and empty containers on ships, perhaps even earlier than they should have due to unpredictable supply times. Thus, freights in the container shipping market reached the highest levels seen since the 2008 global economic crisis, and freight rates in the container shipping market reached their highest levels since the 2008 global economic crisis. At this point, a situation emerged that should be of particular concern to economic theorists. The Baltic Dry Index, previously considered to be the trigger of the development of world trade, increased by 1954.5% in about a year. This indicator, which shows the average freight value of dry bulk ships, is seen as a trigger for production in the world, since the cargoes carried on dry bulk ships are generally raw materials. However, this time, the increase in the Baltic Dry Index followed the increase in the container freight rate, as China continued producing by depleting its war stocks from the third quarter of 2020. In fact, in this period, this increase was also triggered by the fact that containerized products (with a cost advantage in previous years) changed to dry cargo ships for the same reason.

Shipping companies, especially liner shipping companies, turned each of the crises that emerged in the COVID-19 era into an opportunity. Blank sailings applied in the first months of the pandemic reduced costs. This cost reduction was seen as a key to optimizing routes and was applied from time to time throughout the pandemic. Due to the high demand for the end products from the third quarter of 2020, the increased freight rates have whipped the profitability in the market, which was already cost-effective with the help of blank sailing applications. In this period, ship-operating companies were highly effective in the context of time management with the help of digital integration in its business processes. Although business development strategies such as prior container applications in 2021 deepened the problem of port congestion, they reinforced the increase in profitability. In fact, thanks to these port congestions, shippers lined up months in advance to book their products on ships, especially for Christmas shopping. This intense demand turned the market into an open bidding market, and freight rates remained at their highest levels since the 2008 economic crisis. The lack of empty container and empty ship capacity, and the resulting increased freight rates, continued until the end of the 2022 Chinese New Year. The demand situation following the Chinese New Year will determine how the market will behave. However, in this period, the maritime transport market, which uses the acceleration of digitalization in the context of time management, will develop various strategies and take a position according to demand. However, in this period, companies that entered this market by investing in new ships, taking advantage of the increase in freight rates, and that have not experienced a deep crisis such as the 2008 economic crisis, may be waiting for a troubled period.

In this chapter, it was revealed that maritime transport companies were considerably profitable during the COVID-19 pandemic. The main purpose of this research was to present the resilience strategies of maritime transport companies during the COVID-19 pandemic, which was a disaster experienced by global trade for the first time. The strategies of the maritime transport companies against the negative effects of COVID-19 that resulted in this profitability were demonstrated. These strategies, such as blank sailings, prior container applications, cost reduction as a result of slow steaming and teleworking, and time management based on digital integration increased the resiliency of the market against COVID-19 disruptions. It was proposed that these strategies may be investigated with MCDM methods and they can be prioritized to select the most effective one for the profitability or resiliency of the maritime companies in the COVID-19 period. Each of these strategies can also be handled as a criterion and, thanks to MCDM methods, it can be concluded by embedding these criteria in the model which of the alternatives, such as customer satisfaction, profitability, resilience, and cost-effectiveness, are more prominent.

REFERENCES

Al Yami, H., Yang, Z., Riahi, R., Bonsall, S., Wang, J., Wan, C., and Z. Qu. 2017. Analytical strategic safety management in container ports. In *Proceedings of the 4th International Conference on Transportation Information and Safety (ICTIS), Banff, Alberta, 2017*, 184–191.

Attinasi, M.G., Bobasu, A., and R. Gerinovics. 2021. What is driving the recent surge in shipping costs? ECB Economic Bulletin, March. https://www.ecb.europa.eu/pub/economic-bulletin/focus/2021/html/ecb.ebbox 202103_01~8ecbf2b17c.en.html (accessed September 9, 2021).

Baker, J. 2021. Maersk posts record frst-quarter earnings. Lloyd's List, 5 May. https://lloydslist.maritimeintelligence.informa.com/LL1136670/Maersk-posts-record-frst-quarter-earnings (accessed October 3).

Başhan, V. and H. Demirel. 2019. Application of fuzzy DEMATEL technique to assess most common critical operational faults of marine boilers. *Politeknik Dergisi*, 22(3): 545–555.

Başhan, V. and H. Demirel. 2018. Evaluation of critical operational faults of marine diesel generator engines by using DEMATEL method. *Journal of ETA Maritime Science*, 6(2): 119–128.

Başhan, V., Demirel, H., and M. Gul. 2020. A novel risk evaluation approach for frequently encountered risks in ship engine rooms. *Brodogradnja: Teorija I Praksa Brodogradnje I Pomorske Tehnike*, 71(2): 31–54.

Börjeson, N. and M. Boström. 2018. Towards reflexive responsibility in a textile supply chain. *Business Strategy and the Environment*, 27(2): 230–239.

Carvalho, H., Cruz-Machado, V., and J. G. Tavares. 2012. A mapping framework for assessing supply chain resilience. *International Journal of Logistics Systems and Management*, 12(3): 354–373.

Cengiz, H. and E. Turan. 2021. Business impact of Covid-19 pandemic on global maritime industry. *Journal of Naval Sciences and Engineering*, 17(1): 43–75.

Chang, C. H., Xu, J., and D. P. Song. 2014. An analysis of safety and security risks in container shipping operations: a case study of Taiwan. *Safety Science*, 63: 168–178.

Chen, H., Cullinane, K., and N. Liu. 2017. Developing a model for measuring the resilience of a port hinterland container transportation network. *Transportation Research Part E: Logistics and Transportation Review*, 97: 282–301.

Cullinane, K. and H. Haralambides. 2021. Global trends in maritime and port economics: the COVID-19 pandemic and beyond. *Maritime Economics & Logistics*, 23: 369–380.

de Sá, M. M., de Souza Miguel, P. L., de Brito, R. P., and S. C. F. Pereira. 2019. Supply chain resilience: the whole is not the sum of the parts. *International Journal of Operations & Production Management*, 40(1): 92–115.

Ding, J. F. 2009. Fuzzy MCDM approach for selecting strategic partner: an empirical study of a container shipping company in Taiwan. *International Journal of Innovative Computing, Information and Control*, 5(4): 1055–1068.

Ding, J. F. and C. C. Chou. 2012. A fuzzy MCDM model to evaluate investment risk of location selection for container terminals. *WSEAS Transactions on Information Science and Applications*, 9(10): 295–304.

Dui, H., Zheng, X., and S. Wu. 2021. Resilience analysis of maritime transportation systems based on importance measures. *Reliability Engineering & System Safety*, 209: 107461.

El Baz, J. and S. Ruel. 2021. Can supply chain risk management practices mitigate the disruption impacts on supply chains' resilience and robustness? Evidence from an empirical survey in a COVID-19 outbreak era. *International Journal of Production Economics*, 233: 107972.

Gerber, M. and R. von Solms. 2005. Management of risk in the information age. *Computers & Security*, 24: 16–30.

Grida, M., Mohamed, R., and A. N. H. Zaied. 2020. Evaluate the impact of COVID-19 prevention policies on supply chain aspects under uncertainty. *Transportation Research Interdisciplinary Perspectives*, 8: 1–8.

Gurning, S. and S. Cahoon. 2011. Analysis of multi-mitigation scenarios on maritime disruptions. *Maritime Policy & Management*, 38(3): 251–268.

Humphreys, R. M., Dumitrescu, A., Biju, N. O., and Y. Y. Lam. 2020. COVID-19 and the maritime and logistics sector in Africa. Working paper, World Bank Group. https://openknowledge.worldbank.org/handle/10986/34080 (accessed August 24, 2021).

IMO. 2015. Frequently asked questions on maritime security. Working paper, International Maritime Organization. http://www.imo.org/OurWork/Security/Guide_to_Maritime_Security/Pages/FAQ.aspx (accessed September 1, 2021).

Jamshidiantehrani, M., Ahmadzadeh, A., Rahimisadr, M., and M. Abdolmohammadi. 2020. Identifying and prioritizing the factors affecting the agility of the supply chain of pharmaceutical company using multi-criteria decision-making methods in COVID-19 pandemic. *Systematic Reviews in Pharmacy*, 11(7): 371–381.

Jensen, L. 2015. Challenges in maritime cyber-resilience. *Technology Innovation Management Review*, 5(4): 35–39.

Kamal, M. R., Chowdhury, M. A. F., and M. M. Hosain. 2021. Stock market reactions of maritime shipping industry in the time of COVID-19 pandemic crisis: an empirical investigation. *Maritime Policy & Management*, 1–16. doi.org/10.1080/03088839.2021.1954255.

Kamalahmadi, M. and M. M. Parast. 2016. A review of the literature on the principles of enterprise and supply chain resilience: major findings and directions for future research. *International Journal of Production Economics*, 171: 116–133.

Kharisma, S. A. and R. Ardi. 2020. Supply chain risk assessment of generic medicine in Indonesia using DEMATEL-based ANP (DANP). In *Proceedings of the 2020 IEEE International Conference on Industrial Engineering and Engineering Management (IEEM), Singapore, December 14–17,* 716–720.

Khorram, S., and M. A. Khalegh. 2020. A novel hybrid MCDM approach to evaluate ports' dredging project criteria based on intuitionistic fuzzy DEMATEL and GOWPA. *WMU Journal of Maritime Affairs*, 19(1): 95–124.

Koyuncu, K., Tavacioğlu, L., Gökmen, N., and U. Ç. Arican. 2021. Forecasting COVID-19 impact on RWI/ISL container throughput index by using SARIMA models. *Maritime Policy & Management,* 48(8): 1096–1108.

Kuhn, K., Bicakci, S., and S. A. Shaikh. 2021. COVID-19 digitization in maritime: understanding cyber risks. *WMU Journal of Maritime Affairs*, 20: 1–22.

Kumar, S. and A. Jolly. 2021. Consequences of COVID-19 pandemic on global maritime trade industry. *International Maritime Health*, 72(1): 82–83.

Lam, J. S. L. and X. Bai. 2016. A quality function deployment approach to improve maritime supply chain resilience. *Transportation Research Part E: Logistics and Transportation Review*, 92: 16–27.

Maliszewska, M., Mattoo, A., and D. Van Der Mensbrugghe. 2020. The potential impact of COVID-19 on GDP and trade: A preliminary assessment. World Bank Policy Research, 9211.

Mańkowska, M., Pluciński, M., Kotowska, I., and L. Filina-Dawidowicz. 2021. Seaports during the COVID-19 pandemic: the terminal operators' tactical responses to disruptions in maritime supply chains. *Energies*, 14(14): 4339. https://doi.org/10.3390/en14144339.

Mansouri, M., Mostashari, A., and R. Nilchiani. 2009. A decision analysis framework for resilience strategies in maritime systems. In *Proceeding of the Incose International Symposium, Singapore, July 19–23*, 1406–1427.

Manuj, I. and J. T. Mentzer. 2008. Global supply chain risk management strategies. *International Journal of Physical Distribution & Logistics Management*, 38(3): 192–223.

Menhat, M. N., Zaideen, I. M. M., Yusuf, Y., Salleh, N. H. M., Zamri, M. A., and J. Jeevan. 2021. The impact of Covid-19 pandemic: a review on maritime sectors in Malaysia. *Ocean & Coastal Management*, 209. doi.org/10.1016/j.ocecoaman.2021.105638.

Millefiori, L. M., Braca, P., Zissis, D., Spiliopoulos, G., Marano, S., Willett, P. K., and S. Carniel. 2021. COVID-19 impact on global maritime mobility. *Scientific Reports*, 11(1): 1–16.

Narasimha, P. T., Jena, P. R., and R. Majhi. 2021. Impact of COVID-19 on the Indian seaport transportation and maritime supply chain. *Transport Policy*, 110: 191–203.

Notteboom, T. E. and Haralambides, H. E. 2020. Port management and governance in a post-COVID-19 era: quo vadis? *Maritime Economics & Logistics*, 22(3): 329–352.

Öztürk, O. and Turan, E. 2020. Examination of the impact of the Covid-19 outbreak on sea freight transportation. In *Proceeding of the 2nd International Symposium of Engineering Applications on Civil Engineering and Earth Sciences (IEACES), Karabuk, September 23–25*, 40–47.

Paul, A., Shukla, N., Paul, S. K., and A. Trianni. 2021. Sustainable supply chain management and multi-criteria decision-making methods: a systematic review. *Sustainability*, 13(13): 7104. doi.org/10.3390/su13137104.

Piñeiro, L. C., Mejia, M. Q., and F. Ballini. 2021. Beyond COVID-19: the future of maritime transport. *WMU Journal of Maritime Affairs*, 20: 127–133.

Praharsi, Y., Jami'in, M. A., Suhardjito, G., and H. M. Wee. 2021. The application of Lean Six Sigma and supply chain resilience in maritime industry during the era of COVID-19. *International Journal of Lean Six Sigma*, 12(4): 800–834.

Rahman, R. 2021. The creative ways retailers are trying to sidestep shipping delays, from Home Depot flying in power tools to Coca-Cola chartering bulk vessels used for coal. Business Insider, October 8. www.businessinsider.com/creative-ways-retailers-sidestep-shipping-delays-port-backup-2021-9 (accessed October 8, 2021).

Rodbundith, T. S. and A. Sopadang. 2021. Evaluation of factors affecting air cargo terminal operation performance during COVID-19. In *Proceeding of the 11th Annual International Conference on Industrial Engineering and Operations Management, IEOM, Singapore, March 7–11*, 6687–6689.

Russon, M-A. 2021. The cost of the Suez Canal blockage. BBC News, March 29. www.bbc.com/news/business-56559073 (accessed August 24, 2021).

Schröder-Hinrichs, J. U., Praetorius, G., Graziano, A., Kataria, A., and M. Baldauf. 2016. Introducing the concept of resilience into maritime safety. In *Proceeding of the 6th Symposium on Resilience Engineering, Lisbon, Portugal, June 22–25*, 176–182.

Sharma, R., Shishodia, A., Kamble, S., Gunasekaran, A., and A. Belhadi. 2020. Agriculture supply chain risks and COVID-19: mitigation strategies and implications for the practitioners. *International Journal of Logistics Research and Applications*. doi.org/10.1080/13675567.2020.1830049.

Shehzad, K., Sarfraz, M., and S. G. M. Shah. 2020. The impact of COVID-19 as a necessary evil on air pollution in India during the lockdown. *Environmental Pollution*, 266(1): 115080. doi.org/10.1016/j.envpol.2020.115080.

Shen, J. H., Liu, C. P., Chang, K. Y., and Y. W. Chen. 2021. Ship deficiency data of port state control to identify hidden risk of target ship. *Journal of Marine Science and Engineering*, 9(10): 1120. doi.org/10.3390/jmse9101120.

Sodhi, M. S., Son, B. G., and C. S. Tang. 2012. Researchers' perspectives on supply chain risk management. *Production and operations management*, 21(1): 1–13.

Soni, U., Jain, V., and S. Kumar. 2014. Measuring supply chain resilience using a deterministic modeling approach. *Computers & Industrial Engineering*, 74: 11–25.

Spekman, R. E., and E. W. Davis. 2004. Risky business: expanding the discussion on risk and the extended enterprise. *International Journal of Physical Distribution & Logistics Management*, 34(5): 414–433.

Stannard, S. 2020. COVID-19 in the maritime setting: the challenges, regulations and the international response. *International Maritime Health*, 71(2): 85–90.

Tatenhove, J. P. 2021. COVID-19 and European maritime futures: different pathways to deal with the pandemic. *Maritime Studies*, 20(1): 63–74.

Thunstrom, L., Newbold, S., Finnoff, D., Ashworth, M., and J. F. Shogren. 2020. The benefits and costs of flattening the curve for COVID-19. *SSRN*. 3561934.

Tummala, R. and T. Schoenherr. 2011. Assessing and managing risks using the supply chain risk management process (SCRMP). *Supply Chain Management*, 16(6): 474–483.

UNCTAD. 2020. *Review of Maritime Transport 2020*. New York: United Nations Publications.

Verschuur, J., Koks, E. E., and J. W. Hall. 2021. Global economic impacts of COVID-19 lock-down measures stand out in high-frequency shipping data. *PLoS ONE*, 16(4): e0248818. doi.org/10.1371/journal.pone.0248818.

Vilko, J. P. and J. M. Hallikas. 2012. Risk assessment in multimodal supply chains. *International Journal of Production Economics*, 140(2): 586–595.

Wieland, A. and C. F. Durach. 2021. Two perspectives on supply chain resilience. *Journal of Business Logistics*, 42: 315–322.

Wong, C. W., Lirn, T. C., Yang, C. C. and K. C. Shang. 2020. Supply chain and external conditions under which supply chain resilience pays: An organizational information processing theorization. *International Journal of Production Economics*, 226: 107610. doi.org/10.1016/j.ijpe.2019.107610Get.

World Bank. 2021. COVID-19: impact and response. World Bank. https://openknowledge.worldbank.org/bitstream/handle/10986/33477/211565-ch01.pdf (accessed September 1, 2021).

World Economic Forum. 2013. *Building Resilience in Supply Chains*. Geneva: World Economic Forum.

Yam, L. 2021. Technology will help maritime transport navigate through the pandemic and beyond. https://blogs.worldbank.org/transport/technology-will-help-maritime-transportnavigate-through-pandemic-and-beyond (accessed September 9, 2021).

Yan, X. P., Wan, C. P., Zhang, D., and Z. L. Yang. 2017. Safety management of waterway congestions under dynamic risk conditions—a case study of the Yangtze River. *Applied Soft Computing*, 59: 115–128.

Yang, C. C. and W. L. Hsu. 2018. Evaluating the impact of security management practices on resilience capability in maritime firms—a relational perspective. *Transportation Research Part A: Policy and Practice*, 110: 220–233.

Yazır, D., Şahin, B., Yip, T. L., and P. H. Tseng. 2020. Effects of COVID-19 on maritime industry: a review. *International Maritime Health*, 71(4): 253–264.

Zavitsas, K., Zis, T., and M. G. Bell. 2018. The impact of flexible environmental policy on maritime supply chain resilience. *Transport Policy*, 72: 116–128.

6 Assessment and Reinforcement of Airport Climate Resilience with a BWM-Enhanced Hoshin Kanri Method

Turan Paksoy,[1] Enes Demiralay,[2] and Engin Hasan Çopur[3]

[1]Department of Aviation Management, Necmettin Erbakan University, Konya, Turkey

[2]Department Of Industrial Engineering, Konya Technical University, Konya, Turkey

[3]Department Of Astronautical Engineering, Necmettin Erbakan University, Konya, Turkey

CONTENTS

6.1 Introduction 71
6.2 A General Structure for an Airport Resilience System 74
6.3 A Brief Background of the Hoshin Kanri Method and BWM 77
 6.3.1 Hoshin Kanri 77
 6.3.2 BWM 79
6.4 A Novel Approach: BWM-Enhanced Hoshin Kanri Method 80
6.5 A Case Study: Strategy Development for Airport Climate Resilience 83
6.6 Conclusion 89
References 90

6.1 INTRODUCTION

Climate change is now regarded as the most dangerous threat to all living things globally. All continents across the world are profoundly affected by the effects of climate change, and it is forecasted that these adverse effects will become more drastic and more frequent in the coming decades. According to the results of the scientific observations made on and above Earth's surface, climate change makes its presence felt in global warming, warming oceans, decreased snow cover, declining Arctic sea ice, a sea-level rise, extreme events, etc. (Shepherd et al., 2018, 2019; Zscheischler

DOI: 10.1201/9781003212904-6

et al., 2018). Some of these global issues can bring daily life to a standstill. Many events that have occurred in past years are cited as examples of how badly they can hit international trade, the production of industrial and agricultural goods, and the transport of goods and passengers of all kinds. Inevitably, the aviation sector lies at the heart of this highly elaborate transportation network system. To put it more precisely, the International Civil Aviation Organization (ICAO) Annual Report 2018 provides a statistical analysis that reveals how rapidly the aviation sector is growing. According to figures in this report, 4.3 billion passengers traveled by aircraft in 2018, and 58 billion tons of goods were sent by air freight. Before the COVID-19 pandemic period, it was expected that the total number of air passengers would rise to 10 billion passengers by 2040 (ICAO, 2018). Hence, it would not be surprising that crises caused by flight cancellations and delays greatly affect air transportation (Greer, Rakas, and Horvath, 2020). This malfunction can produce some unintended and disastrous consequences.

The US economy faced heavy losses, estimated to be between nearly $33 billion (Ball et al., 2010) and $41 billion (Joint Economic Committee 2008) due to flight delays in 2007. In February 2010, heavy snowstorms hit the Eastern coast of the US badly, leading to the cancelation of more than 20,000 pre-planned flights. These extreme meteorological events lasted the whole month; at its peak, 23% of scheduled flights had to be canceled. Although it was impossible to evaluate the actual losses, it was estimated they ranged from $80 million to $100 million (Dunn and Wilkinson, 2016). Another unfavorable weather condition is atmospheric turbulence. Based on in-flight turbulence statistics (FAA, 2017), 234 aviation accidents involving death and injury were recorded in the USA, and airlines had to pay million dollars in compensation for the damages and lost working days of flight attendants (Sharman and Lane, 2016). These examples can be multiplied, but they show how important it is to be prepared for the worst and reveal a poor grasp of this fact. Despite not being directly related to climate change, the most recent case is the COVID-19 pandemic, which has negatively affected all aspects of life, such as education, health care, economy, and society. Considering the global issues arising from this infectious disease, it is evident that disaster preparedness must be strengthened to take the required actions rapidly and efficiently at all levels.

In light of the present situation, airports must show remarkable resilience in the face of climate change-induced disasters. However, stating that any action against the impacts of these disasters on the airport services has not been taken is very harsh criticism. The reason is that airports and aviation organizations periodically introduce comprehensive planning processes, such as master plans or climate change risk assessments and adaptation plans, to describe short- and long-term strategies, which also cover climate change-induced issues (Adelaide Airport, 2019; Sydney Airport, 2019a; Brisbane Airport, 2020; National Academies of Sciences, Engineering, and Medicine, 2019; OECD, 2018). Even though it may be uncertain whether the strategies for dealing with climate change are sufficient or implemented successfully, airport administrations are at least aware that they have to prevent the negative impacts of climate change-induced extreme events on vital airport operations (Burbidge, 2016).

Airport climate resilience is a new concept developed to meet the needs of airports arising from climate change, and is defined by the ICAO as "one that has taken measures to adjust for the challenges that climate change and drastic weather bring." This definition may seem unclear and need further interpretation. To make it more understandable for everyone, three components form resilience capacity (Francis and Bekera, 2014; Vugrin and Camphouse, 2011). The first component is the absorptive capacity to reduce the effects of extreme meteorological events. The second component is recovery and vital capacity to return the system to its normal state. The last component, called adaptive capacity, reflects the adaptation of a system to unintended and unpredictable conditions.

Some studies have focused on airport climate resilience. One recent study using this performance measure system proposed a resilience metric to measure the recovery speed when severe weather events hit airports in China (Zhou and Chen, 2020). Another case study focused on the impacts of flood hazards at Faleolo International Airport due to heavy rainfalls induced by climate change and suggested constructing new drainage systems and retrofitting existing airport systems to alleviate flooding (Wu, Taniguchi, and Tajima, 2021). A review by Poo et al. (2018) states that the most harmful impacts on airport and seaport operations are due to sea-level rise and flooding. Despite this, it claimed that more research is needed to ensure airport resilience.

In line with previous studies, to sustain airport resilience, the preparedness of three resilience components to respond to meteorological emergencies at airports is very important. This aim can be achieved by developing the required tools for monitoring and controlling the resilience-related strategic plan, including long- and short-term strategies (Ferrulli, 2016). Therefore, in the next section, possible long- and short-term strategies for airport climate resilience are introduced, as well as the climate change-induced risks for airports and the possible actions that can be taken to minimize these risks.

This chapter proposes an original approach based on the best–worst method (BWM) and Hoshin Kanri method. By using measurable scores for the first time in the Hoshin Kanri X matrix, the weights of the actions to be taken to achieve long-term goals with different degrees of importance were determined. A case study on the assessment and reinforcement of climate resilience of an airport has been prepared. Unlike Hoshin Kanri studies in the literature, an updated Hoshin Kanri X matrix has been proposed for the results to be more effective and controllable. The proposed approach's strategic plans in the Hoshin Kanri X matrix can be shaped according to various real-life problems and used to solve all kinds of strategy distribution problems. The benefits of the BWM method have been demonstrated in the literature via studies on multiple problems. However, an approach in which a lean management tool is integrated with the BWM method has been studied for the first time.

Every job, every service, and every business has multiple long-, medium-, and short-term goals. These goals may not be of equal importance. Achieving some plans may be more important than others. For this reason, it is necessary to determine the degree of importance of the goals. While determining the degree of importance, evaluations can be made on more than one criterion. In such cases, multi-criteria decision-making (MCDM) methods are used to determine the degree of importance. Several MCDM methods can be used to determine their severity. When the literature research on MCDM methods was conducted, it was observed that the BWM method had the following advantages over other MCDM methods: i. Rezaei (2015) has proved that the consistency ratio of the BWM method gives better results than matrix-based MCDM methods. Also, Ali (2020) compared analytic hierarchy process (AHP) and BWM methods, and concluded that BWM achieved better results with a better consistency ratio. ii. Guo (2017) concluded that fewer data comparisons are made while determining the criterion weights in the BWM method, and therefore less processing is done.

Hoshin Kanri is a strategy deployment tool that helps businesses determine the most appropriate way to achieve their long-, medium-, and short-term strategies. Hoshin Kanri is not as popular as other lean management tools. However, holding the high potential for impact at a business or organization's strategic, tactical, and operational levels has been made it Hoshin Kanri attractive. For these reasons, BWM and Hoshin Kanri have been used in this chapter.

In section 6.2., a general framework of airport climate resilience is introduced. This framework includes four main components: long-term strategies; short-term strategies; climate change-induced risks; and actions required to satisfy the predefined strategy to deal with the adverse effects of climate change on airport operations. Section 6.3. gives a brief introduction to the Hoshin Kanri method from theoretical and practical points of view. Section 6.4. proposes a novel decision-making method, namely BWM-enhanced Hoshin Kanri. In section 6.5., a case study of the creation of a climate-resilient airport is presented to demonstrate the applicability of the proposed approach. The conclusion is given in section 6.6.

6.2 A GENERAL STRUCTURE FOR AN AIRPORT RESILIENCE SYSTEM

This section provides a brief projection of the current state of airport climate resilience. First, the primary climate stressors that severely impact airport operations are identified. According to the literature, the common stressors are temperature, precipitation, wind, and humidity. These stressors produce serious risks for airport operations. Therefore, before describing the actions required to reduce the effects of climate change-induced disasters, climate change-related risks are identified and linked with the climate stressors. The climate stressors and their risks for airports are given in Table 6.1. The climate stressors and risk definitions to be used in this chapter have been determined with the help of the study entitled "Climate Resilience and Benefit–Cost Analysis: A Handbook for Airports" (National Academies of Sciences, Engineering, and Medicine, 2019).

Having described the climate stressors that pose severe risks to airport operations, a strategic approach must be taken to monitor, evaluate, and eliminate the impacts of the climate change-induced risks. Therefore, strategies to develop or strengthen airport climate resilience are needed. Considering the long- and short-term requirements to satisfy high preparedness for climate change-induced disasters, possible long- and

TABLE 6.1
Climate Stressors and Risk Descriptions

Temperature	Increased heating, ventilation, and air conditioning demand and duration
	Damage to pavements
	Electrical support system failure
	Health problems
	Delays and cancellations of flights
	Increased need for maintenance and repair
	Increased risk of fire
Precipitation	Flood damage to buildings, structures, and underground infrastructure
	Increased contaminations on runways and taxiways
	Overloaded flooding and drainage systems
	Disruptions to snow and ice removal operations
	Accidents
Wind	Delays and cancellations of flights
	Inappropriate flight routes due to changes in prevailing wind
	Damage of structures and external equipment
	Increased need for maintenance and repair
Humidity	Operational delays
	Isolation materials damage
	Malfunction of climatization equipment

short-term strategies are introduced (Verdijk, 2020; Baglin, 2012; Sydney Airport, 2019b; Heathrow Airport, 2016; Gresham and Marchi, 2015).

Long-Term Strategies: i) To develop coherent policies systems, and processes; ii) to improve the effectiveness of the coordination and successful implementation of the activities needed to develop a great resilience to the climate change-related effects; iii) to enhance the resource capacity to ensure the continuity of climate change activities; iv) to raise the awareness of the importance of the climate-resilient airport; and v) to increase the participation level of all stakeholders, employees, suppliers, business partners, and customers.

Short-Term Strategies: i) Information sharing (to improve information sharing between internal and external stakeholders to enhance transparency, develop effective relationships between suppliers, business partners, and customers, and to develop innovative methods of communication to raise the awareness); ii) monitoring (to develop new methods to monitor and assess the effectiveness of the new regulations, to provide new instruments for monitoring and controlling resources effectively for reduced greenhouse gas emission, and to monitor the quality of infrastructure and maintenance operations on a regular basis); iii) organizational management (to construct an organizational structure to achieve the desired goals leading to a climate-resilient airport, to update operational and organizational processes according to requirements to ensure resilience to the adverse effects of the climate change); iv) capital investment (to manage the capital investments required for additional human, technology, and premises resources to handle vulnerabilities to the climate change issue); and v) in-service training (to prepare staff to deal with climate change-based problems in operational and organizational processes).

Now, actions needed to successfully implement the short- and long-term strategies must be performed. A set of actions are given below that could serve as appropriate ways or tools to deal with the impacts of climate change-induced disasters.

Actions: i) Development of adaptive procedures and resources (review the safety and health regulations for airport employees and passengers; describe areas including high fire risks; assess fire safety regulations and operations; assess and improve operation plans to reduce the effects of extreme snow and rain; realign runway directions; and strengthen management guidelines by preparing, updating, and recommending management policies and procedures); ii) enhancement of effective robustness (review new technologies for disaster-management and control; development of new suppliers for new technologies supporting effective robustness; coordinating and ensuring inbound and outbound flows of goods in support of airport resilience; process testing and continuous improvement; managing budgets and payments; analyzing possible impacts on infrastructure systems and plan required actions to eliminate or reduce the adverse effects of such impacts; and assess the possible hotspots in extreme weather events and take reasonable precautions to eliminate the hotspots); iii) installation of effective systems (recalculate the required length of runways for take-offs and build longer runways if needed; modify the allowable load limit for pavements; replace or rebuild pavements by using new materials to resist extreme conditions; design new methods to monitor and evaluate the effectiveness and readiness of all components of the business; modify current infrastructure systems by installing advanced technology; strengthen building structures; increase the capability of heating, cooling, and ventilation systems; test and analyze the quality of electrical support systems such as lighting and navigational aid systems in extreme weather events; increase the flexibility of rescheduling or re-routing; assess the maintenance and repair operation plans and improve the repair and maintenance systems if needed; and robust modeling of wind direction and strength regimes to predict changes in prevailing direction and strength); iv) establishment of effective management and communication tools (improve the relationships between external parties to shape the organization's vision; manage business reviews effectively by using IT technologies and Customer Relationship Management (CRM) programs; organize meetings regularly with stakeholders to share important information and build robust relationships; organize regular meetings with managers to improve the processes continuously; reorganize activities to control resources effectively; understand and satisfy stakeholders' individual needs and respond to stakeholders' complaints quickly and efficiently); and v) raising awareness (train employees continuously to help retain their knowledge and awareness of the planned and modified actions for dealing with the impacts of climate change; assess the job requirements and modify the job descriptions of all positions; consider climate change, and develop professional and technical knowledge by organizing events (webinars, educational workshops, seminars, etc.) with the support of key experts; develop personal networks and join professional societies; train employees in climate change by preparing posters, brochures, handouts, and all kinds of visual material; manage the preparation and implementation of the necessary activity plans to keep employees constantly resilient against natural disasters and extreme events).

The actions mentioned earlier are not identical, and each action leads decision-makers in different directions in terms of resilience strategy. Therefore, a challenging question is raised that is of incredible interest to managers who feel the pressure of implementing and managing appropriate actions to deal with the impacts of climate change: Which action is the most relevant to satisfy in long-term strategies? The novel decision-making method proposed in the next section has been developed to answer this critical question.

6.3 A BRIEF BACKGROUND OF THE HOSHIN KANRI METHOD AND BWM

6.3.1 Hoshin Kanri

Hoshin means "direction" and Kanri means "management." When they are used together, the meaning is "How do we manage our direction." Hoshin Kanri has also referred to in the literature as "management by policy," "Hoshin planning," and "policy deployment." However, "strategy deployment" is the most accurate interpretation of Hoshin Kanri. Hoshin Kanri has various descriptions in the literature. The most common two descriptions used for Hoshin Kanri are: i) "a systematic approach that unites the whole organization's operational activities with its strategic targets" (Akao, 1991); and ii) "a method of developing plans, goals, controls, and regions of development based on the previous policies and evaluation of the last years' performance" (Dale, 1990).

Hoshin Kanri is a management and planning technique developed to ensure that strategic plans are considered and implemented at all levels in an organization. However, Hoshin Kanri is not a strategic planning tool; it is an accomplishment tool to deploy an existing strategic plan throughout all levels of employee, depending on the organization's goals (Roberts and Tennant, 2003). The essential aims of Hoshin Kanri are to determine the best ways to achieve an organizations' long-term goals, enable implementation to all levels of organization effectively, and review the key performance indicators (KPIs) and processes while considering strategic plans. Hoshin Kanri intends to guide employees in the same direction by eliminating the waste arising from inconsistent administration and poor communication (Andrés-López, González-Requena, and Sanz-Lobera, 2015; Leth et al., 2019).

Hoshin Kanri was developed by companies in Japan in the 1960s, and its origin is unclear (Melander et al., 2016). The most authoritative text of Hoshin Kanri was that edited by Akao (1991). By the mid-1980s, a few western companies had started implementing their versions of Hoshin Kanri, such as "Hoshin planning" (Bank of America), "management by policy" (Donnelly), "policy deployment" (Proctor & Gamble), "strategy or goal deployment" (Caradoc), "strategy into action" (Unilever), and "managing for results" (Xerox Corporation; Witcher, 2018; Dennis Beecroft, 1999). Several organizations suppose that their "world-class performance" and success are also dependent on the use of planning approaches based on the Hoshin Kanri principles of the "prioritization of strategic goals" (Marsden, 1998). The variety observed in the literature on the structure and principles of Hoshin Kanri shows that there is no consensus. Therefore, it may confuse researchers interested in it (Giordani da Silveira et al., 2018).

Thürer et al. (2018) defined the Hoshin planning process as the cascading of goals from the top down, starting from the purposes of the top management and progressing to the lowest organizational level. However, Hoshin planning is not just a top-down process. The Hoshin planning mechanism is built with catchball and plan–do–check–act (PDCA) tools to ensure continuous improvement based on the Kaizen philosophy. Catchball is a technique that creates feedback loops by providing a two-way flow of information between each level, from senior management to operational level employees (Manos, 2010). When an organization directs its employees to the goals set by top managers without feedback from the lower levels, it may face the risks of involuntary work and faulty production. However, when a strategy is developed by setting realistic targets agreed on by employees at all levels, it is seen that the ownership and motivation of employees increase, and processes are carried out more efficiently by managers. The PDCA tool is a process-improvement tool developed by Deming (1952), and is also called the Deming cycle. The PDCA tool improves next month's plans by monthly reviews. As the use of the PDCA tool increases, so does the probability of success of processes through continuous improvement.

There are various methods for applying Hoshin Kanri; the X matrix is one of them. It provides a strategic framework for organizations by linking long-term goals with annual goals, tactical improvements, and KPIs. The X matrix includes strategic components that are generally given as long-term objectives, annual objectives, priorities and activities, and KPIs (Yakubiv et al., 2020). The strategy deployment starts from the long-term objectives and progresses clockwise. The strategic components must converge from strategical to operational levels, but the names of features can be chosen by practitioners arbitrarily according to the missions and goals of companies.

In recent years, it has been seen that Hoshin Kanri has been used to develop strategies in different sectors. However, no study has yet been found in the literature in the aviation industry, and this study is pioneering. An offshore oil and gas construction case study implemented Hoshin Kanri on a megaproject. The implementation of Hoshin Kanri in this mega project provided a 10% reduction in project delivery time, \$1 million in savings from the risk incentive account, and a daily revenue potential of \$2.8 million due to the delivery of the project before the deadline (Leth et al., 2019). A forecast of the variability in the rate of development of biofuel production up to 2025 has been made by analyzing the current trends in the development of biofuel engineering in Ukraine. A strategy for developing agro-based biofuel production in Ukraine has been proposed using an approach based on the Hoshin Kanri matrix (Yakubiv et al., 2020). With the increasing demands for ceramic tiles, the floor and wall coverings industry is constantly growing. For Peruvian companies to be ahead of countries with low production costs, such as China and Brazil, which have large shares in the international market, they need to develop an effective strategy to reduce costs without changing their product quality. A strategy for this was proposed using the Hoshin Kanri method (Telles Rodriguez et al., 2021).

6.3.2 BWM

BWM is a MCDM method developed by Rezaei in 2015, used to determine criterion (alternative) weights. A comparison system is created with two paired comparison vectors, comparing the best criterion (alternative) with other criteria (alternatives) and the other criteria (alternatives) with the worst criterion (alternative). Optimal criterion (alternative) weights and the consistency ratio of the comparison system are calculated through the optimization model created using the comparison system. The BWM method is preferred over MCDM methods using a pairwise comparison matrix. It makes a consistent decision-making process by eliminating the possible fixation bias of the decision-makers. The BWM is comprised of the following five steps (Rezaei, 2015):

Step 1. Determine a set of decision criteria.

Step 2. Determine the best (e.g., most essential, most ambitious) criterion and the worst (e.g., least essential, least ambitious) criterion.

Step 3. Determine the priority of the best criterion over all the other criteria using a scale. The resulting best-to-others vector would be:

$$A_B = \left(a_{B1}, a_{B2}, \ldots, a_{Bn}\right) \tag{6.1}$$

where a_{Bj} indicates the decision of the best criterion C_B over criterion C_j. It is clear that $a_{BB} = 1$.

Step 4. Determine the priority of all the criteria over the worst criterion using a scale. The resulting others-to-worst vector would be:

$$A_W = \left(a_{1W}, a_{2W}, \ldots, a_{nW}\right) \tag{6.2}$$

where a_{jW} indicates the preference of criterion C_j over the worst criterion C_W. It is clear that $a_{WW} = 1$.

Step 5. Find the optimal weights.

$$\left(w_1^*, w_2^*, \ldots, w_n^*\right) \tag{6.3}$$

The optimal weight for the criteria is the one where, for each pair of w_B / w_j and w_j / w_W, we have $w_B / w_j = a_{Bj}$ and $w_j / w_W = a_{jW}$. To satisfy these conditions for all j, we should find a solution where the maximum absolute differences $\left|\frac{w_B}{w_j} - a_{Bj}\right|$ and $\left|\frac{w_j}{w_W} - a_{jW}\right|$ for all j are minimized. Considering the non-negativity and sum condition for the weights, the following problem results:

$$\begin{gathered}\min \max_j \left\{ \left|\frac{w_B}{w_j} - a_{Bj}\right|, \left|\frac{w_j}{w_W} - a_{jW}\right| \right\} \\ \sum_j w_j = 1 \\ w_j \geq 0, \text{ for all } j\end{gathered} \tag{6.4}$$

Problem (6.4) can be transferred to the following problem:

$$\begin{gathered}\min \xi \\ \left|\frac{w_B}{w_j} - a_{Bj}\right| \leq \xi \text{ for all } j \\ \left|\frac{w_j}{w_W} - a_{jW}\right| \leq \xi \text{ for all } j \\ \sum_j w_j = 1 \\ w_j \geq 0, \text{ for all } j\end{gathered} \tag{6.5}$$

Solving Eq. (6.5), the optimal weights $\left(w_1^*, w_2^*, \ldots, w_n^*\right)$ and ξ^* are obtained.

6.4 A NOVEL APPROACH: BWM-ENHANCED HOSHIN KANRI METHOD

This section offers a systematic way to deploy the strategy from the long-term to an action level with the help of a novel approach based on Hoshin Kanri and BWM methods. The framework for the proposed approach is shown in Figure 6.1.

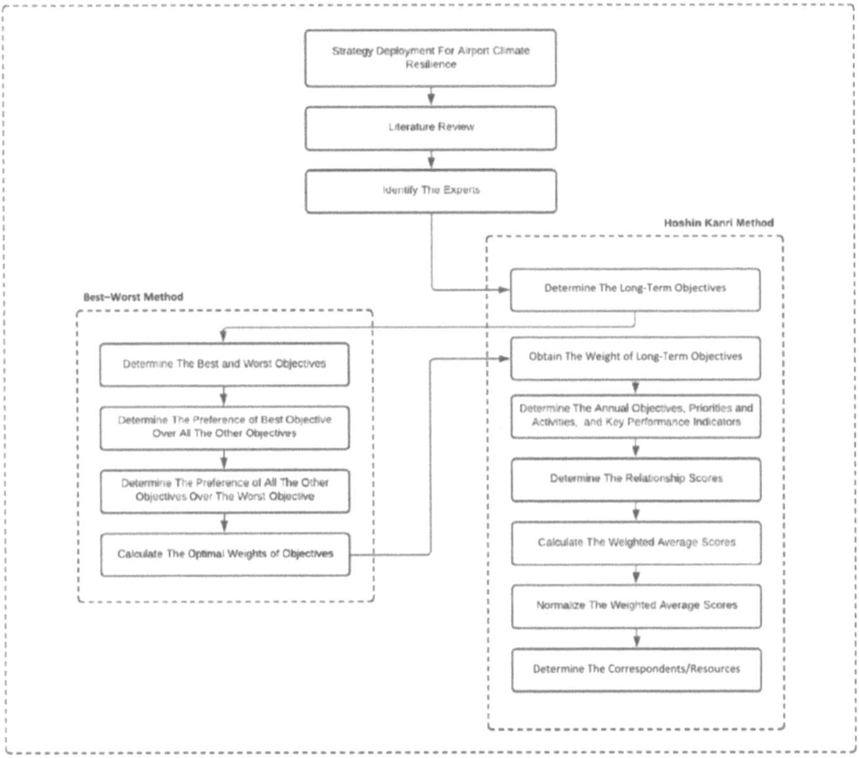

FIGURE 6.1 The framework of the proposed approach.

The proposed X matrix starts from south (S) to determine long-term objectives and ends at southeast (SE) to determine correspondents. It operates in the clockwise direction, as shown in Figure 6.2.

Step 1. Determine the "long-term objectives" (i) and write down in S matrix $(i = 1,\ldots,m)$.
Step 2. Calculate the weight (w_i) of "long-term objectives" by using BWM.
Step 3. Determine the "annual objectives" (j) and write down in W matrix $(j = 1,\ldots,n)$.
Step 4. Determine the relationship scores (SW_{ij}) between "long-term objectives" and "annual objectives" with respect to a Likert scale (1 – very weak; 2 – weak; 3 – fair; 4 – strong; 5 – very strong) and write down in southwest (SW) matrix.
Step 5. Calculate the weighted average scores (AO_j) of "annual objectives" by sum of multiplication of relationship scores (SW_{ij}) with weights of "long-term objectives" (w_i).

$$AO_j = \sum_{i=1}^{m} w_i * SW_{ij}, \qquad \forall j \tag{6.6}$$

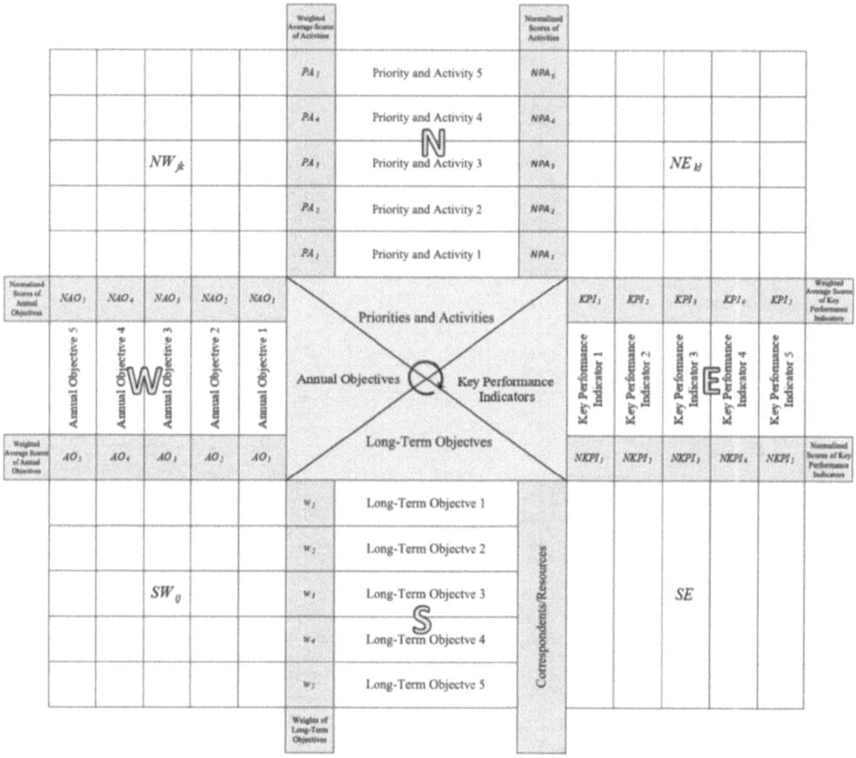

FIGURE 6.2 The proposed Hoshin Kanri X matrix enhanced by BWM.

Step 6. Normalize the weighted average scores (NAO_j) in Eq. (6.2) given as:

$$NAO_j = AO_j / \sum_{j=1}^{n} AO_j, \qquad \forall j \tag{6.7}$$

Step 7. Determine the "priorities and activities" (k) and write down in N matrix $(k = 1,\ldots,p)$.

Step 8. Determine the relationship scores (NW_{jk}) between "annual objectives" and "priorities and activities" with respect to the Likert scale and write down in northwest (NW) matrix.

Step 9. Calculate the weighted average scores (PA_k) of "priorities and activities" by sum of multiplication of relationship scores (NW_{jk}) with normalized score of annual objectives (NAO_j).

$$PA_k = \sum_{j=1}^{n} NAO_j * NW_{jk}, \qquad \forall k \tag{6.8}$$

Step 10. Normalize the weighted average scores (NPA_k) by Eq. (6.9) given as:

$$NPA_k = PA_k / \sum_{k=1}^{p} PA_k, \qquad \forall k \tag{6.9}$$

Step 11. Determine the "key performance indicators" (l) and put them in E matrix $(l = 1,\ldots,r)$.

Step 12. Determine the relationship scores (NE_{kl}) between "priorities and activities" and "key performance indicators" with respect to Likert scale and write down in NE matrix.

Step 13. Calculate the weighted average scores (KPI_l) of "key performance indicators" by sum of multiplication of relationship scores (NE_{kl}) with normalized score of "priorities and activities" (NPA_k).

$$KPI_l = \sum_{k=1}^{p} NPA_k * NE_{kl}, \qquad \forall l \tag{6.10}$$

Step 14. Normalize the weighted average scores ($NKPI_l$) by Eq. (6.11) given as:

$$NKPI_l = KPI_l / \sum_{l=1}^{r} KPI_l, \qquad \forall l \tag{6.11}$$

Step 15. Determine the "correspondents/resources" and match them with the "key performance indicators" on SE matrix.

6.5 A CASE STUDY: STRATEGY DEPLOYMENT FOR AIRPORT CLIMATE RESILIENCE

Climate change, and natural disasters as their inevitable consequences, are among the most urgent problems our world and humanity faces. Climate change-induced disasters affect the pace of life in many ways. For example, flight operations may be interrupted depending on the strength of impacts. These interruptions may prevent aid deliveries to regions that natural disasters hit. Delays in the aviation sector cause hefty financial losses. As stated in the section 6.1., a loss of $33 billion was incurred in the USA in 2007 due to flight delays. Airports need to continue operations and activities without severe disruptions. Therefore, airport management should determine new policies and strategies, considering the risks caused by climate change. In this section, a strategy deployment case study for airport climate resilience has been prepared using the proposed approach: Hoshin Kanri X matrix enhanced by BWM.

Step 1. Determine the "long-term strategies" (see Figure 6.3) and write them down in the S matrix.

Step 2. Calculate the weight of "long-term strategies" (w_i) using the BWM according to best-to-others and others-to worst matrices shown in Table 6.2.

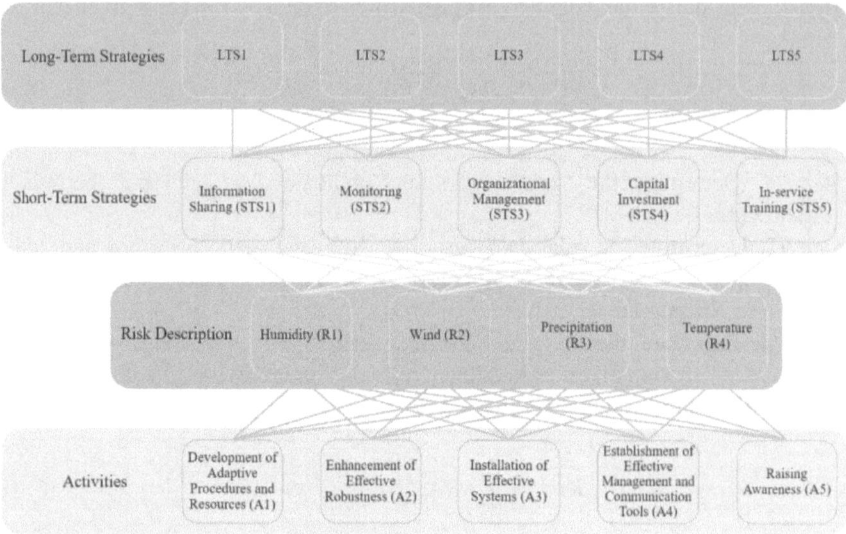

LTS1: To develop coherent policies, systems, and processes
LTS2: To improve the effectiveness of the coordination and successful implementation of the activities needed to develop a great resilience to the climate change-related effects
LTS3: To enhance the resource capacity to ensure the continuity of climate change activities
LTS4: To raise the awareness of the climate-resilient airport
LTS5: To increase the participation level of all stakeholders, employees, suppliers, business partners, and customers

FIGURE 6.3 Hierarchical decision structure of the case study.

TABLE 6.2
Best-to-Others and Others-to-Worst Pairwise Comparison Vectors

	LTS_1	LTS_2	LTS_3	LTS_4	LTS_5
Best	1	3	2	5	4
Worst	5	3	4	1	2

In order to determine the effect levels of the actions that will minimize or eliminate the effects of climate risk stressors that may cause disasters that may occur due to climate, it is necessary for decision-makers to determine the weights of long-term strategies. As decision-makers, three academics who are experts in aviation meteorology, civil aviation security, and airport equipment and activities, respectively, were chosen. The experts created best to others and others to worst comparison vectors to apply the BWM method.

After the determination of pairwise comparison vectors, the criteria weights in Table 6.3 were obtained with the GAMS package software.

Step 3. Determine the "short-term strategies" (see Figure 6.3) and write them down in the W matrix.

TABLE 6.3
Long-Term Strategy Weights

	LTS_1	LTS_2	LTS_3	LTS_4	LTS_5
w_i	0.41	0.16	0.24	0.07	0.12

TABLE 6.4
Relationship Scores Between "Long-Term Strategies" and "Short-Term Strategies"

	STS_1	STS_2	STS_3	STS_4	STS_5
LTS_1	4	3	5	5	5
LTS_2	5	5	5	3	4
LTS_3	5	2	5	5	4
LTS_4	5	4	2	2	5
LTS_5	5	4	4	2	5

TABLE 6.5
Weighted Average Scores of "Short-Term Strategies"

	j=1	j=2	j=3	j=4	j=5
STS_j	4.59	3.27	4.67	4.11	4.60

Step 4. Determine the relationship scores (SW_{ij}) between "long-term strategies" and "short-term strategies" with respect to the Likert scale and write them down in the SW matrix as shown in Table 6.4.

Step 5. Calculate the weighted average scores of "short-term strategies" (STS_j) (see Table 6.5) by the sum of multiplication of relationship scores (SW_{ij}) with the weights of "long-term strategies" (w_i) using Eq. (6.6).

$$\begin{aligned} STS_1 &= w_1{*}SW_{11} + w_2{*}SW_{21} + \ldots + w_5{*}SW_{51} \\ &= 0.41{*}4 + 0.16{*}5 + \ldots + 0.12{*}5 = 4.59 \end{aligned} \tag{6.12}$$

Step 6. Normalize the weighted average scores of the "short-term strategies" ($NSTS_j$) (see Table 6.6) using Eq. (6.7) given as:

$$NSTS_1 = STS_1 / \sum_{j=1}^{5} STS_j = 4.59 / 4.59 + 3.27 + \ldots 4.60 = 0.22 \tag{6.13}$$

TABLE 6.6
Normalized Scores of "Short-Term Strategies"

	j=1	*j*=2	*j*=3	*j*=4	*j*=5
$NSTS_j$	0.22	0.15	0.22	0.19	0.22

TABLE 6.7
Relationship Scores Between "Short-Term Strategies" and "Risk Descriptions"

	R_1	R_2	R_3	R_4
STS_1	5	5	1	1
STS_2	5	5	5	5
STS_3	4	4	4	4
STS_4	5	5	5	5
STS_5	5	5	4	4

TABLE 6.8
Weighted Average Scores of "Risks Descriptions"

	k=1	*k*=2	*k*=3	*k*=4
R_k	4.78	4.78	3.70	3.70

Step 7. Determine the "risk description" (see Figure 6.3) and write it down in the N matrix.

Step 8. Determine the relationship scores (NW_{jk}) between "short-term strategies" and "risk descriptions" with respect to the Likert scale and write them down in NW matrix as shown in Table 6.7.

Step 9. Calculate the weighted average scores of "risk descriptions" (R_k) (see Table 6.8) by the sum of multiplication of relationship scores (NW_{jk}) with normalized score of "short-term strategies" ($NSTS_j$) using Eq. (6.8).

$$\begin{aligned} R_1 &= NSTS_1 * NW_{11} + NSTS_2 * NW_{21} + \ldots + NSTS_5 * NW_{51} \\ &= 0.22*5 + 0.15*5 + \ldots + 0.22*5 = 4.78 \end{aligned} \tag{6.14}$$

Step 10. Normalize the weighted average scores of "risk descriptions" (NR_k) (see Table 6.9) by Eq. (6.9) given as:

$$NR_1 = R_1 / \sum_{k=1}^{4} R_k = 4.78 / 4.78 + 4.78 + \ldots 3.70 = 0.28 \tag{6.15}$$

TABLE 6.9
Normalized Scores of "Risk Descriptions"

	k=1	k=2	k=3	k=4
NR_k	0.28	0.28	0.22	0.22

TABLE 6.10
Relationship Scores Between "Risk Descriptions" and "Actions"

	A_1	A_2	A_3	A_4	A_5
R_1	5	5	5	4	1
R_2	5	5	4	4	1
R_3	5	5	4	4	3
R_4	4	5	5	3	1

TABLE 6.11
Weighted Average Scores of "Actions"

	l=1	l=2	l=3	l=4	l=5
A_l	4.78	5.00	4.50	3.78	1.44

Step 11. Determine the "actions" (see Figure 6.3) and write them down in the E matrix.

Step 12. Determine the relationship scores (NE_{kl}) between "risk descriptions" and "actions" with respect to the Likert scale and write them down in the NE matrix as shown in Table 6.10.

Step 13. Calculate the weighted average scores of "actions" (A_l) (see Table 6.11) by the sum of multiplication of relationship scores (NE_{kl}) with normalized score of "risk descriptions" (NR_k) using Eq. (6.10).

$$\begin{aligned} A_1 &= NR_1 * NE_{11} + NR_2 * NE_{21} + \ldots + NR_4 * NE_{41} \\ &= 0.28*5 + 0.28*5 + \ldots + 0.22*4 = 4.78 \end{aligned} \tag{6.16}$$

Step 14. Normalize the weighted average scores of "actions" (NA_l) (see Table 6.12) by Eq. (6.11) given as:

$$NA_1 = A_1 / \sum_{l=1}^{5} A_l = 4.78 / 4.78 + 5.00 + \ldots 1.44 = 0.25 \tag{6.17}$$

Step 15. Determine the "correspondents/resources" (see Figure 6.4) and match them with the "actions" in the SE matrix.

TABLE 6.12
Normalized Scores of "Actions"

	l=1	l=2	l=3	l=4	l=5
NA_l	0.25	0.26	0.23	0.19	0.07

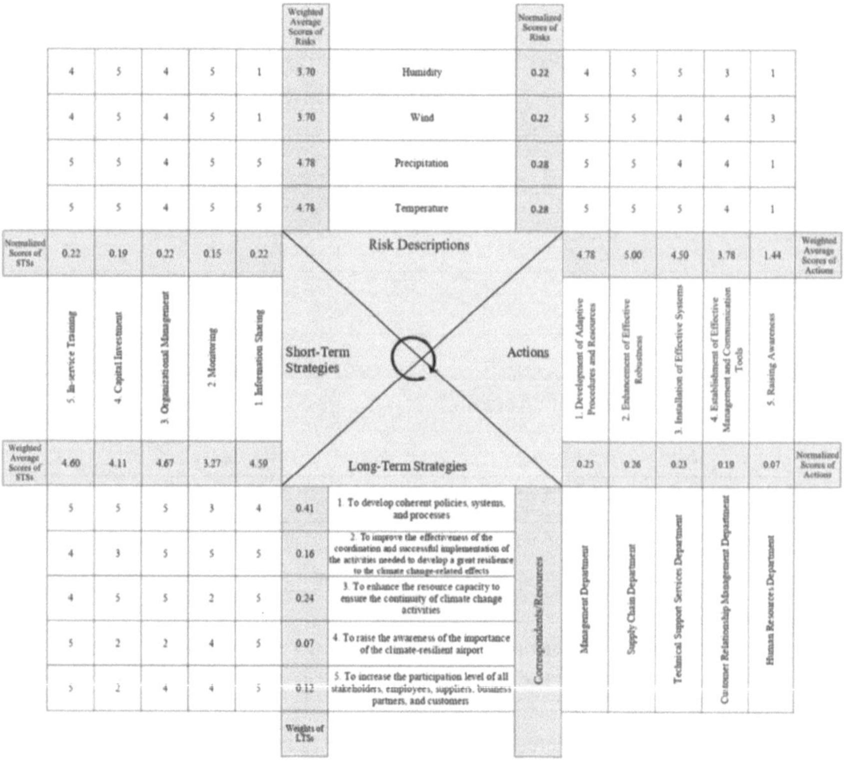

FIGURE 6.4 Hoshin Kanri X matrix for the case study.

The case study of the proposed approach has been prepared by applying a step-by-step approach to create a climate-resilient airport against the problems and delays that may occur as a result of the effects of climate change. Figure 6.4 shows the correlations between long-term strategies, short-term strategies, risk descriptions, and actions determined by the airport management. Based on the correlations, the significance level of each action in realizing the long-term strategy has been determined. Then, convenient actions were assigned to airport units.

As a result of the evaluations made by the decision-makers, the most important long-term strategy with the application of the BWM method was "To develop coherent policies, systems, and processes" (41% weight). "To raise the awareness of the importance of the climate-resilient airport" was at the bottom of the priority list (7% weight).

By using the weights of the long-term strategies evaluated by decision-makers in line with the geographic conditions and climate stressors in the region where the airport is located, the importance levels of the actions to be taken to ensure climate resilience were calculated as follows: "enhancement of effective robustness" (26%); "development of adaptive procedures and resources" (25%); "installation of effective systems" (23%); "establishment of effective management and communication tools" (19%); and "raising awareness" (7%).

6.6 CONCLUSION

The effects of climate change on nature are increasing significantly. In different parts of the world, the impact of climate change can be observed as an increase in average temperature, an increase in the number of storms, the melting of glaciers, changing seasons, sea-level rise, or extreme weather conditions. In recent years, the increasing effects of climate change have negatively affected modern life. Climate change can affect societies by impacting on socio-cultural life and natural resources (US EPA, 2017). For example, due to the increasing average temperatures, freshwater resources in lakes are being depleted by evaporation, affecting society. Some societies, plant and animal species, and industrial sectors are more affected by climate change than others. The effects of climate change on the aviation sector may result in flight delays and airport closures, and the costs that may arise as a consequence; shortages may occur due to the inability to ship critical goods. In the short term, it is hoped that the effects of climate change on the aviation sector will not cause more severe consequences than extra costs. However, in the medium and long term, it is estimated that the aviation industry will be affected by the consequences of potential climate change such as sea-level rise, increased storm intensity, temperature change, precipitation change, changes in wind, changing ice conditions, desertification, and changes in biodiversity (European Union, European Civil Aviation Conference, and EUROCONTROL, 2010). In this study, what kind of climate change resiliency strategy will be adopted by top management in airports, which is one of the main elements of the aviation industry, has been determined by using the Hoshin Kanri method. By associating the long-term strategies, which were determined according to the opinions of employees from top management to the operational level, with short-term strategies, risk descriptions, and actions, the level of importance of each action in achieving the long-term strategy has been determined. Since, according to airport top management, each long-term strategy does not have the same significance, the importance levels of the long-term strategies were calculated using the BWM method. When the aim is to implement a long-term strategy "To develop the coherent policies, systems, and processes," the action with the highest importance is calculated to be "Enhancement of effective robustness." When the weights of long-term strategies change according to the evaluation of the decision-makers, the degree of importance of the actions to be taken to ensure climate resilience also changes. In addition, when the risks to be assessed on the Hoshin Kanri X matrix change according to different geographic regions, the degree of importance of the actions to be taken to ensure climate resilience also changes. Temperature, precipitation, wind, and humidity were determined to be climate stressors that may cause risk in this study. However, these climate stressors vary according to the geographic conditions of the airports' locations. While the climate stressor in some regions may be precipitation, it may be a volcanic threat in another region. For this reason, it is crucial to choose climate stressors that are common to all geographic regions. In future studies, dynamic assessment and reinforcement of airport climate resilience can be carried out, considering the geographic conditions of each region where airports are located.

REFERENCES

Adelaide Airport. 2019. "Adelaide Airport Master Plan." www.adelaideairport.com.au/corporate/community/adelaide-airport-master-plan/.

Akao, Y. 1991. *Hoshin Kanri: Policy Deployment for Successful TQM*. Productivity Press.

Ali, S. S., T. Paksoy, B. Torğul, and R. Kaur. "Reverse Logistics Optimization of An Industrial Air Conditioner Manufacturing Company For Designing Sustainable Supply Chain: A Fuzzy Hybrid Multi-Criteria Decision-Making Approach." *Wireless Network* 26: 5759–5782.

Andrés-López, E., I. González-Requena, and A. Sanz-Lobera. 2015. "Lean Service: Reassessment of Lean Manufacturing for Service Activities." *Procedia Engineering* 132 (January): 23–30. doi: 10.1016/J.PROENG.2015.12.463.

Baglin, C. 2012. "Airport Climate Adaptation and Resilience." www.national-academies.org/trb/bookstore.

Ball, M., C. Barnhart, M. Dresner, M. Hansen, K. Neels, A. Odoni, E. Peterson, et al. 2010. "Total Delay Impact Study A Comprehensive Assessment of the Costs and Impacts of Flight Delay in the United States." Report no. 6234. Federal Aviation Administration Air Traffic Organization Strategy and Performance Business Unit.

Brisbane Airport. 2020. "Sustainable Development I Brisbane Airport." www.bne.com.au/corporate/community-and-environment/environment-and-sustainability/sustainable-development.

Burbidge, R. 2016. "Adapting European Airports to a Changing Climate." *Transportation Research Procedia* 14: 14–23. doi: 10.1016/J.TRPRO.2016.05.036.

Dale, B. G. 1990. "Policy Deployment." *The TQM Magazine* 2(6). doi: 10.1108/EUM0000000003052.

Deming, W. 1952. *Elementary Principles of the Statistical Control of Quality: A Series of Lectures*. Nippon Kagaku Gijutsu Remmei, Tokyo.

Dennis Beecroft, G. 2016. "The Role of Quality In Strategic Management." *Management Decision* 37(6): 499–502. doi: 10.1108/00251749910278005.

Dunn, S. and S. M. Wilkinson. 2016. "Increasing the Resilience of Air Traffic Networks Using a Network Graph Theory Approach." *Transportation Research Part E: Logistics and Transportation Review* 90: 39–50. doi: 10.1016/J.TRE.2015.09.011.

European Union, European Civil Aviation Conference, and EUROCONTROL. 2010. "Effects of Climate Change on Aviation Safety." *ICAO High-Level Safety Conference 2010*. www.icao.int/Meetings/AMC/HLSC/Dowload%20all%20documents%20at%20once/Forms/AllItems.aspx.

FAA. 2017. "Turbulence: Staying Safe." www.faa.gov/travelers/fly_safe/turbulence/.

Ferrulli, P. 2016. "Resilient Architectural Design: Considerations in the Design of Airports to Withstand Climate Change Effects." *Climate Change Management*: 381–93.

Francis, R. and B. Bekera. 2014. "A Metric and Frameworks for Resilience Analysis of Engineered and Infrastructure Systems." *Reliability Engineering & System Safety* 121: 90–103. doi: 10.1016/J.RESS.2013.07.004.

Giordani da Silveira, W., E. Pinheiro de Lima, F. Deschamps, and S. E. Gouvea da Costa. 2018. "Identification of Guidelines for Hoshin Kanri Initiatives." *International Journal of Productivity and Performance Management* 67: 85–110.

Greer, F., J. Rakas, and A. Horvath. 2020. "Airports and Environmental Sustainability: A Comprehensive Review." *Environmental Research Letters* 15(10): 103007. doi: 10.1088/1748-9326/ABB42A.

Gresham, Smith and Partners, and R. Marchi. 2015. "Climate Change Adaptation Planning: Risk Assessment for Airports." *Transportation Research Board*. doi: 10.17226/23461.

Guo, S. and H. Zhao. "Fuzzy Best-Worst Multi-Criteria Decision-Making Method and its Applications." *Knowledge-Based Systems* 121: 23–31. doi: 10.1016/j.knosys.2017.01.010.

Heathrow Airport. 2016. "Climate Change Adaptation and Resilience Progress Report." https://assets.publishing.service.gov.uk/government/uploads/system/uploads/attachment_data/file/566147/climate-adrep-heathrow.pdf

ICAO. 2018. "The World of Air Transport in 2018." www.icao.int/annual-report-2018/Pages/the-world-of-air-transport-in-2018.aspx.

Joint Economic Committee. 2008. "Flight Delays Cost Passengers, Airlines and the U.S. Economy Billion." http://online.wsj.com/public/resources/documents/jecreport05222008.pdf

Leth, S., J. Lerche, H. H. Neve, and S. Wandahl. 2019. "Lessons for Construction: Delivering Mega Project on Time through Culture and Hoshin Kanri." *27th Annual Conference of the International Group for Lean Construction, IGLC 2019*. The International Group for Lean Construction, 1011–22. doi: 10.24928/2019/0139.

Manos, A. 2010. "Hoshin Promotion." *Six Sigma Forum Magazine* 7–14.

Marsden, N. 1998. "The Use of Hoshin Kanri Planning and Deployment Systems in the Service Sector: An Exploration." *Total Quality Management* 9(4–5): 167–71. doi: 10.1080/0954412988820.

Melander, A., M. Löfving, D. Andersson, F. Elgh, and M. Thulin. 2016. "Introducing the Hoshin Kanri Strategic Management System in Manufacturing SMEs." *Management Decision* 54(10): 2507–23. doi: 10.1108/MD-03-2016-0148.

National Academies of Sciences, Engineering, and Medicine. 2019. *Climate Resilience and Benefit–Cost Analysis: A Handbook for Airports*. https://nap.nationalacademies.org/catalog/25497/climate-resilience-and-benefit-cost-analysis-a-handbook-for-airports

OECD. 2018. "Climate-Resilient Infrastructure." https://www.oecd-ilibrary.org/deliver/4fdf9eaf-en.pdf?itemId=%2Fcontent%2Fpaper%2F4fdf9eaf-en&mimeType=pdf

Poo, M. C. P., Z. Yang, D. Dimitriu, and Z. Qu. 2018. "Review on Seaport and Airport Adaptation to Climate Change: A Case on Sea Level Rise and Flooding." *Marine Technology Society Journal* 52(2): 23–33. doi: 10.4031/MTSJ.52.2.4.

Rezaei, J. 2015. "Best-Worst Multi-Criteria Decision-Making Method." Omega (United Kingdom) 53: 49–57. doi: 10.1016/j.omega.2014.11.009.

Roberts, P. and C. Tennant. 2003. "Application of the Hoshin Kanri Methodology at a Higher Education Establishment in the UK." *The TQM Magazine* 15(2): 82–87. doi: 10.1108/09544780310461080.

Sharman, R. and T. Lane. 2016. "Aviation Turbulence: Processes, Detection, Prediction." *Aviation Turbulence: Processes, Detection, Prediction* January: 1–523. doi: 10.1007/978-3-319-23630-8.

Shepherd, A., E. Ivins, E. Rignot, B. Smith, M. van den Broeke, I. Velicogna, P. Whitehouse, et al. 2019. "Mass Balance of the Greenland Ice Sheet from 1992 to 2018." *Nature* 579(7798): 233–39. doi: 10.1038/s41586-019-1855-2.

Shepherd, A., E. Ivins, E. Rignot, B. Smith, M. Van Den Broeke, I. Velicogna, P. Whitehouse, et al. 2018. "Mass Balance of the Antarctic Ice Sheet from 1992 to 2017." *Nature* 558(7709): 219–22. doi: 10.1038/s41586-018-0179-y.

Sydney Airport. 2019a. "Master Plan 2039 – Overview." www.sydneyairport.com.au/corporate/planning-and-projects/master-plan.

Sydney Airport. 2019b. "Sydney Airport Environment Strategy 2019–2024." https://assets.ctfassets.net/v228i5y5k0x4/cxZTrfYEhY0GQuKBVb1b2/a0caa7b4c46aebad0b69c59793123eff/Sydney_Airport_Environmental_Strategy_2019-2024_F.pdf

Telles Rodriguez, A., C. Vasquez Calampa, E. Altamirano, and C. Del Carpio. 2021. "Management Model for Improving the Quality of Glazing by Applying Lean Manufacturing Tools in a Ceramics Plant." *ACM International Conference Proceeding Series*, January. Association for Computing Machinery, 105–11. doi: 10.1145/3447432.3447443.

Thürer, M., T. Maschek, L. Fredendall, P. Gianiodis, M. Stevenson, and J. Deuse. 2018. "On the Integration of Manufacturing Strategy: Deconstructing Hoshin Kanri." *Management Research Review* 42(3): 412–26. doi: 10.1108/MRR-04-2018-0178.

US EPA. 2017. "Climate Impacts on Society." https://19january2017snapshot.epa.gov/climate-impacts/climate-impacts-society_.html.

Verdijk, P. F. M. 2020. "Climate-Resilient Airports: An Airport Climate Resilience Assessment Scan From a Theoretical Concept to an Operational Method." https://essay.utwente.nl/85464/.

Vugrin, E. D. and R. C. Camphouse. 2011. "Infrastructure Resilience Assessment through Control Design." *International Journal of Critical Infrastructures* 7(3): 243–60. doi: 10.1504/IJCIS.2011.042994.

Witcher, B., V. S. Chau, and P. Harding. 2008. "Dynamic Capabilities: Top Executive Audits and Hoshin Kanri at Nissan South Africa." *International Journal of Operations and Production Management* 28(6): 540–561. doi: 10.1108/01443570810875359.

Wu, L., K. Taniguchi, and Y. Tajima. 2021. "Impact of Climate Change on Flood Hazard at Airports on Pacific Islands: A Case Study of Faleolo International Airport, Samoa." *Journal of Disaster Research* 16(3): 351–62. doi: 10.20965/JDR.2021.P0351.

Yakubiv, V., I. Boryshkevych, I. Piatnychuk, N. Iwaszczuk, and A. Iwaszczuk. 2020. "Strategy for the Development of Bioenergy Based on Agriculture: Case for Ukraine." *International Journal of Renewable Energy Research* 10(3). www.ijrer.org/ijrer/index.php/ijrer/article/view/10867/pdf.

Zhou, L. and Z. Chen. 2020. "Measuring the Performance of Airport Resilience to Severe Weather Events." *Transportation Research Part D: Transport and Environment* 83(June): 102362. doi: 10.1016/J.TRD.2020.102362.

Zscheischler, J., S. Westra, B. J. J. M. van den Hurk, S. I. Seneviratne, P. J. Ward, A. Pitman, A. AghaKouchak, et al. 2018. "Future Climate Risk from Compound Events." *Nature Climate Change* 8(6): 469–77. doi: 10.1038/s41558-018-0156-3.

7 The Evaluation of Social Vulnerability Factors for Pandemics Based on Fuzzy DEMATEL and the Delphi Method

Sukran Seker
Yildiz Technical University

CONTENTS

7.1 Introduction 93
7.2 Literature Review 94
7.3 Fuzzy DEMATEL Approach Based on the Delphi Method 95
7.4 Case Study 98
7.5 Conclusion 103
References 103

7.1 INTRODUCTION

Pandemics affect a wide geographical area when an epidemic spreads to more than one continent. The novel SARS-CoV-2 (COVID-19) virus was reported first in a Chinese city in late 2019. COVID-19 has spread rapidly to most countries and regions worldwide. As of the end of August 2020, approximately 25 million cases and 840,000 deaths had been recorded globally (OECD, 2020). However, pandemics have been seen throughout history and they have a significant impact on especially socially vulnerable communities. The socio-economic conditions of societies have revealed urban vulnerability against pandemics. While pandemics came to the fore with the COVID-19 crisis, it has been seen that pandemics seriously affect society. Certain segments of the population are in a more vulnerable position in pandemics, since the crisis threatens the livelihood, health, and well-being of people (Calderón-Larrañaga, 2020).

Social vulnerability refers to the social, economic, demographic, cultural, geographical, and historical characteristics that affect the vulnerability of societies to disasters. Therefore, there are concerns about the impacts of pandemics on the poorest populations, especially in low- and middle-income countries. While these

DOI: 10.1201/9781003212904-7

people have limited access to health care when infection occurs, they have difficulty in adopting preventive measures (such as social isolation; Souza et al., 2020). Therefore, social vulnerability shows how social, economic, and political factors intensify or reduce the effects of disasters in a society. Social categories and inequalities are also manifested in the quality and disposal of resources, such as health care or emergency services. Accordingly, socially vulnerable groups are at greater risk during disasters and pandemics (Thomas et al., 2020). However, there are some common themes or factors that affect social vulnerability such as age, gender, race and ethnicity, socio-economic status, populations with special needs, language barriers, type and infrastructure of housing, income inequality, and community capital (Rifat and Liu, 2021).

The aim of this chapter is to examine the social vulnerability factors in pandemics. To achieve this aim, a new conceptual framework for social vulnerability factors influencing pandemics is proposed. A hybrid study based on the Delphi method and DEMATEL (decision-making trial and evaluation laboratory), proposed by Gabus and Fontela (1972), under a fuzzy environment, is proposed to recognize and evaluate social vulnerability factors in pandemics. Since many real-world problems consist of imprecise and uncertain information, fuzzy sets integrate with the DEMATEL method. Thus, the proposed method was used to prepare a vulnerability map using social vulnerability factors and taking necessary measures against pandemics. Accordingly, while gender, demographic factors, socio-economic factors, public resources, etc., are determined as social vulnerability factors for pandemics, a modified Delphi method presented by (Sung (2001) and Gumus (2009)) and fuzzy DEMATEL was applied to represent the interrelations between social vulnerability factors with visual representation and ranking of social vulnerability factors. Thus, government, policymakers, and practitioners can regulate the necessary measures to fight against pandemics.

The chapter is constructed as follows: In the second section, there is a literature review about disaster management based on social vulnerability factors. Some preliminaries for the suggested method are presented in Section 7.3. To represent the interdependence relationship among vulnerability factors, the proposed approach is applied in Section 7.4. Lastly, concluding remarks are presented in Section 7.5.

7.2 LITERATURE REVIEW

Social vulnerability assessment is critical for disaster mitigation and resilience. Therefore, the social vulnerability concept has been extensively studied in the disaster-management literature. In this section, studies of social vulnerability assessment specific to disaster-management (pandemics, natural disasters, etc.) are discussed.

In 1918, the flu epidemic showed variation in mortality according to the effect of socio-economic factors in different geographical regions (Mamelund, 2006). In the 2009 H1N1 influenza pandemic, there was a threefold difference in mortality between the poorest and wealthiest populations in Great Britain (Rutter et al., 2012).

Furthermore, Waitzberg et al. (2020) evaluated Israel's answer to the COVID-19 pandemic. In Israel, the ultra-Orthodox Jewish and the Arab people are minor groups, comprising 12% and 20% of the total population, respectively. Due to the spread of the COVID-19 epidemic in these groups, the government took special measures adapted to this population. Calderón-Larrañaga et al. (2020) proposed a conceptual framework developed for older people that addressed biological and social differences in the COVID-19 pandemic, and showed potential differences for avoidance, treatment, diagnosis, and/or long-term outcomes. Bwire et al. (2020) highlighted biological differences in the immune systems of men and women that can affect the ability to fight off many infections, including COVID-19. In general, women are more resilient to infections than men due to sex differences and coronavirus receptors (ACE2), and having different lifestyles, such as lower levels of smoking and alcohol consumption in women. Lakhani (2020) demonstrated that older adults (65+ years of age) with disabilities are vulnerable to COVID-19 and need palliative care. The Lancet (2020) showed that socially vulnerable groups are not only the elderly, those with health problems and co-morbidities, or homeless people, but also people from various socio-economic groups who are struggling financially, mentally, or physically to cope with the crisis. Roy et al. (2020) showed the importance of appropriate identification and treatment of patients with COVID-19 infection. He also showed that rapid recognition and appropriate treatment of patients with other pathologies, including bacterial infection, is critical.

The fuzzy DEMATEL method used in this chapter has been applied extensively in many vulnerability assessments in disaster-management studies. For example, Hiete et al. (2012) suggested a fuzzy DEMATEL method to develop a compound indicator in disaster resilience for Germany. Accordingly, while urban German areas and southern Germany had high levels of resilience, rural areas, particularly eastern Germany, had a low level of resilience. Chukwuma et al. (2021) used GIS (geographic information system) and the integration of interval value fuzzy rough number (IVFRN), DEMATEL, and analytical network process (ANP) to evaluate flood hazard vulnerability. Khazai et al. (2013) focused on the disaster vulnerability of regions. DEMATEL methodology was applied to examine direct and indirect dependencies within social and industrial vulnerability factors by considering both social and economic factors. Hosseini et al. (2021) assessed flood vulnerability at the basin level. DEMATEL, ANP, and fuzzy theories are integrated for flood vulnerability assessment. Trivedi (2018) suggested DEMATEL to provide a comprehensive understanding of the complex relationships among all dimensions considered for shelter layout planning for effective disaster planning. Mao et al. (2019) presented the gray rough DEMATEL method to assess the causal relationship among ice disaster risk factors and determine the importance weight of risk factors for ice disaster prevention and management.

In this chapter, a new conceptual framework for social vulnerability factors influencing pandemics is proposed. The interaction between the most critical social vulnerabilities during a pandemic are examined to recognize and evaluate the social vulnerability factors of pandemics by applying a fuzzy DEMATEL approach based on the Delphi method.

7.3 FUZZY DEMATEL APPROACH BASED ON THE DELPHI METHOD

Multi-criteria decision-making (MCDM) is a tool used to evaluate potential alternatives considering multiple and often conflicting criteria. Fuzzy MCDM methods are applied for real-world problems where there are uncertain and incomplete data for the solution.

Fuzzy sets introduced by Zadeh (1965) are used in this study, since today's decision-making problems involve uncertainty and hesitancy due to the opinions of experts. Fuzzy systems are needed because decision problems are complex and, in many cases, problems are modeled through linguistic knowledge based on fuzzy-based linguistic approaches. While the proposed method shows the relationships between social factors, it also reveals which factors are more vulnerable to disasters.

The DEMATEL method is an inclusive method used in solving complex problems. Thus, the model is used to measure the interaction level and the direction of influence of the criteria. It is applied to analyze the causal relationship among real-world factors by constructing a structural model (Trivedi, 2018). Fuzzy DEMATEL based on a modified Delphi method was used in this study to handle vague and subjective nature of human opinions by facilitating the decision-making process. The stepwise proposed method is shown in Figure 7.1.

Fuzzy DEMATEL based on a modified Delphi method includes the following steps:

Step 1. The factors related to the problem are determined and a fuzzy direct relation matrix is constructed using pairwise comparison based on the modified Delphi method. The Delphi method is applied, as it is used extensively to collect the opinions of anonymous experts in the literature. In the Delphi method anonymous experts make decisions in consensus by following some procedures (Gumus, 2009). Once the factors associated with the problem are determined using the modified Delphi method (Sung (2001) and Gumus (2009)), to construct a direct relation matrix, decision-makers (DMs) compare the factors in terms of verbal

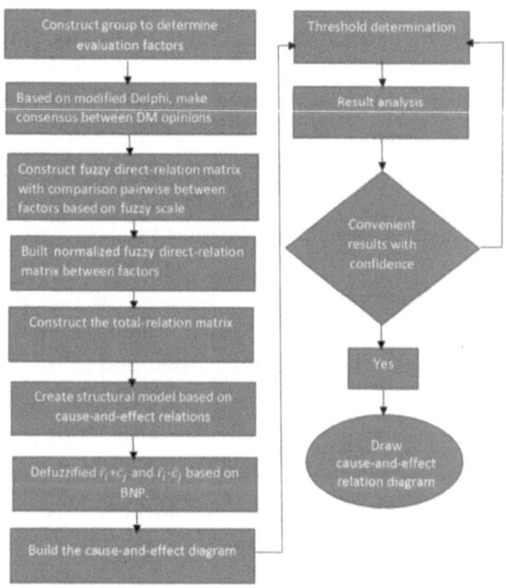

FIGURE 7.1 The proposed method.

variables of fuzzy sets. The pair-wise comparison is obtained with consensus based on linguistics variables represented in a scale presented by Akyuz and Celik (2015). The initial fuzzy direct relation matrix considering fuzzy pair-wise influenced relationships between the vulnerable factors is constructed as in Equation (7.1).

$$\tilde{X} = \left[\tilde{x}_{ij}\right]_{mxm} = \begin{bmatrix} \tilde{x}_{11}\ \tilde{x}_{12} & \cdots & \tilde{x}_{1m} \\ \vdots & \ddots & \vdots \\ \tilde{x}_{m1}\tilde{x}_{m2} & \cdots & \tilde{x}_{mm} \end{bmatrix} \tag{7.1}$$

where

$$\tilde{x}_{ij} = (l_{ij}, m_{ij}, u_{ij}) \tag{7.2}$$

Step 2. The normalized fuzzy direct relation matrix is established. The normalized fuzzy direct-relation matrix is obtained by applying Equations (7.3–7.5).

$$\gamma = max\left(\sum_{j=1}^{n} u_{ij}\right) \tag{7.3}$$

The linear scale conversion is applied to translate the factors into comparable scales.

$$\tilde{Y} = \left[\tilde{y}_{ij}\right]_{mxm} = \begin{bmatrix} \tilde{y}_{11}\ \tilde{y}_{12} & \cdots & \tilde{y}_{1m} \\ \vdots & \ddots & \vdots \\ \tilde{y}_{m1}\tilde{y}_{m2} & \cdots & \tilde{y}_{mm} \end{bmatrix} \tag{7.4}$$

where

$$\tilde{y}_{ij} = \frac{\tilde{x}_{ij}}{\gamma} = \left(\frac{l_{ij}}{\gamma}, \frac{m_{ij}}{\gamma}, \frac{u_{ij}}{\gamma}\right) \tag{7.5}$$

Step 3. Construct the total relation matrix. Total relation matrix $\tilde{T}$ is obtained using Equations (7.6–7.11).

$$\tilde{T} = \lim_{w \to \infty}(\tilde{Y} + \tilde{Y}^2 + \ldots.. + \tilde{Y}^w) \tag{7.6}$$

$$\tilde{T} = \begin{bmatrix} \tilde{t}_{11}\ \tilde{t}_{12} & \cdots & \tilde{t}_{1m} \\ \vdots & \ddots & \vdots \\ \tilde{t}_{m1}\tilde{t}_{m2} & \cdots & \tilde{t}_{mm} \end{bmatrix} \tag{7.7}$$

where

$$\tilde{t}_{ij} = \left(l_{ij}^{*}, m_{ij}^{*}, u_{ij}^{*}\right) \tag{7.8}$$

$$\text{Matrix}\left[l_{ij}^{*}\right] = Y_l x\left(I - Y_l\right)^{-1} \tag{7.9}$$

$$\text{Matrix}\left[m_{ij}^{*}\right] = Y_m x\left(I - Y_m\right)^{-1} \tag{7.10}$$

$$\text{Matrix}\left[u_{ij}^{*}\right] = Y_u x\left(I - Y_u\right)^{-1} \tag{7.11}$$

Step 4. Create the structural model. To construct the structural model, $\tilde{r}_i + \tilde{c}_j$ and $\tilde{r}_i - \tilde{c}_j$ are determined based on total relation matrix.$\tilde{r}_i$ and $\tilde{c}_j$show the row and column sum of matrix $\tilde{T}$, respectively. In addition, $\tilde{r}_i + \tilde{c}_j$ represents the significance of factor I and $\tilde{r}_i$ - $\tilde{c}_j$ represents the net effect of factor i.

Step 5. $\tilde{r}_i + \tilde{c}_j$ and $\tilde{r}_i - \tilde{c}_j$ values are defuzzified by applying the center of area defuzzification technique presented by Ross (2004) to obtain BNP (best non-fuzzy performance) values.

$$BNP_{ij} = \frac{u_{ij} - l_{ij} + m_{ij-} - l_{ij}}{3} + l_{ij} \tag{7.12}$$

Step 6. Draw the cause-and-effect diagram. The cause-and-effect diagram as an easy-to-understand structural model shows the complex relationships of the factors. While the horizontal axis ($\tilde{r}_i + \tilde{c}_j$) shows the importance value of the factor as a "prominence", the vertical axis ($\tilde{r}_i - \tilde{c}_j$) represents the extent of the influence as a "relation". Accordingly, if the $\tilde{r}_i - \tilde{c}_j$ axis is positive, the factor is among the cause group. If the $\tilde{r}_i - \tilde{c}_j$ axis is negative, the factor is among the effect group.

7.4 CASE STUDY

The emerging COVID-19 data show that the health and economic impacts of the pandemic have affected individuals not only biologically, but also socially. For example, social factors such as population density, ethnic diversity, urban/rural environment, education level, lifestyle, and living conditions related to the region are associated with different health and well-being outcomes. The effects of these social differences been shown in pandemics and affect societies at different levels.

In this study, a fuzzy DEMATEL method was applied to show the causal relationship between socially vulnerable factors with respect to causal and effect relations. Social vulnerability factors were determined using components of the Social Vulnerability Index from the Johns Hopkins Center for Systems Science and Engineering database and also from the literature. The social vulnerability factors for pandemics were determined as in Table 7.1.

TABLE 7.1
Social Vulnerability Factors for Pandemics

Main Criteria	Sub-Criteria
Socio-economic factors	Unemployment (V1)
	Income (V2)
	Education (V3)
Household composition	Age 65 years or older (V4)
	Age 17 years or younger (V5)
Ethnicity	Ethnic diversity (V6)
	Lifestyle and living conditions (V7)
People	Age (V8)
	Gender (V9)
	Population density (V10)

TABLE 7.2
Linguistic Assessment of the Decision-Makers' Opinions via Consensus

Code	V1	V2	V3	V4	V5	V6	V7	V8	V9	V10
V1	No	VH	H	No	No	H	VH	No	No	H
V2	H	No	VH	VH	No	H	VH	H	L	VH
V3	H	H	No	H	VH	VH	H	H	H	VH
V4	VH	H	H	No	VH	H	H	H	L	H
V5	H	H	H	VH	No	H	L	H	VL	H
V6	VH	H	H	H	H	No	VH	No	VL	No
V7	H	H	H	H	H	L	No	H	L	VH
V8	H	H	H	VH	H	H	H	No	L	H
V9	L	VL	L	H	H	L	H	L	No	L
V10	VH	H	H	H	VL	H	H	VH	H	No

The social vulnerability factors for pandemics were evaluated to determine cause and effect group in drawing a roadmap for governments.

The social vulnerability factors were shown to the DMs in order to assess the relations among each pair of social vulnerability factors as given in Table 7.1. Then, five university DMs from health sociology reached consensus to evaluate the relationships among social vulnerable factors using a fuzzy linguistic scale presented by Akyuz and Celik (2015). Table 7.2 represents the DMs' linguistic assessments arrived at via consensus. Once the initial direct-fuzzy matrix was established using Equation (7.1), the normalized direct-relation matrix was built by performing Equations (7.3–7.5). The normalized direct-relation matrix is illustrated in Table 7.3. Then, using Equations (7.6–7.11), the total-relation fuzzy matrix was obtained and is shown in Table 7.4. The defuzzified threshold values of total-relation fuzzy matrix were obtained, as well

TABLE 7.3
Normalized Initial Direct-Relation Fuzzy Matrix

Code	V1			V2			V3			V4				V9			V10		
V1	(0.00	0.00	0.03)	(0.08	0.11	0.11)	(0.06	0.08	0.11)	(0.00	0.00	0.03)	…	80.00	0.00	0.03)	(0.06	0.08	0.11)
V2	(0.06	0.08	0.11)	(0.00	0.00	0.03)	(0.08	0.11	0.11)	(0.08	0.11	0.11)	…	(0.03	0.06	0.08)	(0.08	0.11	0.11)
V3	(0.06	0.08	0.11)	(0.06	0.08	0.11)	(0.00	0.00	0.03)	(0.06	0.08	0.11)	…	(0.06	0.08	0.11)	(0.08	0.11	0.11)
V4	(0.08	0.11	0.11)	(0.06	0.08	0.11)	(0.06	0.08	0.11)	(0.00	0.00	0.03)	…	(0.03	0.06	0.08)	(0.06	0.08	0.11)
.																			
.																			
V9	(0.03	0.06	0.08)	(0.00	0.03	0.06)	(0.03	0.06	0.08)	(0.06	0.08	0.11)	…	(0.00	0.00	0.03)	(0.03	0.06	0.08)
V10	(0.08	0.11	0.11)	(0.06	0.08	0.11)	(0.06	0.08	0.11)	(0.06	0.08	0.11)	…	(0.06	0.08	0.11)	(0.00	0.00	0.03)

TABLE 7.4
Total-Relation Fuzzy Matrix

Code	V1			V2			V3			V4				V9			V10		
V1	(0.04	0.13	0.87)	(0.11	0.22	0.93)	(0.09	0.20	0.95)	(0.04	0.12	0.82)	…	(0.02	0.08	0.65)	(0.09	0.20	0.88)
V2	(0.11	0.28	1.20)	(0.05	0.19	1.10)	(0.13	0.29	1.20)	(0.13	0.28	1.13)	…	(0.05	0.17	0.89)	(0.13	0.29	1.11)
V3	(0.12	0.30	1.33)	(0.11	0.28	1.30)	(0.06	0.21	1.25)	(0.11	0.28	1.25)	…	(0.08	0.20	1.00)	(0.13	0.30	1.23)
V4	(0.14	0.31	1.30)	(0.10	0.27	1.27)	(0.11	0.27	1.30)	(0.05	0.19	1.15)	…	(0.05	0.17	0.95)	(0.11	0.26	1.20)
.																			
.																			
V9	(0.06	0.20	1.09)	(0.03	0.17	1.04)	(0.06	0.19	1.09)	(0.08	0.21	1.05)	…	(0.01	0.09	0.77)	(0.06	0.19	1.01)
V10	(0.13	0.30	1.26)	(0.10	0.26	1.24)	(0.10	0.26	1.26)	(0.10	0.26	1.19)	…	(0.08	0.19	0.95)	(0.05	0.18	1.09)

TABLE 7.5
Crisp Values of r_i, c_j, $r_i + c_j$, and $r_i - c_j$

Code	r_i	c_j	r_i+c_j	r_i-c_j
V1	3.54	5.19	8.73	-1.65
V2	4.79	4.99	9.78	-0.19
V3	5.30	5.10	10.40	0.20
V4	5.12	4.84	9.95	0.28
V5	4.80	4.06	8.86	0.74
V6	4.09	4.94	9.03	-0.85
V7	4.95	5.18	10.13	-0.23
V8	5.09	4.20	9.28	0.89
V9	4.15	3.55	7.70	0.60
V10	4.96	4.75	9.71	0.20

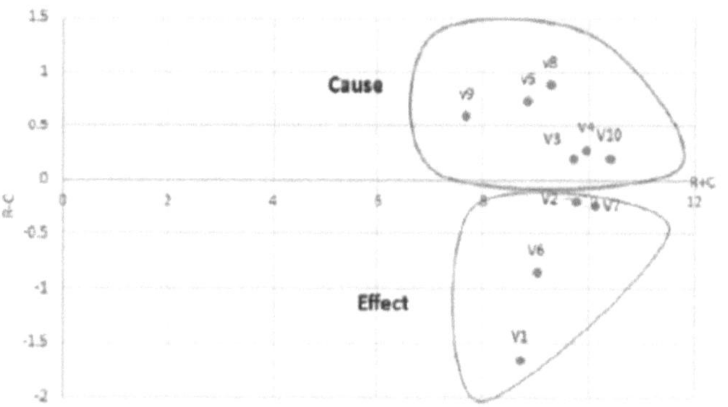

FIGURE 7.2 Cause-and-effect relation diagram.

as fuzzy values of $\tilde{r}_i$, $\tilde{c}_j$, $\tilde{r}_i$-$\tilde{c}_j$, and $\tilde{r}_i + \tilde{c}_j$. Using Equation (7.12), crisp values of $\tilde{r}_i$, $\tilde{c}_j$, $\tilde{r}_i - \tilde{c}_j$, and $\tilde{r}_i + \tilde{c}_j$ were calculated to draw the cause-and-effect relation diagram, as given in Table 7.5. The cause-and-effect relation diagram is displayed in Figure 7.2.

In order to assess the vulnerability of pandemics based on social factors, we the status of vulnerable factors must be identified. Accordingly, age (V8) and age 17 years or younger (V5) have the most significant impact on pandemic vulnerability since they have the highest $\tilde{r}_i - \tilde{c}_j$ values. In other words, the factors of age and age 17 years or younger have a substantial influence on vulnerable populations during pandemics. These are consistent since chronic diseases increase with age, and those aged over 60 years are prone to unhealthy conditions like lung or heart disease, diabetes, or conditions that disturb the immune system. People younger than 17 years of age are vulnerable to coronavirus infection due to their lifestyle, and they can spread the virus. However, influential effect factors can easily be affected by other factors.

Education (V3) and lifestyle and living conditions (V7) lead to serious consequences in dealing with pandemics. Furthermore, their influenced impact index (*cj*) has the biggest value, indicating that the degree of their influence is important in a whole. While lack of education is known as a multi-dimensional social disadvantage, having good lifestyle and living conditions provide increased advantages for employment, often due to more stable employment conditions. Thus, economically vulnerable social groups do not have health insurance, and therefore can suffer more from pandemic and economic crises. As a result, to reduce the risks of people of different ethnic backgrounds and struggle with unemployment due to socio-economic disadvantages, governments and policymakers should suggest solutions with different strategies, learning from previous pandemics.

7.5 CONCLUSION

The pandemics negatively affect people's lives, since they cause economic, social, and political disruptions in a society. Social vulnerability is determined by the social, economic, demographic, cultural, geographical, and historical characteristics that determine an individual's or society's ability to cope with or respond to disasters. Socially vulnerable people are especially at risk during public health emergencies due to demographic, socio-economic, and household compositions.

A vulnerability assessment addresses whom or what is vulnerable, along with how it is vulnerable. In this chapter, social vulnerability factors in a pandemic were evaluated. Fuzzy DEMATEL on the basis of Delphi method was used to evaluate the interconnected relationships between social vulnerability factors against pandemics. A visual structural model was constructed to determine the most critical factors. According to the results, age, age 17 years or younger, and gender are the most significant social vulnerability factors that require great caution for the spread of a pandemic due to having the highest $r_i–c_j$ values (0.89, 0.74, and 0.60, respectively) among all factors in the cause group. Thus, they have more impact in dealing with pandemics and they have to be taken into consideration more. However, lifestyle and living conditions, and education lead to serious consequences in dealing with pandemics since they have the highest $ri+cj$ values (10.13 and 9.95, respectively) among the effect factor group. In addition, their influenced impact index (*cj*) had the biggest value (5.8 and 5.10, respectively). Thus, lifestyle and living conditions and education lead to serious consequences in dealing with pandemics and they must be taken into consideration, firstly to reduce the effect of pandemics. It is necessary for policymakers to consider them extensively, since they can lead to serious consequences. For further studies, different MCDM methods can be integrated with the proposed method. In addition, the methods can be extended with different fuzzy extensions.

REFERENCES

Akyuz, E. & Celik, E. 2015. A fuzzy DEMATEL method to evaluate critical operational hazards during gas freeing process in crude oil tankers. *Journal of Loss Prevention in the Process Industries*, *38*, 243–253.

Bwire, G. M. 2020. Coronavirus: why men are more vulnerable to Covid-19 than women? *SN Comprehensive Clinical Medicine*, *2*(7), 874–876.

Calderón-Larrañaga, A., Dekhtyar, S., Vetrano, D. L., Bellander, T., & Fratiglioni, L. 2020. COVID-19: risk accumulation among biologically and socially vulnerable older populations. *Ageing Research Reviews*, *63*, 101149.

Chukwuma, E. C., Okonkwo, C. C., Ojediran, J. O., Anizoba, D. C., Ubah, J. I., & Nwachukwu, C. P. 2021. A GIS based flood vulnerability modelling of Anambra State using an integrated IVFRN-DEMATEL-ANP model. *Heliyon*, *7*(9), e08048.

de Souza, C. D. F., Machado, M. F., & do Carmo, R. F. 2020. Human development, social vulnerability and COVID-19 in Brazil: a study of the social determinants of health. *Infectious Diseases of Poverty*, *9*(1), 1–10.

Gabus, A. & Fontela, E. 1972. World problems, an invitation to further thought within the framework of DEMATEL. *Battelle Geneva Research Center, Geneva, Switzerland*, 1–8.

Gumus AT. 2009. Evaluation of hazardous waste transportation firms by using a two-step fuzzy-AHP and TOPSIS methodology. *Expert Systems with Applications*, *36*(2), 4067–4074.

Hiete, M., Merz, M., Comes, T., & Schultmann, F. 2012. Trapezoidal fuzzy DEMATEL method to analyze and correct for relations between variables in a composite indicator for disaster resilience. *OR Spectrum*, *34*(4), 971–995.

Hosseini, F. S., Sigaroodi, S. K., Salajegheh, A., Moghaddamnia, A., & Choubin, B. 2021. Towards a flood vulnerability assessment of watershed using integration of decision-making trial and evaluation laboratory, analytical network process, and fuzzy theories. *Environmental Science and Pollution Research*, *28*(44), 62487–62498.

Khazai, B., Merz, M., Schulz, C., & Borst, D. 2013. An integrated indicator framework for spatial assessment of industrial and social vulnerability to indirect disaster losses. *Natural Hazards*, *67*(2), 145–167.

Lakhani, A. 2020. Which Melbourne metropolitan areas are vulnerable to COVID-19 based on age, disability, and access to health services? Using spatial analysis to identify service gaps and inform delivery. *Journal of Pain and Symptom Management*, *60*(1), e41–e44.

Lancet, T. 2020. Redefining vulnerability in the era of COVID-19. *Lancet*, *395*(10230), 1089.

Mamelund, S. E. 2006. A socially neutral disease? Individual social class, household wealth and mortality from Spanish influenza in two socially contrasting parishes in Kristiania 1918–19. *Social Science & Medicine*, *62*(4), 923–940.

Mao, W., Wang, W., Luo, D., & Sun, H. 2019. Analyzing interactions between risk factors for ice disaster in Ning-Meng reach of Yellow River based on grey rough DEMATEL method. *Natural Hazards*, *97*(3), 1025–1049.

OECD. 2020. The impact of the COVID-19 pandemic on jobs and incomes in G20 economies. https://www.ilo.org/global/about-the-ilo/how-the-ilo-works/multilateral-system/g20/reports/WCMS_756331/lang--en/index.htm

Rifat, S. A. A., & Liu, W. 2021. One year into the pandemic: the impacts of social vulnerability on COVID-19 outcomes and urban–rural differences in the conterminous United States. *International Journal of Environmental Health Research*. doi: 10.1080/09603123.2021.1979196

Ross, T. J. 2004. *Fuzzy logic with engineering applications*. New York, John Wiley & Sons.

Roy, N. B., Telfer, P., Eleftheriou, P., de la Fuente, J., Drasar, E., Shah, F., ... & National Haemoglobinopathy Panel. (2020). Protecting vulnerable patients with inherited anaemias from unnecessary death during the COVID-19 pandemic. *British Journal of Haematology*, *189*(4), 635–639.

Rutter, P. D., Mytton, O. T., Mak, M., & Donaldson, L. J. 2012. Socio-economic disparities in mortality due to pandemic influenza in England. *International Journal of Public Health*, *57*(4), 745–750.

Sung W. C. 2001. Application of Delphi method, a qualitative and quantitative analysis, to the healthcare management. *Journal of Healthcare Management*, 2(2), 11–19.

Thomas, D. S., Jang, S., & Scandlyn, J. 2020. The CHASMS conceptual model of cascading disasters and social vulnerability: The COVID-19 case example. *International Journal of Disaster Risk Reduction*, *51*, 101828.

Trivedi, A. 2018. A multi-criteria decision approach based on DEMATEL to assess determinants of shelter site selection in disaster response. *International Journal of Disaster Risk Reduction*, *31*, 722–728.

Waitzberg, R., Davidovitch, N., Leibner, G., Penn, N., & Brammli-Greenberg, S. 2020. Israel's response to the COVID-19 pandemic: tailoring measures for vulnerable cultural minority populations. *International Journal for Equity in Health*, *19*(1), 1–5.

Zadeh L. H. 1965. Fuzzy sets. Information and Control, *8*, 338–353.

8 Drought Vulnerability Assessment Based on IVIF AHP and IVIF WASPAS

A Case Study in Turkey

Esra Ilbahar
Yildiz Technical University

CONTENTS

8.1 Introduction 107
8.2 Literature Review 108
8.3 Methodology 110
  8.3.1 Intuitionistic Fuzzy Sets 110
  8.3.2 Interval-Valued Intuitionistic Fuzzy AHP 111
  8.3.3 Interval-Valued Intuitionistic Fuzzy WASPAS (IVIF–WASPAS) 112
8.4 Application and Discussion 113
8.5 Conclusion 115
References 118

8.1 INTRODUCTION

Drought has many direct and indirect effects, as water, an indispensable natural resource, is used both for the daily needs of the population and for production. (Azevedo Reis et al., 2020). Drought, a natural disaster, harms agricultural production and negatively affects the economy, society, and the environment as a result (Pei et al., 2019; Sivakumar et al., 2021, Mokarram et al., 2021). That is why droughts can be characterized as meteorological, agricultural, hydrological, and socio-economic types. Meteorological drought is described as lack of rainfall and the length of the dry period, whereas agricultural drought is considered to be damage to crops due to a deficiency in soil moisture and lack of rainfall. Hydrological drought is defined as the effect of lack of rainfall on streamflow, and reservoir and lake levels, whereas socio-economic drought is described as the effects of drought on the supply and demand balance of some economic goods (Dabanli, 2018; Mun et al., 2020; Karamouz et al., 2015; https://drought.unl.edu/Education/DroughtIn-depth/TypesofDrought.aspx).

DOI: 10.1201/9781003212904-8

Rainfall has reduced by 20% in the last 25 years in the Mediterranean Basin, where Turkey is located (https://wwftr.awsassets.panda.org/downloads/wwf__turkiyenin_yarinlari_projesi_final_raporu.pdf?3420/turkiyeninyarinlarraporu). In this regard, the effects of global climate change are estimated mostly as drought and, consequently, loss of productivity and income in agriculture, an increase in forest fires, and a loss of biodiversity. It is estimated by the European Commission that the impact of drought conditions in the EU-15 countries on the general economy over the last 30 years has been €85 billion (https://wwftr.awsassets.panda.org/downloads/wwf__turkiyenin_yarinlari_projesi_final_raporu.pdf?3420/turkiyeninyarinlarraporu). According to the January 2021 Global Drought Observatory report (https://edo.jrc.ec.europa.eu/documents/news/GDODroughtNews202102_Turkey.pdf), the total monthly rainfall, snow, and hail in Turkey continued to be below average for most of the past 18 months, and there are some concerns regarding water supply. It was stated that the current drought in Turkey has started to cause crop damage in some cities such as Edirne, Izmir, and Konya (https://edo.jrc.ec.europa.eu/documents/news/GDODroughtNews202102_Turkey.pdf). As vulnerability to drought, based on various factors, differs from region to region, some regions will probably be affected more by changes in temperature, precipitation, soil structure, slope, and lack of groundwater, etc. That is why the evaluation of different cities is crucial to prioritize drought vulnerability and, consequently, in taking appropriate prevention or mitigation actions (Pei et al., 2019; Sivakumar et al., 2021).

As there are many criteria to be considered in the evaluation of different cities in terms of drought vulnerability, multi-criteria decision-making (MCDM) methods should be used to find a solution to the problem considered. Analytic hierarchy process (AHP), one of the most widely used MCDM methods, is preferred to determine factor weights in cases where there is a hierarchical structure. Weighted aggregated sum product assessment (WASPAS) is able to provide a more comprehensive assessment than the weighted sum method and weighted product methods since it combines both of these methods (Mardani et al., 2017). Both AHP and WASPAS are extended with intuitionistic fuzzy sets (IFSs) and are used in the literature since intuitionistic fuzzy sets are able to express uncertainty better by enabling both membership and non-membership degrees to be assigned. In this study, interval-valued intuitionistic fuzzy AHP was utilized to identify the weights of assessment criteria, whereas the interval-valued intuitionistic fuzzy WASPAS method was used to evaluate different cities with respect to drought vulnerability. The rest of the chapter is constructed as follows: The preliminaries on IFSs and the steps of the methods employed in this study are summarized in Section 8.2. The application of the proposed method and related discussion are provided in Section 8.3. Finally, concluding remarks and suggestions for future studies are given in Section 8.4.

8.2 LITERATURE REVIEW

Vulnerability to natural disasters has been widely studied in the literature. Studies have been conducted on the vulnerability to different kinds of disasters such as flood (Tanir et al., 2021; Papilloud and Keiler, 2021; Desalegn and Mulu, 2021; Yang et al., 2021; Figueiredo et al., 2021; Chukwuma et al., 2021; Shreevastav et al., 2021), earthquake (Alam and Haque, 2021; Jena et al., 2021; Jena et al., 2020; Izquierdo-Horna and Kahhat, 2020; Guo and Kapucu, 2020), landslide (Kaynia et al., 2008; Pham et al., 2020; de Assis Dias et al., 2020; Bera et al., 2020; Alemayo and Eritro, 2021), volcano (Maharani et al., 2016; Hizbaron et al., 2018; Reyes-Hardy et al., 2021), wildfire (Andersen and Sugg, 2019; Paveglio et al., 2018; Oliveira et al., 2018; Vaiciulyte et al., 2019; Palaiologou et al., 2019), and tsunami (Voulgaris and Murayama, 2014; Nayak et al., 2014; Santos et al., 2014; Mishra et al., 2014) in order to take the necessary measures and prepare effective plans against natural disasters. With increasing environmental concerns, drought vulnerability began to come to the fore in recent years.

Drought risk assessment has been studied in the literature by considering several factors; the factors used for drought assessment in the literature are summarized in Table 8.1.

In the study of Hoque et al. (2021) fuzzy logic and geospatial techniques were employed to evaluate the risks of agricultural drought in Australia. It was revealed that drought risk is significantly affected by the ability to incorporate mitigation measures. In the study of Karamouz et al. (2015), drought vulnerability of East Azerbaijan province was assessed by considering the factors of precipitation, temperature, land use, groundwater level, solar radiation, and slope. The region maps obtained from GIS (geographic information system) for these factors were combined to form a total vulnerability map, while the weights of these factors were determined using the AHP method. Azevedo Reis et al. (2020) also used GIS along with MCDM to obtain a drought index by taking climatological, economic, social, and water management factors into consideration. Both AHP and GIS were utilized by Sivakumar et al. (2021) to make a drought vulnerability assessment in India. Dabanli (2018) investigated the drought risk for provinces in Turkey by calculating a drought risk index based on vulnerability and hazard. It was revealed that one province has a high drought risk, six provinces have moderate drought risk, and the drought risk of the other provinces is low. Sun et al. (2019) utilized a comprehensive gray relational analysis and max-comprehensive gray relational analysis clustering to determine the main factors for regional agricultural drought vulnerability in China. The main factors influencing drought were identified as effective irrigation area, agricultural population

TABLE 8.1
Factors Used for Drought Assessment in the Literature

Authors	Temperature	Rainfall	Slope	Elevation	Population	Land use
Hoque et al.	✓	✓	✓	✓	✓	✓
Sivakumar et al.	✓	✓	✓		✓	✓
Guo et al.	✓	✓			✓	✓
Azevedo Reis et al.		✓			✓	
Sun et al.					✓	✓
Dabanli		✓			✓	✓
Palchaudhuri & Biswas	✓	✓	✓		✓	✓
Karamouz et al.	✓	✓	✓			✓

percentage, the income of farmers, and agricultural machinery power. Palchaudhuri and Biswas (2016) also used GIS along with AHP to evaluate drought risk in India. In their study, AHP was used to determine the weights of assessment criteria and it was found that annual rainfall and monthly rainfall are the most influential climatic factors on drought. Moreover, it was revealed that land use has much more impact on drought than soil texture. In the study of Guo et al. (2021), the factor weights were calculated by the entropy weight method. Then, different provinces in China were assessed with respect to their agricultural vulnerability to drought by utilizing the contribution model. As can be seen by examining these studies, different methods have been utilized throughout the literature to evaluate drought risk and vulnerability. However, an integration of WASPAS, AHP, and interval-valued IFSs was used for the first time in this assessment. The advantage of using IFSs is that they enable decision-makers to assign both membership and non-membership values to express their judgments. When it is difficult to express uncertainty with single values, interval-valued IFSs are preferred to better define uncertainty in an assessment process. Therefore, prioritization based on the methodology utilized in this study can better represent uncertainty in the assessment process and, consequently, provide more comprehensive results for decision-makers to be able to take the necessary preventive actions.

8.3 METHODOLOGY

In the methodology, an integration of IFSs, AHP, and WASPAS was utilized, because WASPAS is, because of its ease of use, one of the techniques that has started to be preferred in recent years, whereas AHP is one of the most commonly used MCDM methods in the literature where a hierarchical structure exists. Extension of these methods with IFSs enables this methodology to better define uncertainty by using both membership and non-membership values. In this section, preliminaries on IFSs, the steps of interval-valued intuitionistic fuzzy (IVIF) AHP, and IVIF WASPAS methods are provided.

8.3.1 Intuitionistic Fuzzy Sets

IFSs, one of the extensions of ordinary fuzzy sets (Zadeh, 1965), were introduced by Atanassov (1986). Different from ordinary fuzzy sets, IFSs express uncertainty via both membership and non-membership values whose sum is equal to, at most, 1 (Atanassov, 1986).

Definition. Let interval-valued IFS be represented as follows (Atanassov, 1999):

$$\tilde{A} \cong \left\{ \left\langle x, \left[\mu_L(x), \mu_U(x) \right], \left[v_L(x), v_U(x) \right] \right\rangle, x \in X \right\} \tag{8.1}$$

where $\mu_A(x) = \left[\mu_L(x), \mu_U(x) \right] \subset [0,1]$ and $v_A(x) = \left[v_L(x), v_U(x) \right] \subset [0,1]$, then the following equation is satisfied:

$$0 \leq \mu_U(x) + v_U(x) \leq 1 \tag{8.2}$$

Moreover, the indeterminacy interval, $\pi_A(x) = [\pi_L(x), \pi_U(x)]$, is calculated as follows (Atanassov, 1999; Wu et al., 2013):

$$\pi_L(x) = 1 - \mu_U(x) - v_U(x) \tag{8.3}$$

$$\pi_U(x) = 1 - \mu_L(x) - v_L(x) \tag{8.4}$$

Some operations on interval-valued intuitionistic fuzzy numbers (IFNs) are defined as follows (Atanassov, 1999; Zavadskas et al., 2014; Oztaysi et al., 2017):

$$\tilde{A}_1 \oplus \tilde{A}_2 = \left(\left[\mu_1^L + \mu_2^L - \mu_1^L\mu_2^L, \mu_1^U + \mu_2^U - \mu_1^U\mu_2^U\right], \left[v_1^L v_2^L, v_1^U v_2^U\right]\right) \tag{8.5}$$

$$\tilde{A}_1 \otimes \tilde{A}_2 = \left(\left[\mu_1^L\mu_2^L, \mu_1^U\mu_2^U\right], \left[v_1^L + v_2^L - v_1^L v_2^L, v_1^U + v_2^U - v_1^U v_2^U\right]\right) \tag{8.6}$$

$$\lambda\tilde{A} = \left(1 - (1 - \mu_L)^\lambda, 1 - (1 - \mu_U)^\lambda], \left[v_L^\lambda v_U^\lambda\right]\right) \tag{8.7}$$

$$\tilde{A}^\lambda = \left(\left[\mu_L^\lambda \mu_U^\lambda\right]\left[1 - (1 - v_L)^\lambda, 1 - (1 - v_U)^\lambda\right]\right] \tag{8.8}$$

$$\tilde{A}_1 / \tilde{A}_2 = \left(\left[\min\left(\mu_1^L, \mu_2^L\right), \min\left(\mu_1^U \mu_2^U\right)\right], \max\left(v_1^L, v_2^L\right), \left[\max\left(v_1^U, v_2^U\right)\right]\right) \tag{8.9}$$

$$d(\tilde{A}) = \frac{\mu_L + \mu_U + (1 - v_L) + (1 - v_U) + \mu_L\mu_U - \sqrt{(1 - v_L)(1 - v_U)}}{4} \tag{8.10}$$

8.3.2 Interval-Valued Intuitionistic Fuzzy AHP

The steps of IVIF–AHP, introduced by Wu et al. (2013), can be summarized as follows:

Step 1. Pairwise comparisons of assessment criteria are obtained via the linguistic scale provided in the study of Wu et al. (2013) and then these terms are expressed in terms of their corresponding interval-valued IFNs.

Step 2. The pairwise comparison matrix with these interval-valued IFNs is converted to a score judgment matrix, $\tilde{s}_{ij}$, by utilizing Eq. (8.11).

$$\tilde{s}_{ij} = \tilde{\mu}_{ij} - \tilde{\vartheta}_{ij} = [\mu_{ij}^L - \vartheta_{ij}^U, \mu_{ij}^U - \vartheta_{ij}^L], \text{for all i, j,}\ldots\text{, n} \tag{8.11}$$

Step 3. The interval multiplicative matrix, $\tilde{A}$, is obtained from the score judgment matrix with Eq. (8.12).

$$\tilde{A} = 10^{\tilde{s}} \tag{8.12}$$

Step 4. Then, the priority vector of interval multiplicative matrix is obtained by using Eq. (8.13):

$$\tilde{w}_i = \frac{\sum_{j=1}^{n} \tilde{a}_{ij}}{\sum_{i=1}^{n} \sum_{j=1}^{n} \tilde{a}_{ij}} = \left[\frac{\sum_{j=1}^{n} \tilde{a}_{ij}^{-}}{\sum_{i=1}^{n} \sum_{j=1}^{n} \tilde{a}_{ij}^{+}}, \frac{\sum_{j=1}^{n} \tilde{a}_{ij}^{+}}{\sum_{i=1}^{n} \sum_{j=1}^{n} \tilde{a}_{ij}^{-}} \right] \tag{8.13}$$

Step 5. The comparison of each $\tilde{w}_i$ interval number with others is carried out and a possibility degree matrix, $P = (p_{ij})_{nxn}$, is formed by utilizing Eq. (8.14). When $a = [a^-, a^+]$ and $b = [b^-, b^+]$ are interval numbers, the possibility degree of $a \geq b$ is calculated as follows:

$$p(a \geq b) = \frac{\min\left\{L_a + L_b, \max\left(a^+ - b^-, 0\right)\right\}}{L_a + L_b} \tag{8.14}$$

where $L_a = a^+ - a^-$ and $L_b = b^+ - b^-$.

Step 6. Prioritization of $P = (p_{ij})_{nxn}$ to obtain weights is performed by utilizing Eq. (8.15).

$$w_i = \frac{1}{n}\left[\sum_{j=1}^{n} p_{ij} + \frac{n}{2} - 1\right] \tag{8.15}$$

8.3.3 Interval-Valued Intuitionistic Fuzzy WASPAS (IVIF–WASPAS)

WASPAS, proposed by Zavadskas et al. (2012), is a MCDM method that facilitates the contribution of both the weighted sum model (WSM) and the weighted product model (WPM) for a comprehensive assessment.

IVIF–WASPAS is the integration of IFSs and the WASPAS method (Zavadskas et al., 2014). The steps of this method are summarized as follows:

Step 1. Assessments are collected in linguistic form and converted to interval-valued IFNs through the scale introduced by Zavadskas et al. (2014).

Step 2. After aggregation, in order to normalize the values, either Eq. (8.16) or Eq. (8.17) is used, depending on the criteria type.

$$\tilde{r}_{ij} = \frac{\tilde{x}_{ij}}{\max_i \tilde{x}_{ij}} \tag{8.16}$$

$$\tilde{r}_{ij} = \frac{\min_i \tilde{x}_{ij}}{\tilde{x}_{ij}} \tag{8.17}$$

where $\max_i \tilde{x}_{ij}$ = $([max_i\, \mu_{ij}^L, max_i \mu_{ij}^U],\ [min_i\, \vartheta_{ij}^L\ \ min_i \vartheta_{ij}^U])$ and $\min_i \tilde{x}_{ij}$ = $([min_i\, \mu_{ij}^L, min_i \mu_{ij}^U],\ [max_i\, \vartheta_{ij}^L\ \ max_i \vartheta_{ij}^U])$. Then, in order to perform division operation for interval-valued IFNs, Eq. (8.9) is used.

Step 3. In order to calculate weighted sum values, $\tilde{Q}_i^{(1)}$, as given in Eq. (8.18), operations for interval-valued IFNs provided in Eqs. (8.5)–(8.7) are used.

$$\tilde{Q}_i^{(1)} = \sum_{j=1}^{n} \tilde{r}_{ij} w_j \tag{8.18}$$

Step 4. In order to calculate weighted product values, $\tilde{Q}_i^{(2)}$, as given in Eq. (8.19), operations for interval-valued IFNs provided in Eqs. (8.6)–(8.8) are used.

$$\tilde{Q}_i^{(2)} = \prod_{j=1}^{n} \tilde{r}_{ij}^{w_j} \tag{8.19}$$

Step 5. Then, weighted sum values, $\tilde{Q}_i^{(1)}$, and weighted product values, $\tilde{Q}_i^{(2)}$, values are combined as given in Eq. (8.20) by using Eqs. (8.5)–(8.7) in order to calculate the total relative score of each alternative, $\tilde{Q}_i$.

$$\tilde{Q}_i = \lambda\, \tilde{Q}_i^{(1)} + (1 - \lambda)\tilde{Q}_i^{(2)} \tag{8.20}$$

Step 7. Finally, defuzzification of $\tilde{Q}_i$ values is performed by utilizing Eq. (8.10) and the alternative with the highest score is identified as the best alternative.

8.4 APPLICATION AND DISCUSSION

The conditions such as changes in temperature, precipitation, soil structure, slope, and lack of groundwater will probably have more influence on some regions. For this reason, in this study, the aim was to assess cities with respect to their drought vulnerability to be able to put appropriate preventive measures in place. In this context, eight assessment criteria were determined for drought vulnerability assessment and categorized under three main headings. The necessary data for this analysis were obtained from information published by official institutions, publications, and decision-makers. IVIF–AHP was applied to identify criteria weights after four pairwise comparison matrices were formed, as provided in Tables 8.2–8.5.

After criteria weights were obtained, as given in Table 8.6, thirteen cities in the Central Anatolia Region of Turkey were evaluated with respect to these criteria and it is provided in Table 8.7. By applying the steps of the IVIF–WASPAS method, $\lambda\, \tilde{Q}_i^{(1)}$ and $(1 - \lambda)\tilde{Q}_i^{(2)}$ values were calculated as presented in Table 8.8. After the sum of the values presented in Table 8.8 were calculated using Eq. (8.5), Eq. (8.10) was used to obtain the defuzzied values in Table 8.9.

TABLE 8.2
Pairwise Comparison of Main Criteria

C	C1	C2	C3
C1	E	VH	H
C2	VL	E	L
C3	L	H	E

TABLE 8.3
Pairwise Comparison of C1 Criteria

C1	C11	C12	C13
C11	E	H	L
C12	L	E	VL
C13	H	VH	E

TABLE 8.4
Pairwise Comparison of C2 Criteria

C2	C21	C22
C21	E	L
C22	H	E

TABLE 8.5
Pairwise Comparison of C3 Criteria

C3	C31	C32	C33
C31	E	VL	L
C32	VH	E	H
C33	H	L	E

As can be seen from Table 8.9, the most vulnerable city identified was Konya, followed by Ankara, Niğde, Kırıkkale, and Kırşehir. The findings show some similarities with previous research and reports. In the study of Dabanli (2018), Konya was also identified as the most vulnerable city in terms of drought, followed by Ankara. Moreover, it was stated by the January 2021 Global Drought Observatory report that there was crop damages in Konya, and Ankara is one of the cities in the Central

TABLE 8.6
Criteria Weights Calculated with IVIF AHP

Criteria		Weights
C1	Climate-related criteria	0.440
C11	Average temperature	0.147
C12	Average maximum temperature	0.100
C13	Annual rainfall	0.194
C2	Topography-related criteria	0.227
C21	Elevation	0.100
C22	Slope	0.126
C3	Population and land use-related criteria	0.333
C31	Agricultural land areas	0.089
C32	Irrigated land areas	0.134
C33	Population	0.111

Anatolia Region affected by drought (https://edo.jrc.ec.europa.eu/documents/news/GDODroughtNews202102_Turkey.pdf).

A sensitivity analysis was performed to validate the reliability of the outputs. In the analysis, eight cases were evaluated by setting the weight of only one of the criteria equal to 0.125 each time while adjusting the other weights proportionally. Figure 8.1 depicts the change in the prioritization of the cities with respect to drought vulnerability. In this figure, the y-axis represents the ranking of 13 cities, while the x-axis represents the different cases where changes are made to the criteria weights.

Konya was identified as the most vulnerable city in all cases, followed by Ankara, Niğde, and Kırıkkale. It was also seen that the rankings of Kayseri and Çankırı are relatively more sensitive to the change in weights, and ranks 9 and 10 vary between these cities. Moreover, relatively sharper changes in the ranks of Sivas and Aksaray occurred when the changes were made to the weights of criteria C11 (average temperature) and C31 (agricultural land areas).

8.5 CONCLUSION

Drought is a detrimental natural disaster and has come to the fore due to the increasing effects of climate change. As it is influenced by various factors such as temperature, rainfall, and land use, cities' vulnerability to drought varies. In order to make effective plans and take necessary measures, prioritizing cities with respect to drought vulnerability is required. This prioritization can be made effectively with MCDM methods since several factors affecting drought vulnerability exist. That is why, in this study, after the weights of evaluation criteria were identified with interval-valued intuitionistic fuzzy AHP, the drought vulnerability of different cities was compared using the IVIF–WASPAS method. As a result of the analysis, Konya has been found to be more vulnerable to a drought, followed by Ankara, Niğde, Kırıkkale, and Kırşehir. Therefore, these cities should be given priority while making the necessary plans

TABLE 8.7
Assessment of Cities in Central Anatolia Region

	Ankara	Konya	Kayseri	Eskişehir	Sivas	Kırıkkale	Aksaray	Karaman	Kırşehir	Niğde	Nevşehir	Yozgat	Çankırı
C11	H	H	M	H	L	VH	H	H	H	H	M	L	H
C12	H	H	H	H	L	VH	VH	VH	H	H	M	VL	H
C13	VL	VVL	M	H	VL	ML	L	VVL	VVL	L	M	ML	VVL
C21	M	H	H	L	VH	L	M	H	M	VH	VH	VH	L
C22	H	M	H	H	VH	H	M	M	H	H	H	H	H
C31	H	VH	M	M	M	L	L	L	L	L	L	M	L
C32	M	VH	M	M	M	M	M	M	M	M	M	M	M
C33	VVH	H	M	L	L	VL	VL	VL	VL	VL	VL	VL	VL

TABLE 8.8
The Contribution of Weighted Sum, $\tilde{Q}_i^{(1)}$, and Weighted Product, $\tilde{Q}_i^{(2)}$, Values

Cities	$\lambda\tilde{Q}_i^{(1)}$	$(1-\lambda)\tilde{Q}_i^{(2)}$
Ankara	[0.343, 0.403], [0.451, 0.563]	[0, 0.269], [0.67, 0.714]
Konya	[0.308, 0.385], [0, 0.569]	[0, 0.267], [0.664, 0.713]
Kayseri	[0.259, 0.329], [0.509, 0.632]	[0, 0.24], [0.692, 0.74]
Eskişehir	[0.226, 0.3], [0.529, 0.659]	[0, 0.206], [0.711, 0.77]
Sivas	[0.215, 0.296], [0, 0.656]	[0, 0.188], [0.723, 0.786]
Kırıkkale	[0.235, 0.319], [0, 0.631]	[0, 0.183], [0.721, 0.788]
Aksaray	[0.22, 0.295], [0, 0.663]	[0, 0.188], [0.725, 0.785]
Karaman	[0.241, 0.319], [0, 0.637]	[0, 0.197], [0.714, 0.776]
Kırşehir	[0.225, 0.298], [0.533, 0.662]	[0, 0.193], [0.72, 0.781]
Niğde	[0.261, 0.341], [0, 0.612]	[0, 0.205], [0.704, 0.767]
Nevşehir	[0.22, 0.295], [0, 0.662]	[0, 0.188], [0.725, 0.786]
Yozgat	[0.188, 0.265], [0, 0.691]	[0, 0.156], [0.747, 0.814]
Çankırı	[0.205, 0.28], [0.547, 0.679]	[0, 0.177], [0.73, 0.795]

TABLE 8.9
Total Relative Score of Each Alternative and the Corresponding Defuzzified Values

Cities	$\tilde{Q}_i = \lambda\tilde{Q}_i^{(1)} + (1-\lambda)\tilde{Q}_i^{(2)}$	Defuzzified $\tilde{Q}_i$ values
Ankara	[0.343, 0.563], [0.303, 0.402]	0.437
Konya	[0.308, 0.55], [0, 0.406]	0.462
Kayseri	[0.259, 0.49], [0.353, 0.468]	0.367
Eskişehir	[0.226, 0.444], [0.376, 0.507]	0.333
Sivas	[0.215, 0.428], [0, 0.516]	0.381
Kırıkkale	[0.235, 0.444], [0, 0.497]	0.394
Aksaray	[0.22, 0.428], [0, 0.521]	0.382
Karaman	[0.241, 0.453], [0.381, 0.494]	0.342
Kırşehir	[0.225, 0.433], [0, 0.516]	0.386
Niğde	[0.261, 0.476], [0, 0.47]	0.416
Nevşehir	[0.22, 0.427], [0, 0.52]	0.382
Yozgat	[0.188, 0.38], [0.408, 0.563]	0.290
Çankırı	[0.205, 0.407], [0, 0.539]	0.369

and carrying out some mitigation. Moreover, a sensitivity analysis was performed by adjusting criteria weights and it was shown that the findings are reliable.

For future studies, this assessment can be applied to other regions as well. Moreover, instead of WASPAS, other MCDM methods such as TOPSIS, EDAS, and VIKOR can be integrated into the solution procedure. To validate the findings, other fuzzy set extensions such as type 2 fuzzy sets and Pythagorean fuzzy sets can also be used.

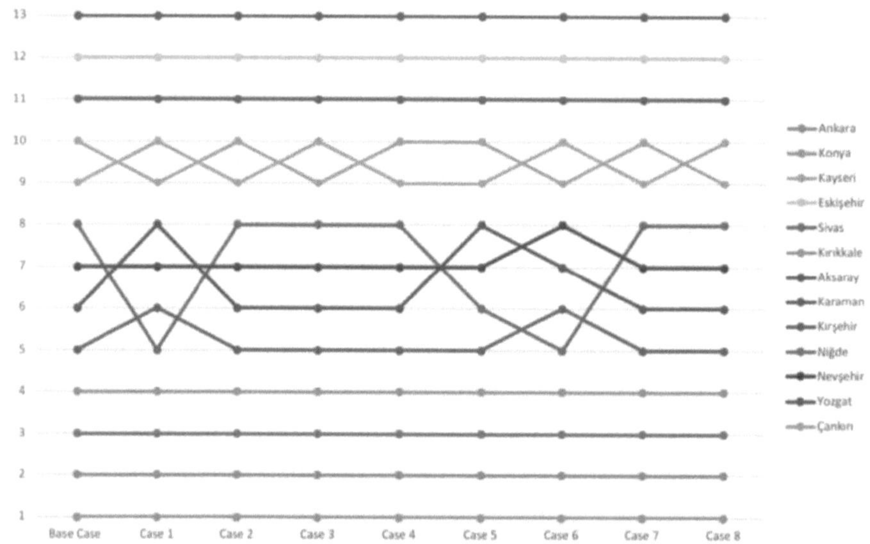

FIGURE 8.1 Change in the prioritization of cities with different criteria weights.

REFERENCES

Alam, M. S. & Haque, S. M. 2021. Multi-dimensional earthquake vulnerability assessment of residential neighborhoods of Mymensingh City, Bangladesh: A spatial multi-criteria analysis-based approach. *Journal of Urban Management* 11:37–58.

Alemayo, G. G. & Eritro, T. H. 2021. Landslide vulnerability of the Debre Sina-Armania road section, Central Ethiopia: Insights from geophysical investigations. *Journal of African Earth Sciences* 184:104383.

Andersen, L. M. & Sugg, M. M. 2019. Geographic multi-criteria evaluation and validation: A case study of wildfire vulnerability in Western North Carolina, USA following the 2016 wildfires. *International Journal of Disaster Risk Reduction* 39:101123.

Atanassov, K. T. 1999. Interval valued intuitionistic fuzzy sets. In *Intuitionistic Fuzzy Sets* (pp. 139–177). Physica, Heidelberg.

Atanassov, K. T. 1986. Intuitionistic fuzzy sets. *Fuzzy sets and Systems* 20:87–96.

Azevedo Reis, G., de Souza Filho, F. A., Nelson, D. R., Rocha, R. V., & da Silva, S. M. O. 2020. Development of a drought vulnerability index using MCDM and GIS: study case in São Paulo and Ceará, Brazil. *Natural Hazards* 104(2):1781–1799.

Bera, S., Guru, B., & Oommen, T. 2020. Indicator-based approach for assigning physical vulnerability of the houses to landslide hazard in the Himalayan region of India. *International Journal of Disaster Risk Reduction* 50:101891.

Chukwuma, E. C., Okonkwo, C. C., Ojediran, J. O., Anizoba, D. C., Ubah, J. I., & Nwachukwu, C. P. 2021. A GIS based flood vulnerability modelling of Anambra State using an integrated IVFRN-DEMATEL-ANP model. *Heliyon* 7(9):e08048.

Dabanli, I. 2018. Drought hazard, vulnerability, and risk assessment in Turkey. *Arabian Journal of Geosciences* 11(18):1–12.

de Assis Dias, M. C., Saito, S. M., & dos Santos Alvalá, R. C. et al. 2020. Vulnerability index related to populations at-risk for landslides in the Brazilian Early Warning System (BEWS). *International Journal of Disaster Risk Reduction* 49:101742.

Desalegn, H. & Mulu, A. 2021. Flood vulnerability assessment using GIS at Fetam watershed, upper Abbay basin, Ethiopia. *Heliyon* 7(1):e05865.

Figueiredo, R., Romão, X., & Paupério, E. 2021. Component-based flood vulnerability modelling for cultural heritage buildings. *International Journal of Disaster Risk Reduction* 61:102323.

Guo, H., Chen, J., & Pan, C. 2021. Assessment on agricultural drought vulnerability and spatial heterogeneity study in China. *International Journal of Environmental Research and Public Health* 18(9):4449.

Guo, X. & Kapucu, N. 2020. Assessing social vulnerability to earthquake disaster using rough analytic hierarchy process method: A case study of Hanzhong City, China. *Safety Science* 125:104625.

Hizbaron, D. R., Hadmoko, D. S., & Mei, E. T. W., et al. 2018. Towards measurable resilience: Mapping the vulnerability of at-risk community at Kelud Volcano, Indonesia. *Applied Geography* 97:212–227.

Hoque, M. A. A., Pradhan, B., Ahmed, N., & Sohel, M. S. I. 2021. Agricultural drought risk assessment of Northern New South Wales, Australia using geospatial techniques. *Science of the Total Environment* 756:143600.

Izquierdo-Horna, L., & Kahhat, R. 2020. An interdisciplinary approach to identify zones vulnerable to earthquakes. *International Journal of Disaster Risk Reduction* 48:101592.

Jena, R., Naik, S. P., Pradhan, B., Beydoun, G., Park, H. J., & Alamri, A. 2021. Earthquake vulnerability assessment for the Indian subcontinent using the Long Short-Term Memory model (LSTM). *International Journal of Disaster Risk Reduction* 66:102642.

Jena, R., Pradhan, B., & Beydoun, G. 2020. Earthquake vulnerability assessment in Northern Sumatra province by using a multi-criteria decision-making model. *International Journal of Disaster Risk Reduction* 46:101518.

Karamouz, M., Zeynolabedin, A., & Olyaei, M. A. 2015. Mapping regional drought vulnerability: A case study. *International Archives of the Photogrammetry, Remote Sensing & Spatial Information Sciences* 40:369–377.

Kaynia, A. M., Papathoma-Köhle, M., Neuhäuser, B., Ratzinger, K., Wenzel, H., & Medina-Cetina, Z. 2008. Probabilistic assessment of vulnerability to landslide: application to the village of Lichtenstein, Baden-Württemberg, Germany. *Engineering Geology* 101(1–2):33–48.

Maharani, Y. N., Lee, S., & Ki, S. J. 2016. Social vulnerability at a local level around the Merapi volcano. *International Journal of Disaster Risk Reduction* 20:63–77.

Mardani, A., Nilashi, M., & Zakuan, N., et al. 2017. A systematic review and meta-analysis of SWARA and WASPAS methods: Theory and applications with recent fuzzy developments. *Applied Soft Computing* 57:265–292.

Mishra, P., Usha, T., & Ramanamurthy, M. V. 2014. Evaluation of tsunami vulnerability along northeast coast of India. *Continental Shelf Research* 79:16–22.

Mokarram, M., Pourghasemi, H. R., Hu, M., & Zhang, H. 2021. Determining and forecasting drought susceptibility in southwestern Iran using multi-criteria decision-making (MCDM) coupled with CA-Markov model. *Science of the Total Environment* 781:146703.

Mun, Y. S., Nam, W. H., Jeon, M. G., Bang, N. K., & Kim, T. 2020. Assessment of vulnerability to drought disaster in agricultural reservoirs in South Korea. *Atmosphere* 11(11):1244.

Nayak, S., Reddy, M. H. O., Madhavi, R., & Dutta, S. C. 2014. Assessing tsunami vulnerability of structures designed for seismic loading. *International journal of Disaster Risk Reduction* 7:28–38.

Oliveira, S., Félix, F., Nunes, A., Lourenço, L., Laneve, G., & Sebastián-López, A. 2018. Mapping wildfire vulnerability in Mediterranean Europe. Testing a stepwise approach for operational purposes. *Journal of Environmental Management* 206:158–169.

Oztaysi, B., Onar, S. C., Goztepe, K., & Kahraman, C. 2017. Evaluation of research proposals for grant funding using interval-valued intuitionistic fuzzy sets. *Soft Computing* 21(5):1203–1218.

Palaiologou, P., Ager, A. A., Nielsen-Pincus, M., Evers, C. R., & Day, M. A. 2019. Social vulnerability to large wildfires in the western USA. *Landscape and Urban Planning* 189:99–116.

Palchaudhuri, M., & Biswas, S. 2016. Application of AHP with GIS in drought risk assessment for Puruliya district, India. *Natural Hazards* 84(3):1905–1920.

Papilloud, T. & Keiler, M. 2021. Vulnerability patterns of road network to extreme floods based on accessibility measures. *Transportation Research Part D: Transport and Environment* 100:103045.

Paveglio, T. B., Edgeley, C. M., & Stasiewicz, A. M. 2018. Assessing influences on social vulnerability to wildfire using surveys, spatial data and wildfire simulations. *Journal of Environmental Management* 213:425–439.

Pei, W., Fu, Q., Liu, D., Li, T., Cheng, K., & Cui, S. 2019. A novel method for agricultural drought risk assessment. *Water Resources Management* 33(6): 2033–2047.

Pham, N. T. T., Nong, D., Sathyan, A. R., & Garschagen, M. 2020. Vulnerability assessment of households to flash floods and landslides in the poor upland regions of Vietnam. *Climate Risk Management* 28:100215.

Reyes-Hardy, M. P., Barraza, F. A., Birke, J. P. S., Cáceres, A. E., & Pizarro, M. I. 2021. GIS-based volcanic hazards, vulnerability and risks assessment of the Guallatiri Volcano, Arica y Parinacota Region, Chile. *Journal of South American Earth Sciences* 109:103262.

Santos, A., Tavares, A. O., & Emidio, A. 2014. Comparative tsunami vulnerability assessment of an urban area: An analysis of Setúbal city, Portugal. *Applied Geography* 55:19–29.

Shreevastav, B. B., Tiwari, K. R., Mandal, R. A., & Nepal, A. 2021. Assessing flood vulnerability on livelihood of the local community: A case from southern Bagmati corridor of Nepal. *Progress in Disaster Science* 12:100199.

Sivakumar, V. L., Krishnappa, R. R., & Nallanathel, M. 2021. Drought vulnerability assessment and mapping using Multi-Criteria decision making (MCDM) and application of Analytic Hierarchy process (AHP) for Namakkal District, Tamilnadu, India. *Materials Today: Proceedings* 43:1592–1599.

Sun, H., Dang, Y., & Mao, W. 2019. Identifying key factors of regional agricultural drought vulnerability using a panel data grey combined method. *Natural Hazards* 98(2):621–642.

Tanir, T., Sumi, S. J., & de Lima, A. D. S., et al. 2021. Multi-scale comparison of urban socio-economic vulnerability in the Washington, DC Metropolitan Region resulting from compound flooding. *International Journal of Disaster Risk Reduction* 61:102362.

Vaiciulyte, S., Galea, E. R., Veeraswamy, A., & Hulse, L. M. 2019. Island vulnerability and resilience to wildfires: A case study of Corsica. *International Journal of Disaster Risk Reduction* 40:101272.

Voulgaris, G. & Murayama, Y. 2014. Tsunami vulnerability assessment in the southern boso peninsula, Japan. *International Journal of Disaster Risk Reduction* 10:190–200.

Wu, J., Huang, H. B., & Cao, Q. W. 2013. Research on AHP with interval-valued intuitionistic fuzzy sets and its application in multi-criteria decision making problems. *Applied Mathematical Modelling* 37(24): 9898–9906.

Yang, Y., Guo, H., Wang, D., Ke, X., Li, S., & Huang, S. 2021. Flood vulnerability and resilience assessment in China based on super-efficiency DEA and SBM-DEA methods. *Journal of Hydrology* 600:126470.

Zadeh, L. A. 1965. Fuzzy sets. *Information and Control* 8:338–353.

Zavadskas, E. K., Antucheviciene, J., Hajiagha, S. H. R., & Hashemi, S. S. 2014. Extension of weighted aggregated sum product assessment with interval-valued intuitionistic fuzzy numbers (WASPAS-IVIF). *Applied Soft Computing* 24:1013–1021.

Zavadskas, E. K., Turskis, Z., Antucheviciene, J., & Zakarevicius, A. 2012. Optimization of weighted aggregated sum product assessment. *Elektronika ir Elektrotechnika* 122(6):3–6.

9 Evaluation of the Performance of the Health Systems of 12 OECD Countries in Terms of Preparedness for Pandemics

Esra Ayhan[1] *and Serap Pelin Türkoğlu*[2]
[1]Ankara University
[2]Ankara Yildirim Beyazit University

CONTENTS

9.1 Introduction 123
9.2 Method and Data 125
 9.2.1 CRITIC Method 125
 9.2.2 WASPAS Method 127
9.3 Results 128
9.4 Conclusion 132
References 133

9.1 INTRODUCTION

A pandemic is an epidemic that occurs globally or over a wide area, and it generally affects many people (Qui et al., 2017). In the 21st century, there have been three pandemics in effect all over the world: H1N1 (2009), MERS-CoV (2012), and COVID-19 (2019).

H1N1 is a pandemic that began to spread with a new strain of influenza virus in April 2009. Although this new strain of the virus is often referred to as swine flu, it is also called Mexican flu, swine-centered flu, North American flu, and H1N1 flu. When the course of the disease to date is examined, it has been seen that children and young people are more affected. The more prominent symptoms are cough, fever, sore throat, headache, muscle aches and runny nose, diarrhea, and nausea and vomiting. In total, 14,286 people have died worldwide due to this pandemic (Wikipedia List of Epidemics, 2021).

DOI: 10.1201/9781003212904-9

MERS-CoV is a contagious respiratory disease caused by a novel coronavirus, first identified in Saudi Arabia in 2012. The MERS-CoV virus is suspected to have originated from camels. However, it is not known exactly where and how the disease is transmitted to humans. The disease can be transmitted from person to person by close contact, but, for now, it has not spread rapidly. Human-to-human transmission occurs between family members, patients in hospitals, and healthcare workers. Common features of respiratory diseases such as fever, chills, headache, dizziness, sore throat, dry cough, shortness of breath, and muscle pain usually appear as the first complaints. Gastrointestinal symptoms such as vomiting, and diarrhea may also be seen. The disease is more severe in those with weakened immune systems, elderly people, and those with cancer, diabetes, and chronic lung disease. Worldwide, 941 people have died from MERS-CoV (Wikipedia List of Epidemics, 2021).

The COVID-19, or coronavirus, pandemic is a virus outbreak that emerged on 1 December 2019, in Wuhan, the capital of the Hubei region of China. Symptoms of the disease include high fever, coughing, difficulty in breathing, and the loss of sense of smell and taste; the virus can cause death. People with chronic diseases, those over 60 years of age, and healthcare workers are the people most affected by this virus. As of October 2021, the total number of confirmed cases worldwide was 242,688,319 and the total loss of life was 4,932,928 (World Health Organization Covid-19 Dashboard, 2021).

Global pandemics have had significant negative impacts on society and economies. They have also caused political and social disruption (Qui et al., 2017). Therefore, it is important for countries to be prepared for a pandemic.

In the literature, there have performance analysis studies of health systems of Organisation for Economic Co-operation and Development (OECD) countries. Cetin and Bahce (2016) analyzed the efficiency of the health sectors of 34 OECD countries. The data envelopment analysis method was used to analyze the various inputs (the number of patient beds, number of doctors, and health expenditure per capita) and outputs (infant mortality rate and life expectancy at birth) of the study. Samut and Jafri (2016) examined the efficiency of hospitals in 29 OECD countries from 2000 to 2010. They applied a two-stage analysis. The first stage was measuring the hospitals' efficiency using data envelopment analysis (DEA); the second stage was panel tobit analysis. Karagiannis (2016) used a non-parametric two-stage DEA approach to measure the efficiency of health systems in OECD countries. In the first step of the approach, the DEA method was used to determine the output-oriented technical efficiency score for each country. Second, tobit analysis was used to regress first-stage DEA scores with environmental variables (GDP per capita using, dietary habits, alcohol consumption, etc.). Yiğit (2019) used the TOPSIS method to analyze the performance of OECD countries' health performance. Expenditure of GDP on health (%), expenditure on health per capita (US$), life expectancy of the total population at birth (years), infant mortality deaths per 1000 live births, potential years of life lost per 100,000 females, and potential years of life lost per 100,000 males were the study variables. Pekkaya and Dökmen (2019) used both DEA methods and multi-criteria decision-making (MCDM) methods for evaluating the performance of OECD countries' healthcare systems. They analyzed two output and eight input variables belonging to 35 OECD countries for 2010–2016. They calculated criterion/variable importance levels with DEA, entropy, and CRITICS methods, and health performance rankings were carried out using CCR (output-oriented DEA model), TOPSIS, and grey relational analysis (GRA) methods. Yüksel (2021) compared the health inefficiencies of OECD countries with DEA between the years 2014 and 2018.

Pandemics are one of the most important health problems that countries have faced for centuries. Within the scope of combating pandemics, countries need to develop national and international strategies, and allocate sufficient resources to their health systems. Health indicators are used to compare the health systems of countries, and the comparative evaluations guide policymakers and decision-makers. When we examine the literature, no study has evaluated the performance of health systems in OECD countries in terms of preparedness for pandemics with multi-criteria decision methods. The main aim of this study was to investigate the status of the health systems of OECD countries within the scope of preparedness for pandemics. This study is also important for guiding policymakers in the fight against pandemics.

This chapter is structured as follows. Section 9.2 gives detailed information on the methods and data used for this study. Section 9.3 sets out the discussion with the results of the analysis. The final section presents the conclusions and future recommendations in this area.

9.2 METHOD AND DATA

In the study, CRITIC (Criteria Importance Through Inter-Criteria Correlation) and WASPAS (Weighted Aggregated Sum Product Assessment) methods were used for performance analysis. Analysis was carried out for 12 OECD countries with available data for the years 2009 (start of the H1N1 pandemic), 2012 (start of the MERS-CoV pandemic), and 2019 (start of the COVID-19 pandemic). The evaluated OECD countries and their coding are shown in Table 9.1.

In order to evaluate the preparedness performance of the health systems of the countries examined, five criteria were determined by considering the studies in the literature (Reinhardt et al., 2002; Sinimole, 2012; Samut and Cafrı, 2016; Cetin and Bahce, 2016; Karagiannis, 2016; Yiğit, 2019; Lari and Sefiddashti, 2021; Yüksel, 2021). The criteria and codes used in the study are given in Table 9.2. Values of the criteria are taken from the OECD health database (OECD Statistics, 2021). In the next section, the methods used in the analysis are explained.

9.2.1 CRITIC METHOD

The CRITIC method, developed by Diakoulaki et al. (1995), is used to weight the criteria and is an objective criterion weighting method. In this method, the standard deviation of the criteria and the correlations between the criteria are considered to determine the weights of the criteria.

TABLE 9.1
Evaluated Countries

OECD Countries	Coding
Belgium	A1
Canada	A2
Germany	A3
Iceland	A4
Ireland	A5
Korea	A6
Netherlands	A7
Norway	A8
Portugal	A9
Spain	A10
Sweden	A11
United Kingdom	A12

TABLE 9.2
Criteria Used in Analysis

Criteria	Coding	Definition
Infant mortality/deaths per 1000 live births	C1	The number of deaths of children aged <1 years in a given year per 1000 live births.
Current expenditure on health per capita	C2	The amount of health expenditure per capita of a given country.
Current expenditure on health (% GDP)	C3	The amount of health expenditure as a percentage of the GDP of a given country.
Total hospital beds per 1000 population	C4	The number of hospital beds per 1000 population in a given country.
Number of physicians per 1000 population	C5	The number of physicians per 1000 population in a given country.

The basic steps of the method are given below (Jahan et al., 2012; Babatunde and Ighravwe, 2019; Odu, 2019):

Step 1. A decision matrix is created with criteria and alternatives. The criteria are normalized:

For benefit criteria;

$$r_{ij} = \frac{x_{ij} - x_j^{min}}{x_j^{max} - x_j^{min}}, \quad i = 1,\ldots,m; j = 1,\ldots,n \tag{9.1}$$

i, alternatives; *j* indicates the criteria.

For cost criteria;

$$r_{ij} = \frac{x_j^{max} - x_{ij}}{x_j^{max} - x_j^{min}} \quad i = 1,\ldots,m; j = 1,\ldots,n \tag{9.2}$$

Step 2. The correlation between the criteria is calculated:

$$\rho_{jk} = \frac{\sum_{i=1}^{m}\left(r_{ij} - \overline{r_j}\right)\left(r_{ik} - \overline{r_k}\right)}{\sqrt{\sum_{i=1}^{m}\left(r_{ij} - \overline{r_j}\right)^2 \sum_{i=1}^{m}\left(r_{ik} - \overline{r_k}\right)^2}} \quad j,k = 1,\ldots,n \tag{9.3}$$

Step 3. Calculate the criteria weights:

$$w_j = \frac{c_j}{\sum_{k=1}^{n} c_k} \quad j = 1,\ldots,n \tag{9.4}$$

$$c_j = \sigma_j \sum_{k=1}^{n} (1 - \rho_{jk}) \quad j = 1,\ldots,n \tag{9.5}$$

σ_j, is the standard deviation of the criterion *j* and c_j is the amount of information in the *j* criterion.

The reason why this method was preferred in weighting the criteria in the study is that it is both an objective evaluation method and requires easy mathematical processing steps.

9.2.2 WASPAS Method

The WASPAS method developed by Zavadskas et al. (2012) was created by combining the weighted sum model (WSM) and the weighted product model (WPM) methods. The method follows the steps below (Zavadskas et al., 2012; Chakraborty and Zavadskas, 2014):

Step 1. The decision matrix is created.
Step 2. The decision matrix is normalized according to the benefit and cost criteria:

For benefit criteria;

$$\overline{x}_{ij} = \frac{x_{ij}}{max_i x_{ij}} \quad i = 1,\ldots,m; = 1,\ldots,n \tag{9.6}$$

For cost criteria:

$$\bar{x}_{ij} = \frac{min_i x_{ij}}{x_{ij}} \quad i = 1,\ldots,m; j = 1,\ldots,n \tag{9.7}$$

Step 3. Based on the weighted-sum model method, the overall relative importance of the alternative is calculated as:

$$Q_i^{(1)} = \sum_{j=1}^{n} \bar{x}_{ij} w_j \tag{9.8}$$

Step 4. According to the weighted product model method, the total relative importance of the alternative i is calculated as follows:

$$Q_i^{(2)} = \prod_{j=1}^{n} (\bar{x}_{ij})^{w_j} \tag{9.9}$$

Step 5. For the weighted sum and weighted product models, the weighted common general criterion value is calculated:

$$Q_i = 0.5Q_i^{(1)} + 0.5Q_i^{(2)} \tag{9.10}$$

Step 6. Calculate the total relative importance of the alternatives:

$$Q_i = \lambda Q_i^{(1)} + (1 - \lambda) Q_i^{(2)} \tag{9.11}$$

($\lambda = 0, 0.1, \ldots, 1$).

The determined alternatives are ranked according to their Q_i value, that is, the best alternative should be the highest Q_i value. "λ" is a parameter used in the WASPAS method and takes a value between 0 and 1. When $\lambda = 0$, the WASPAS method turns into WPM, and when $\lambda = 1$, it turns into WSM.

The reason for using the WASPAS method for the pandemic preparedness performance of the health systems of OECD countries is that it is a method that tries to reach the highest estimation accuracy (Tuş and Adalı, 2019), has short calculation steps, and is a reliable MCDM method (Stojić et al., 2018).

9.3 RESULTS

In this study, the importance weights of the performance criteria were calculated with the CRITIC method. To calculate the criterion weights, decision matrices were created for all periods examined. While the C1 criterion is the cost criterion, C2, C3, C4, and C5 are the benefit criteria.

In the next step, each element of the initial decision matrices is normalized. In the normalization process, Equation (9.1) was used in terms of the criteria with the benefit feature, and Equation (9.2) was used in terms of the criteria with the cost feature. In

TABLE 9.3
Criterion Weights

Years	Criteria	C1	C2	C3	C4	C5
2009	wj	0.194	0.149	0.139	0.293	0.222
2012	wj	0.173	0.171	0.159	0.272	0.222
2019	wj	0.196	0.182	0.144	0.223	0.252

TABLE 9.4
Results

	2009		2012		2019	
OECD countries	Q_i	Rank	Q_i	Rank	Q_i	Rank
A1	0.724	3	0.640	3	0.609	3
A2	0.492	12	0.458	12	0.432	12
A3	0.832	1	0.753	1	0.713	1
A4	0.750	2	0.682	2	0.682	2
A5	0.578	8	0.510	8	0.485	8
A6	0.581	7	0.551	7	0.572	7
A7	0.632	5	0.574	6	0.495	6
A8	0.700	4	0.632	4	0.666	4
A9	0.540	10	0.481	10	0.504	10
A10	0.572	9	0.499	9	0.509	9
A11	0.625	6	0.604	5	0.599	5
A12	0.528	11	0.476	11	0.448	11

the last step of the CRITIC method, the standard deviation and importance weight of each performance criterion were calculated by means of Equations (9.3) and (9.4), and the weights of the calculated criteria are given by year in Table 9.3. Considering the weights of the criteria found, the most important criteria for determining performance was total hospital beds per 1000 population, number of physicians per 1000 population, infant mortality, current expenditure on health per capita, and expenditure on health (% GDP) for the years 2009 and 2012. The most important criterion for 2019 was the number of physicians per 1000 population; the ranking of other criteria was the same as in previous years.

In the evaluation of the pandemic preparedness performance of the health systems of OECD countries, the importance weights of the criteria were first calculated. Then, the performance ranking of the countries was obtained with the WASPAS method. The results of the WASPAS method (when λ=0.5) and the performance ranking of the evaluated OECD countries are given in Table 9.4.

According to Table 9.4, the countries with the highest performance were Germany, Iceland, and Belgium. The worst performing countries were Portugal, the UK, and Canada. Considering the performance rankings of the health systems of the evaluated

TABLE 9.5
λ Effect on 2009 Performance Rankings

OECD countries	λ = 0	Rank	λ = 0.1	Rank	λ = 0.2	Rank	λ = 0.3	Rank	λ = 0.4	Rank
A1	0.719	3	0.720	3	0.721	3	0.722	3	0.723	3
A2	0.468	12	0.473	12	0.478	12	0.482	12	0.487	12
A3	0.820	1	0.823	1	0.825	1	0.827	1	0.830	1
A4	0.730	2	0.734	2	0.738	2	0.742	2	0.746	2
A5	0.559	8	0.563	8	0.567	7	0.570	8	0.574	8
A6	0.556	9	0.561	9	0.566	8	0.571	7	0.576	7
A7	0.621	5	0.623	5	0.625	5	0.628	5	0.630	5
A8	0.691	4	0.693	4	0.695	4	0.696	4	0.698	4
A9	0.532	10	0.533	10	0.535	10	0.537	10	0.539	10
A10	0.562	7	0.564	7	0.566	9	0.568	9	0.570	9
A11	0.605	6	0.609	6	0.613	6	0.617	6	0.621	6
A12	0.516	11	0.518	11	0.521	11	0.523	11	0.526	11

OECD countries	λ = 0.6	Rank	λ = 0.7	Rank	λ = 0.8	Rank	λ = 0.9	Rank	λ = 1	Rank
A1	0.725	3	0.726	3	0.727	3	0.729	3	0.730	3
A2	0.496	12	0.501	12	0.506	12	0.510	12	0.515	12
A3	0.834	1	0.836	1	0.839	1	0.841	1	0.843	1
A4	0.753	2	0.757	2	0.761	2	0.765	2	0.769	2
A5	0.581	8	0.585	8	0.589	8	0.592	8	0.596	8
A6	0.587	7	0.592	7	0.597	7	0.602	7	0.607	7
A7	0.635	5	0.637	5	0.639	5	0.642	6	0.644	6
A8	0.702	4	0.704	4	0.706	4	0.708	4	0.709	4
A9	0.542	10	0.544	10	0.546	10	0.548	10	0.549	10
A10	0.574	9	0.576	9	0.578	9	0.581	9	0.583	9
A11	0.630	6	0.634	6	0.638	6	0.642	5	0.646	5
A12	0.531	11	0.533	11	0.535	11	0.538	11	0.540	11

countries within the scope of preparedness for pandemics, the number of hospital beds and physicians in the countries with good performance was high, and health expenditure was high. In low-performing countries, the number of physicians, the number of hospital beds, and health expenditure are all low, while infant mortality rates were high. The results also show how effective the health systems of countries are in preparing for pandemics.

In the WASPAS method, the λ effect is considered to increase the accuracy and efficiency of the ranking of the alternatives. The effect of λ on ranking is calculated using Equation (9.11) and shown in Tables 9.5, 9.6, and 9.7. When the effect of λ on the ranking was examined, similar results were found to the rankings shown in Table 9.4. The ranking of the examined OECD countries varied depending on the selected λ value. Accordingly, for all periods evaluated, Germany and Iceland were the best performing OECD countries. The worst performing country was Canada.

TABLE 9.6
λ Effect on 2012 Performance Rankings

OECD countries	λ = 0	Rank	λ = 0.1	Rank	λ = 0.2	Rank	λ = 0.3	Rank	λ = 0.4	Rank
A1	0.619	3	0.623	3	0.628	3	0.632	3	0.636	3
A2	0.422	12	0.429	12	0.436	12	0.444	12	0.451	12
A3	0.732	1	0.736	1	0.740	1	0.745	1	0.749	1
A4	0.649	2	0.656	2	0.662	2	0.669	2	0.675	2
A5	0.475	9	0.482	9	0.489	8	0.496	8	0.503	8
A6	0.525	7	0.530	7	0.535	7	0.540	7	0.545	7
A7	0.544	6	0.550	6	0.556	6	0.562	6	0.568	6
A8	0.609	4	0.614	4	0.618	4	0.623	4	0.627	4
A9	0.464	10	0.467	10	0.471	10	0.474	10	0.477	10
A10	0.479	8	0.483	8	0.487	9	0.491	9	0.495	9
A11	0.563	5	0.572	5	0.580	5	0.588	5	0.596	5
A12	0.449	11	0.455	11	0.460	11	0.465	11	0.471	11

OECD countries	λ = 0.6	Rank	λ = 0.7	Rank	λ = 0.8	Rank	λ = 0.9	Rank	λ = 1	Rank
A1	0.645	3	0.649	3	0.653	3	0.657	3	0.661	3
A2	0.466	12	0.473	12	0.481	12	0.488	12	0.495	12
A3	0.757	1	0.762	1	0.766	1	0.770	1	0.774	1
A4	0.688	2	0.695	2	0.701	2	0.708	2	0.715	2
A5	0.518	8	0.525	8	0.532	8	0.539	8	0.546	8
A6	0.556	7	0.561	7	0.566	7	0.572	7	0.577	7
A7	0.579	6	0.585	6	0.591	6	0.597	6	0.603	6
A8	0.636	4	0.641	4	0.645	4	0.650	4	0.654	4
A9	0.484	10	0.487	10	0.491	11	0.494	11	0.497	11
A10	0.503	9	0.507	9	0.511	9	0.515	9	0.519	9
A11	0.613	5	0.621	5	0.629	5	0.637	5	0.645	5
A12	0.481	11	0.487	11	0.492	10	0.497	10	0.503	10

TABLE 9.7
λ Effect on 2019 Performance Rankings

OECD countries	λ = 0	Rank	λ = 0.1	Rank	λ = 0.2	Rank	λ = 0.3	Rank	λ = 0.4	Rank
A1	0.585	4	0.590	4	0.595	4	0.599	4	0.604	4
A2	0.395	12	0.402	12	0.410	12	0.417	12	0.424	12
A3	0.691	1	0.695	1	0.700	1	0.704	1	0.709	1
A4	0.636	2	0.646	2	0.655	2	0.664	2	0.673	2
A5	0.465	10	0.469	10	0.473	10	0.477	10	0.481	10
A6	0.550	5	0.554	6	0.559	6	0.563	6	0.567	6
A7	0.465	9	0.471	9	0.477	9	0.483	9	0.489	9

(Continued)

TABLE 9.7 (Continued)
λ Effect on 2019 Performance Rankings

OECD countries	λ = 0	Rank	λ = 0.1	Rank	λ = 0.2	Rank	λ = 0.3	Rank	λ = 0.4	Rank
A8	0.632	3	0.639	3	0.646	3	0.652	3	0.659	3
A9	0.488	7	0.490	8	0.494	8	0.497	8	0.501	8
A10	0.487	8	0.491	7	0.496	7	0.500	7	0.505	7
A11	0.547	6	0.558	5	0.568	5	0.578	5	0.589	5
A12	0.419	11	0.425	11	0.430	11	0.436	11	0.442	11

OECD countries	λ = 0.6	Rank	λ = 0.7	Rank	λ = 0.8	Rank	λ = 0.9	Rank	λ = 1	Rank
A1	0.614	4	0.619	5	0.623	5	0.628	5	0.633	5
A2	0.439	12	0.447	12	0.454	12	0.461	12	0.469	12
A3	0.718	1	0.722	1	0.727	1	0.732	1	0.736	1
A4	0.691	2	0.700	2	0.709	2	0.719	2	0.728	2
A5	0.489	10	0.493	10	0.497	10	0.501	10	0.505	10
A6	0.576	6	0.580	6	0.584	6	0.589	6	0.593	6
A7	0.501	9	0.507	9	0.513	9	0.519	8	0.525	8
A8	0.673	3	0.679	3	0.686	3	0.693	3	0.699	3
A9	0.508	8	0.511	8	0.515	8	0.518	9	0.522	9
A10	0.514	7	0.518	7	0.523	7	0.527	7	0.532	7
A11	0.609	5	0.620	4	0.630	4	0.641	4	0.651	4
A12	0.453	11	0.459	11	0.465	11	0.471	11	0.477	11

9.4 CONCLUSION

The performance of the health systems of 12 OECD countries within the scope of preparedness for pandemics was evaluated with the WASPAS method, an MCDM method, for the years 2009, 2012, and 2019. The CRITIC method was used to determine the criterion weights. Considering the criteria weights, the most important criteria are total hospital beds per 1000 population, the number of physicians per 1000 population, infant mortality, current expenditure on health per capita, and current expenditure on health (% GDP) for 2009 and 2012, respectively. The number of physicians per 1000 population is the most important criterion for 2019.

According to the results of the analysis, the countries that showed the best performance in the scope of preparedness for pandemics were Germany, Iceland, and Belgium. The worst performing countries were Portugal, the UK, and Canada. Considering the results of the study, 12 OECD countries with developed health systems, placing importance on health expenditure, and having more hospital beds and physicians are in a better situation to combat pandemics; pandemics threaten the lives of individuals in 12 OECD countries where health expenditures are low, the number of hospital beds and physicians is low, and the infant mortality rate is high. In this context, within the scope of combating pandemics, countries need to develop national and international strategies, and allocate sufficient resources to their health systems. Countries with a bad performance should revise their health systems.

In future studies, the weights of the criteria can be obtained with other weighting methods and different criteria can be used. With different MCDM methods, alternatives can be ranked, and the results can be compared. Different countries can be included in the analysis in terms of the performance of their health systems.

REFERENCES

Babatunde, M. and D. Ighravwe. 2019. A CRITIC-TOPSIS framework for hybrid renewable energy systems evaluation under techno-economic requirements. *Journal of Project Management* 4, 2:109–126.

Cetin, V. R. and S. Bahce. 2016. Measuring the efficiency of health systems of OECD countries by data envelopment analysis. *Applied Economics* 48, 37:3497–3507.

Chakraborty, S. and E. K. Zavadskas. 2014. Applications of WASPAS method in manufacturing decision making. *Informatica* 25, 1:1–20.

Diakoulaki, D., G. Mavrotas, and L. Papayannakis. 1995. Determining objective weights in multiple criteria problems: the CRITIC method. *Computers Operations Research* 22, 7:763–770.

Jahan, A., F. Mustapha, S. M. Sapuan, M. Y. Ismail, and M. Bahraminasab. 2012. A framework for weighting of criteria in ranking stage of material selection process. *The International Journal of Advanced Manufacturing Technology* 58, 1–4:411–420.

Karagiannis, R. 2016. Technical efficiency evaluation of health care systems in OECD countries. *Greek Economic Outlook* 17, 2.1:58.

Lari, M. S. and S. E. Sefiddashti. 2021. Measuring the efficiency of health systems: a case of mental health in Middle East and North Africa countries. *Iranian Journal of Public Health* 50, 5:1017–1027.

Odu, G. O. 2019. Weighting methods for multi-criteria decision making technique. *Journal of Applied Sciences and Environmental Management* 23, 8:1449–1457.

OECD Statistics. 2021. https://stats.oecd.org/# (accessed October 16, 2021).

Pekkaya, M. and G. Dökmen. 2019. OECD ülkeleri kamu sağlik harcamalarinin ÇKKV yöntemleri ile performans değerlendirmesi. *Uluslararası Yönetim İktisat ve İşletme Dergisi* 15, 4:923–950.

Qiu, W., S. Rutherford, A. Mao, and C. Chu. 2017. The pandemic and its impacts. *Health, Culture and Society* 9:1–11.

Reinhardt, U. E., P. S. Hussey, and G. F. Anderson. 2002. Cross-national comparisons of health systems using OECD data, 1999. *Health Affairs* 21, 3:169–181.

Samut, P. K. and R. Cafrı. 2016. Analysis of the efficiency determinants of health systems in OECD countries by DEA and Panel Tobit. *Social Indicators Research* 129, 1: 113–132.

Sinimole, K. R. 2012. Evaluation of the efficiency of national health systems of the members of World Health Organization. *Leadership in Health Services* 25, 2:139–150.

Stojić, G., Ž. Stević, J. Antuchevičienė, D. Pamučar, and M. Vasiljević. 2018. A novel rough WASPAS approach for supplier selection in a company manufacturing PVC carpentry products. *Information* 9, 5:1–16.

Tuş, A. and E. A. Adalı. 2019. The new combination with CRITIC and WASPAS methods for the time and attendance software selection problem. *Opsearch* 56, 2:528–538.

Wikipedia List of Epidemics, https://en.wikipedia.org/wiki/List_of_epidemics, (accessed October 25, 2021).

World Health Organization Covid-19 Dashboard, https://covid19.who.int/ (accessed October 25, 2021).

Yiğit A. 2019. Performance analysis of OECD countries based on health outcomes and expenditure indicators. *Journal of International Health Sciences and Management* 5, 9:114–123.

Yüksel, O. 2021. Comparison of healthcare system performances in OECD countries. *International Journal of Health Services Research and Policy* 6, 2:251–261.

Zavadskas, E. K., Z. Turskis, J. Antucheviciene, and A. Zakarevicius. 2012. Optimization of weighted aggregated sum product assessment. *Elektronika ir Elektrotechnika* 122, 6:3–6.

10 The Use of VIKOR in Location Selection of Field Hospitals

An Application from Çanakkale, Turkey

Mehtap Kılıç,[1] Miraç Karakoç,[2] and Cüneyt Çalışkan[3]

[1]Ardahan University

[2]Investment and Construction Directorate

[3]University of Health Science

CONTENTS

10.1 Introduction 135
10.2 Methodology 137
    10.2.1 VIKOR Implementation Steps 138
10.3 Implementation 140
10.4 Application Steps of the VIKOR Method 142
10.5 Discussion 145
10.6 Conclusion 150
References 150

10.1 INTRODUCTION

Disasters are phenomena that lead to a great number of lost lives and goods in society. When and where they might occur cannot be predicted, and result in devastating impacts when they do occur (Artan and Özkan 2020; Koçak et al. 2015). Along with their devastating impact, disasters can hinder health needs being met, resulting in the activities of current medical facilities and some organizations coming to a halt (Koçak et al. 2021; Ortiz-Barrios et al. 2020; Koçak 2018). Therefore, mobile medical facilities or mini-hospitals, where medical services are provided to those injured during a disaster, provided temporarily in a prescheduled area before the injured are moved safely to a permanent medical care facility, are needed. These hospitals are called field hospitals (Dursun and Karakoç 2019). Field hospitals are those built in the wake of disasters so that disaster victims are provided with safe medical services

DOI: 10.1201/9781003212904-10

and patients who are temporarily staying at predefined safe zones near hospitals can continue their treatment (Tekin et al. 2017). The World Health Organization and Pan American Health Organization (World Health Organization Pan American Health Organization 2003) define a field hospital as "a mobile, self-contained, self-sufficient health care facility capable of rapid deployment and expansion or contraction to meet immediate emergency requirements for a specified period of time". Field hospitals have been successfully used during both natural and man-made disasters (i.e., war). For instance, field hospitals were used in the Battle of Gallipoli (First World War) Dardanelles war, one of the bloodiest and greatest wars in history, and afforded treatment to tens of thousands of soldiers in the most challenging situations (Karakoç 2020; Koçak et al. 2018). Field hospitals used to be units where the first response was provided and had limited facilities, they are now more complex units that have the necessary equipment (Öztürk and Güneş 2019). Thus, efforts to build a field hospital and provide medical support in a disaster area are one of the fundamental goals in saving more lives. The functioning of field hospitals is different from the functioning of a standard hospital unit. However, the significance of field hospitals is never less during disasters, despite changing treatment functions (Halpern et al. 2003).

It is important to identify where to build a field hospital. It should be built in an area that is both close to the area affected by the disaster and protected as much as possible against post-disaster events (i.e., flood, landslide, and damaged building demolition) (Bıçakçı and Melisnur 2021). In this respect, many parameters should be considered when choosing the location of field hospitals. These parameters are defined by regional requirements, type of disaster, area requirements, and field hospitals (Paul and MacDonald 2016; Vahidnia, Alesheikh, and Alimohammadi 2009; Miç and Antmen 2019; Wu, Lin, and Chen 2007; Vafaei and Oztaysi 2014) (Table 10.1). The presence of parameters involved in building field hospitals and the fact that each alternative to be considered has unique advantages make the decision-making process harder (Uludağ and Doğan 2016). Thus, the study used one of the multi-criteria decision-making (MCDM) methodologies commonly used for decision-making processes; it assesses various alternatives with various criteria for choosing the best possible alternative, and, due to its unique capabilities, helps with solving and analyzing real-time complex problems (Ağaç and Baki 2016). The MCDM methodology used in the study was VIKOR (Vise Kriterijumska Optimizacija Kompromisno Resenje), among many others. The VIKOR methodology provides an arbitrated (the one closest to the ideal solution) ranking by comparing each alternative per their proximity to the ideal alternative under the assumption that each alternative is assessed on the basis of decision-making criteria (Kuru and Akin 2012). Since no previous study of the selection of field hospitals via the VIKOR methodology was not found in the literature, this study also allowed for the emergence of criteria that must be considered for any area under the risk of a disaster.

This study was conducted to guide experts and to define the optimal area for a field hospital to be built in Çanakkale province in the case of a potential disaster via the VIKOR methodology. A literature review of the necessary criteria for building field hospitals was done in the first phase of the study, and some were chosen to be used in the study. In the second phase, a VIKOR analysis was carried out and, as a result, the best alternative area was defined.

TABLE 10.1
Parameters that can Impact the Selection of a Location for a Field Hospital

Size	Variables	Sources
Regional properties	Population size	Paul and MacDonald 2016; Vahidnia, Alesheikh, and Alimohammadi 2009; Miç and Antmen 2019; Wu, Lin, and Chen 2007; Vafaei and Oztaysi 2014
	Population density	Paul and MacDonald 2016; Vahidnia, Alesheikh, and Alimohammadi 2009; Miç and Antmen 2019; Wu, Lin, and Chen 2007
	Geographic properties (protected area, flood, earthquake)	Paul and MacDonald 2016
	Environmental properties	Moradian et al. 2017
	Vulnerability	Maharjan and Hanaoka 2019
Type of disaster	Level of disaster	Paul and MacDonald 2016
	Number/severity of victims	Paul and MacDonald 2016
	Damaged facility	Paul and MacDonald 2016
Area properties	Size (m^2)	Rawls and Turnquist 2010; Miç and Antmen 2019
	Settlement (i.e., rural, urban, metropolitan)	Wu, Lin, and Chen 2007
	Disaster history	Rawls and Turnquist 2010
	Proximity to main roads	Rawls and Turnquist 2010; Vahidnia, Alesheikh, and Alimohammadi 2009; Vafaei and Oztaysi 2014; Louri et al. 2021
	Road network density	Vahidnia, Alesheikh, and Alimohammadi 2009; Miç and Antmen 2019
	Distance to medical facilities	Vahidnia, Alesheikh, and Alimohammadi 2009; Vafaei and Oztaysi 2014
	Accessibility for nodal connections (water, food, medical warehouses, blood centers, disaster containers)	Rawls and Turnquist 2010
	Travel time from areas with high population density to the designated area	Vahidnia, Alesheikh, and Alimohammadi 2009; Miç and Antmen 2019
	Proximity to hospital	Bell et al. 2021; Tekin et al. 2017
Field hospital	Setup time for operation and facilities	Vafaei and Oztaysi 2014
	Bed capacity	Vafaei and Oztaysi 2014

10.2 METHODOLOGY

MCDM is a methodology designed to help decision-makers structure an undefined future by considering a series of events in varied layers (Torkayesh, Malmir, and Rajabi Asadabadi 2021). The VIKOR methodology, one of the MCDM methodologies, was used in this study. Opricovic (1998) suggested that the VIKOR methodology could be used for complex decision-making problems (Akyüz 2012). The methodology means there is multi-criteria optimization and an arbitrated solution, relying on the assumption that an alternative has the highest group benefit and lowest individual regret as the best alternative to be chosen for a decision-making problem (Kabak and Çınar 2020). Any previous study regarding the selection of field hospitals via the VIKOR methodology was not found in the literature. The VIKOR methodology has been used in many studies to define criteria such as the assessment of hospital services (Chang 2014), inpatient admission assessment (Zhang et al. 2016), factors affecting the effectiveness of quality-control circles in a hospital (Bahadori et al. 2018), warehouse site selection (Kutlu Gündoğdu and Kahraman 2019), and medicine provider selection (Han and Chen 2014).

10.2.1 VIKOR Implementation Steps

In this subsection, the application stages of the VIKOR method are outlined in seven steps (Kabak and Çınar,2020).

Step 1: Defining a decision-making problem, and setting goals, criteria, and alternatives (Figure 10.1)
Step 2: Creating a decision matrix

A decision matrix (F) contains the assessments of alternatives, one of which is to be chosen, on the basis of criteria. An assessment on the basis of criterion *j* for the decision alternative *i* is indicated by the f_{ij} element of the decision matrix. The decision matrix appears as provided below:

$$F = \begin{bmatrix} f_{11} & f_{12} \cdots & f_{1n} \\ f_{21} & f_{22} \cdots & f_{2n} \\ \vdots & \vdots \ddots & \vdots \\ f_{m1} & f_{m2} \cdots & f_{mn} \end{bmatrix}$$

Step 3: Creating a normalized decision matrix

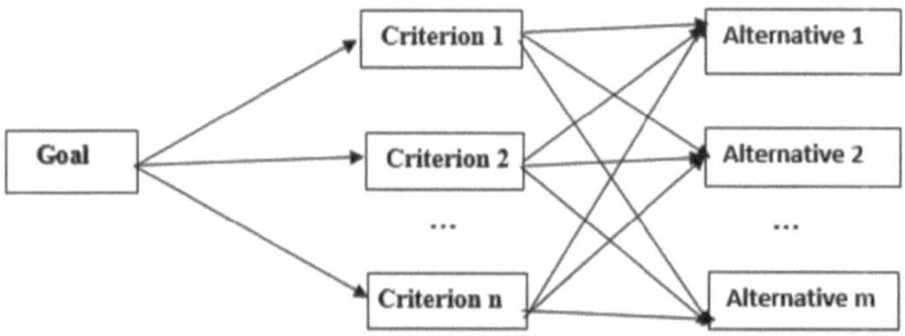

FIGURE 10.1 Hierarchical structure of a multi-criteria decision-making problem. (Kabak and Çınar, 2020.)

The normalized value of alternative *i* on the basis of the criterion *j* for the normalized decision matrix (R) is indicated by r_{ij}. Normalized values are calculated with the help of Equation (10.1) provided below, acknowledging that $f_j^+ = \max_i f_{ij}$ and $f_i^- = \min_i f_{ij}$:

$$r_{ij} = \frac{f_j^+ - f_{ij}}{f_j^+ - f_j^-} \tag{10.1}$$

The normalized values for cost-based criteria are calculated by acknowledging that $f_j^+ = \min_i f_{ij}$ and $f_i^- = \max_i f_{ij}$. A normalized decision matrix appears as provided below:

$$R = \begin{bmatrix} r_{11} & r_{12} \cdots & r_{1n} \\ r_{21} & r_{22} \cdots & r_{2n} \\ \vdots & \vdots \ \ddots & \vdots \\ r_{m1} & r_{m2} \cdots & r_{mn} \end{bmatrix}$$

Step 4: Having a weighted normalized decision matrix

Weighted normalized decision matrix (V) is calculated by multiplying elements of a normalized decision matrix (r_{ij}) with weight values of relevant criteria (w_j). The equality used for the calculation of elements of a weighted normalized decision matrix (v_{ij}) is presented via Equation (10.2), and the appearance of a weighted normalized decision matrix is provided below:

$$v_{ij} = r_{ij} w_j \tag{10.2}$$

$$V = \begin{bmatrix} v_{11} & v_{12} \cdots & v_{1n} \\ v_{21} & v_{22} \cdots & v_{2n} \\ \vdots & \vdots \ \ddots & \vdots \\ v_{m1} & v_{m2} \cdots & v_{mn} \end{bmatrix}$$

Step 5: Calculating group benefit and individual regret benchmark values

Group benefit (S_i) and individual regret benchmark values (R_i) are calculated with the help of equalities provided via Equation (10.3) and Equation (10.4) below.

$$S_i = \sum_{j=1}^{n} w_j \frac{f_j^+ - f_{ij}}{f_j^+ - f_j^-} = \sum_{j=1}^{n} v_{ij} \tag{10.3}$$

$$R_i = \max_j \left(w_j \frac{f_j^+ - f_{ij}}{f_j^+ - f_j^-} \right) = \max_j v_{ij} \tag{10.4}$$

Step 6: Calculating arbitrated benchmark value

Arbitrated benchmark value ($\mathbb{Q}_i$) is calculated for combining group benefit and individual regret benchmark values to be able to decide between alternatives. The arbitrated benchmark value is calculated with the equality provided by the Equation (10.5) below.

$$\mathbb{Q}_i = \vartheta x \frac{S_i - S^+}{S^- - S^+} + (1 - \vartheta) x \frac{R_i - R^+}{R^- - R^+} \tag{10.5}$$

Here: $S_i^+ = \min_i S_i$, $S_i^- = \max_i S_i$, $R_i^+ = \min_i R_i$ and $R_i^- = \max_i R_i$. The θ value, representing the importance of group benefit, is key to the calculation of the arbitrated benchmark value ($\mathbb{Q}_i$). Per the formula, the alternative with the minimum $\mathbb{Q}_i$ value is identified as the best alternative.

Step 7: Validating results

For validation, the results are reviewed to see whether they meet the following two conditions.

- Condition 1: Acceptable advantage
  Let's assume that the alternative with the lowest $\mathbb{Q}_i$value has a value of $\mathbb{Q}(a')$, and the second-best alternative has the second-lowest $\mathbb{Q}_i$ value ($\mathbb{Q}(a'')$), and value D $\mathbb{Q}$ is equal to 1/(m–1). $\mathbb{Q}$ If the condition $(a') - \mathbb{Q}(a'') \geq D\mathbb{Q}$ is met, then the acceptable advantage condition is met.
- Condition 2: Acceptable consistency for decision making
  The alternative with the best $\mathbb{Q}_i$ value must also be the best in terms of group benefit and/or individual regret benchmark values.

When both conditions are met, the alternative with the best $\mathbb{Q}_i$value is defined as the arbitrated solution.

10.3 IMPLEMENTATION

A literature review of the necessary criteria for defining the optimal area in which to build field hospitals in the case of a disaster in Çanakkale, a Turkish province, is provided in Table 10.1. The following criteria were used in this study: (1) Population size, (2) population density, (3) area size (m^2), (4) proximity to main roads, (5) road network density, (6) distance to medical facilities, and (7) proximity to medical warehouses.

Çanakkale is the province of the Dardanelles (Çanakkale Boğazı), which is a connection between Asia and Europe and is named after the city. Being a significant center of population since the early ages, Çanakkale has two transition points enabling connections between Anatolia and Europe, as well as the Mediterranean Sea and the Black Sea ("Governorship of Çanakkale" n.d.; Çalışkan et al. 2014). Çanakkale province was chosen as the study area due to both its geographic location and its history. The study was limited to six neighborhoods in the Center district of Çanakkale. Selected neighborhoods, total population, area, and density information is provided in Table 10.2.

Including the area size and neighborhoods, the list of prospective areas chosen for building field hospitals during disasters as per the study is provided in Table 10.3.

The list of hospitals in the Center district of Çanakkale and the distance of alternative areas to the field hospitals are provided in Table 10.4. Implementing the VIKOR methodology, calculations were made on the basis of the average distance to three hospitals.

The distance of alternative field hospital locations to main arterial roads and the road network density of main arterial roads are provided in Table 10.5. While implementing the VIKOR methodology, the value of distance to the closest main

TABLE 10.2
Neighborhoods in the Center District of Çanakkale: Population, Area, and Density

Neighborhood	Population	Area (m^2)	Density (population/km^2)
Arslanca (İsmetpaşa)	21,319	2,524,500	8,444
Barbaros	56,904	8,299,000	6,856
Cevatpaşa	51,636	6,616,400	7,804
Fevzipaşa	1,928	213,700	9,021
Namık Kemal	1,053	120,800	8,716
Kemalpaşa	1,638	219,800	7,452

TABLE 10.3
The List of All Prospective Areas that can be Used for Selecting Locations for Field Hospitals

Name	Area (m^2)	Neighborhood
City Stadium	14,000	Arslanca (İsmetpaşa)
Friday market	21,500	Namık Kemal
Heliport	8,500	Cevatpaşa
75th Year Park	34,300	Barbaros
Behind special provincial administration	19,900	Barbaros
Next to Kipa Mall	36,100	Barbaros

TABLE 10.4
List of Hospitals in the Center District of Çanakkale and Distance to Alternative Field Hospital Locations (km)

		Alternatives					
Hospital name	**Neighborhood**	**City Stadium**	**Friday Market**	**Heliport**	**75th Year Park**	**Behind Special Provincial Administration**	**Next to Kipa Mall**
Çanakkale Public H.	Barbaros	7.25	6.29	8.86	4.28	3.73	3.22
ÇOMÜ H.	Barbaros	4.17	3.39	5.93	1.89	0.900	0.392
Medicalpark H.	Barbaros	2.37	1.60	3.98	0.523	1.10	1.55

H: Hospital

TABLE 10.5
Distance of Alternative Field Hospital Locations to Main Arterial Roads and Road Network Density

	Distance to Alternative Locations (km)						Road Network Density		
Main Arterial Roads	**City Stadium**	**Friday Market**	**Heliport**	**75th Year Park**	**Behind Special Provincial Administration**	**Next to Kipa Mall**	**Green**	**Yellow**	**Red**
Atatürk Ave.	0	0	0.945	0	0.593	0			
Troya Ave.	1.51	0.790	1.39	0	0.555	1.01			
Beltway	4.08	4.02	3.93	2.04	1.35	0.549			

arterial road was used for calculation. The highest road network density is indicated in red, while the lowest density is indicated in green.

Pharmaceutical warehouses need to be near field hospitals so that necessary materials can be supplied in the case of disasters. The distance to alternative field hospital locations to pharmaceutical warehouses and their neighborhoods is provided in Table 10.6. Implementing the VIKOR methodology, calculations were made on the basis of average distances to pharmaceutical warehouses.

Following the data-collection phase, the VIKOR methodology was implemented. The implementation steps of the methodology are provided in Table 10.7.

10.4 APPLICATION STEPS OF THE VIKOR METHOD

The decision matrix of the problem created in step 2 is given in Table 10.7.

Table 10.7 shows the data used in applying the VIKOR method. The best and worst values were determined and are given at the bottom of the table. In addition, the criteria weights of each criterion are also included.

TABLE 10.6
Distance of Alternative Field Hospital Locations to Pharmaceutical Warehouses (km)

		Alternatives					
Pharmaceutical Warehouse	**Neighborhood**	**City Stadium**	**Friday Market**	**Heliport**	**75th Year Park**	**Behind Special Provincial Administration**	**Next to Kipa Mall**
Selçuk Ecza Deposu	Arslanca (İsmetpaşa)	0.677	0.781	3.41	3.21	3.34	3.79
BEK Ecza Deposu	Arslanca (İsmetpaşa)	2.06	1.79	3.03	3.32	3.90	4.36
Hedef Alliance Ecza Deposu	Cevatpaşa	0.470	1.19	1.69	3.18	3.76	4.22

TABLE 10.7
Identifying the Best and Worst Values with Criteria Weights

Criteria Weights	0.2	0.2	0.23	0.11	0.07	0.07	0.07
Alternatives	**Population Size**	**Population Density**	**Area Size (m^2)**	**Distance to the Closest Main Road**	**Road Network Density**	**Distance to Medical Facilities**	**Proximity to Medical Warehouses**
City Stadium	21,319	8,444	14,000	0	3	4.59	1.06
Friday Market	1,053	8,716	21,500	0	3	3.76	1.25
Heliport	51,636	7,804	8,500	0.945	3	6.25	2.71
75th Year Park	56,904	6,856	34,300	0	2	2.23	3.23
Behind Special Provincial Administration	56,904	6,856	19,900	0.555	2	1.91	3.66
Next to Kipa Mall	56,904	6,856	36,100	0	3	1.72	4.12
Best	**56,904**	**8,716**	**3,6100**	**0**	**2**	**1.72**	**1.06**
Worst	**1,053**	**6,856**	**8,500**	**0.945**	**3**	**6.25**	**4.12**
Criteria Type	beneficial	beneficial	beneficial	n-beneficial	n-beneficial	n-beneficial	n-beneficial

The normalized decision matrix to be applied in step 3 is given in Table 10.8.

The equation used to obtain the normalized decision matrix was: $r_{ij} = \frac{f_j^+ - f_{ij}}{f_j^+ - f_j^-}$

The weighted normalized decision matrix to be applied in step 4 is given in Table 10.9.

The equation used to obtain the weighted normalized decision matrix was: $v_{ij} = r_{ij} w_j$

Group benefit and individual regret criteria that should be applied in step 5 are given in Table 10.10.

The equation used to calculate the group benefit measure was: $S_i = \sum_{j=1}^{n} w_j \frac{f_j^+ - f_{ij}}{f_j^+ - f_j^-} = \sum_{j=1}^{n} v_{ij}$

The equation used to calculate the individual regret criterion was: $R_i = \max_j \left(w_j \frac{f_j^+ - f_{ij}}{f_j^+ - f_j^-} \right) = \max_j v_{ij}$

The consensus criterion calculation to be applied in step 6 is given in Table 10.11.

The equation used to calculate the consensus criterion was: $Q_i = \vartheta \, x \frac{S_i - S^+}{S^- - S^+} + (1 - \vartheta) x \frac{R_i - R^+}{R^- - R^+}$

Rankings according to the consensus criteria values are given in Table 10.12.

Acquired solution values were calculated to validate the results. It was checked whether condition 1 and condition 2 in step 6 were met, according to VIKOR methodology. Calculated values in relation to the validation of the results are provided in Table 10.13.

Where both conditions are met, $Q_j = 0.00$. Where $Q_i = 0.00$, it appeared that the alternative with the lowest Qi value was the City Stadium. As a result of the analyses, decisions-makers decided that City Stadium was the best alternative.

10.5 DISCUSSION

Çanakkale province is at a key point in terms of geographic location and has many past instances of exposure to disasters. Therefore, alternative areas were identified for a field hospital to be built in the Çanakkale and the VIKOR methodology was used to determine the optimal location among them. While identifying these areas, selections were made between the northern and southern sides of the borderline, which was drawn via Sarıçay in the Center district of Çanakkale. City Stadium, the Friday market, and the heliport were identified as areas on the northern side of the Sarıçay, while 75th Year Park, behind the special provincial administration, and next to Kipa Mall were the areas on the southern side of the Sarıçay.

In the studies by Paul and MacDonald (2016), Vahidnia, Alesheikh, and Alimohammadi (2009), Miç and Antmen (2019), and Wu, Lin, and Chen (2007), population size and population density were among the criteria to be considered for the building of field hospitals. Building field hospitals in densely populated areas will enable the maximum number of people to be reached and responded to

TABLE 10.8
Normalized Decision Matrix

	Population Size	Population Density	Area Size(m^2)	Distance to Closest Main Road	Road Network Density	Distance to Medical Facilities	Proximity to Medical Warehouses
City Stadium	0.363	0.854	0.199	0.000	1.000	0.634	0.000
Friday Market	0.000	1.000	0.471	0.000	1.000	0.450	0.062
Heliport	0.906	0.510	0.000	1.000	1.000	1.000	0.539
75th Year Park	1.000	0.000	0.935	0.065	0.000	0.113	0.709
Behind Special Provincial Administration	1.000	0.000	0.413	0.587	0.000	0.042	0.850
Next to Kipa Mall	1.000	0.000	1.000	0.000	1.000	0.000	1.000

TABLE 10.9
Weighted Normalized Decision Matrix

Alternatives	Population Size	Population Density	Area Size (m^2)	Distance to the Closest Main Road	Road Network Density	Distance to Medical Facilities	Proximity to Medical Warehouses
City Stadium	0.127	0.029	0.184	0.000	0.070	0.044	0.000
Friday Market	0.200	0.000	0.122	0.000	0.070	0.032	0.004
Heliport	0.019	0.098	0.230	0.110	0.070	0.070	0.038
75th Year Park	0.000	0.200	0.015	0.000	0.000	0.008	0.050
Behind Special Provincial Administration	0.000	0.200	0.135	0.065	0.000	0.003	0.059
Next to Kipa Mall	0.000	0.200	0.000	0.000	0.070	0.000	0.070

TABLE 10.10
S_i and R_i Values of Alternative Areas and DQ Value

	S_i	R_i
City Stadium	0.455	0.184
Friday Market	0.428	0.200
Heliport	0.635	0.230
75th Year Park	0.273	0.200
Behind Special Provincial Administration	0.462	0.200
Next to Kipa Mall	0.340	0.200
S^+(Min)	0.273	
S^- (Max)	0.635	
R^+(Min)	0.184	
R^- (Max)	0.230	
D(Q) (j=6)	0.2	

TABLE 10.11
Arbitrated Benchmark (Q_j) Values of Alternative Areas

	$Q_j = 0.00$	$Q_j = 0.25$	$Q_j = 0.5$	$Q_j = 0.75$	$Q_j = 1.00$
City Stadium	0.000	0.126	0.252	0.378	0.504
Friday Market	0.345	0.366	0.387	0.407	0.428
Heliport	1.000	1.000	1.000	1.000	1.000
75th Year Park	0.345	0.259	0.173	0.086	0.000
Behind Special Provincial Administration	0.345	0.390	0.434	0.479	0.523
Next to Kipa Mall	0.345	0.306	0.266	0.226	0.186

TABLE 10.12
Ranking Alternative Areas

	$\vartheta = 0.00$	$\vartheta = 0.25$	$\vartheta = 0.5$	$\vartheta = 0.75$	$\vartheta = 1.00$
City Stadium	1	1	2	3	4
Friday Market	2	4	4	4	3
Heliport	6	6	6	6	6
75th Year Park	2	2	1	1	1
Behind Special Provincial Administration	2	5	5	5	5
Next to Kipa Mall	2	3	3	2	2

TABLE 10.13
Validating Results

	$Q_i = 0.00$	$Q_i = 0.25$	$Q_i = 0.5$	$Q_i = 0.75$	$Q_i = 1.00$
Q(A2)	0.345	0.259	0.252	0.226	0.186
Q(A1)	0	0.126	0.173	0.086	0
Q(A2)-Q(A1)	0.345	0.133	0.079	0.14	0.186
DQ=1/(6–1)	0.2	0.2	0.2	0.2	0.2
Condition 1	**Met**	**Not met**	**Not met**	**Not met**	**Not met**
Condition 2	**Met**	**Met**	**Met**	**Met**	**Met**

Condition 1 is met where the difference between the alternative with the best Qi value and the alternative with the second-best Qi value is greater than the DQ value. Q(A2)-Q(A1)>=DQ

Condition 2 is met where the alternative with the lowest Qi value is also the alternative with the lowest benchmark values in terms of group benefit and/or individual regret.

during disasters. The area with the highest population density in the center district of Çanakkale is the Fevzipaşa neighborhood. However, there was no alternative area to be selected from in the Fevzipaşa neighborhood. Subsequent to Fevzipaşa neighborhood, the second most densely populated area is the Namık Kemal neighborhood. Among the selected alternatives, the "Friday market" is located in the area. The area with the highest population is the Barbaros neighborhood. Among the selected alternative areas, the "75th Year Park", "behind the special provincial administration", and "next to Kipa Mall" are located in this area. The area with the lowest population is the Namık Kemal neighborhood. Among the selected alternatives, the "Friday market" is located in this area.

Considering the studies including criteria related to the building of field hospitals, Bıçakçı and Melisnur (2021) highlighted that the area where field hospitals should be built must be close to medical facilities and transportation to these areas must be easy. Having field hospitals close to medical facilities and warehouses to supply medical tools and devices will be efficient in saving the lives of more people by facilitating access to materials, human resources, and medical facilities needed in the response phase. Thus, it was decided that these criteria should be included in this study when choosing the areas to build field hospitals. The average distances of the alternative areas in terms of the criterion of distance to hospitals were defined by averaging out their distances. Therefore, the alternative closest to hospitals was "next to Kipa Mall". The alternative furthest from hospitals was the "heliport". In terms of proximity to medical warehouses, the average of the distances of the alternative areas was calculated. According to average distance results, the alternative area closest to medical warehouses in terms of was "City Stadium". The area furthest from medical warehouses was "next to Kipa Mall".

The studies by Rawls and Turnquist (2010) and Miç and Antmen (2019) found that the area needed to build field hospitals on must be large. Having a large area might allow more people to get medical services more easily and effectively. Efficient use of the opportunities provided by an area helps with the reduction of the negative impact caused by disasters. The alternative with the largest area among the selected alternatives was "next to Kipa Mall". The alternative with the smallest area was the "heliport".

Rawls and Turnquist (2010), Vahidnia, Alesheikh, and Alimohammadi (2009), Vafaei and Oztaysi (2014), and Louri et al. (2021) mentioned that field hospitals must be close to main roads. Having field hospitals close to main roads facilitates the transportation of the people to hospital, easy access to the hospital by the medical teams, and delivery of the required public services to the area as soon as possible. To that effect, the main road used for transportation must be suitable in terms of road network density along with the fact that the field hospital must be close to main roads, as stated by Vahidnia, Alesheikh, and Alimohammadi (2009) and Miç and Antmen (2019). Roads available in terms of road network density will make it convenient for public services to be delivered in the area where the field hospital is located. In the study, "City Stadium", "Friday market", "75th Year Park", and "next to Kipa Mall" were located closer to the main arterial roads than the other areas in terms of distance. The "heliport" and "behind the special provincial administration" are more distant from the main arterial roads than the other selected alternatives. The area with the highest road network density is Atatürk Avenue. The selected alternative areas that are closest to the roads in this area were "City Stadium", "Friday market", "75th Year Park", and "next to Kipa Mall". One of the closest alternatives in terms of distance to main arterial roads, "75th Year Park", is also close to the area marked in yellow in terms of road network density.

10.6 CONCLUSION

Building field hospitals during disasters is important for minimizing the loss of lives resulting from disasters, performing an effective and quick response on site, and minimizing the negative impacts of disasters. The area where field hospitals will be built must be defined accurately so that their efficiency is increased. An accurately selected area will facilitate the post-disaster response and prevent coordination chaos between teams, and a predefined area will ensure preparedness for the time of crisis. The success of disaster crisis management is possible thanks to the plans made during the disaster risk-management phase. Determining the locations of the intervention areas that need to be created will increase speed and efficiency. In this way, the disaster can be managed better. In this study, the VIKOR methodology was used to identify the best of a number of alternative areas for building a field hospital in Çanakkale province. As a result of the VIKOR methodology, the best alternative area in terms of building a field hospital in Çanakkale was selected as "City Stadium". City Stadium is the most suitable place to be chosen for the establishment of a field hospital in case of a disaster that may occur in Çanakkale.

REFERENCES

Ağaç, Gökhan, and Birdoğan Baki. 2016. "Application of Multi-Criteria Decision Making in Health Care Field: A Literature Review." *Hacettepe Sağlık İdaresi Dergisi* 19 (3): 343–63.

Akyüz, Gökhan. 2012. "Bulanık VIKOR Yöntemi İle Tedarikçi Seçimi." *Atatürk Üniversitesi İktisadi ve İdari Bilimler Dergisi* 26 (1): 197–215.

Artan, Taner, and Aydın Olcay Özkan. 2020. "Disasters and Social Work." *Journal of Adem* 1 (1): 47–54.

Bahadori, Mohammadkarim, Ehsan Teymourzadeh, Foad Faizy Bagejan, Ramin Ravangard, Mehdi Raadabadi, and Seyed Mojtaba Hosseini. 2018. "Factors Affecting the Effectiveness of Quality Control Circles in a Hospital Using a Combination of Fuzzy VIKOR and Grey Relational Analysis." *Proceedings of Singapore Healthcare* 27 (3): 180–6. https://doi.org/10.1177/2010105818758088.

Bell, Sue Anne, Lesly A. Dossett, Jesus Cespero, Mayuri Guntupalli, Keith Dickey, Jonathan Eliason, and Dawn Coleman. 2021. "T-Minus 10 Days: The Role of an Academic Medical Institution in Field Hospital Planning." *Prehospital and Disaster Medicine* 36 (3): 338–43. https://doi.org/10.1017/S1049023X21000224.

Bıçakçı, Nurcan, and Nevruz Melisnur. 2021. "Field Hospitals Disasters." *Emerg Aid Disaster Science* 1 (1): 17–21.

Çalışkan, Cüneyt, Aysun Algan, Hüseyin Koçak, Burcu Küçük Biçer, Meltem Şengelen, and Banu Çakir. 2014. "Preparations for Severe Winter Conditions by Emergency Health Personnel in Turkey." *Disaster Medicine and Public Health Preparedness* 8 (2): 170–3. https://doi.org/10.1017/dmp.2014.28.

Chang, Tsung-Han. 2014. "Fuzzy VIKOR Method: A Case Study of the Hospital Service Evaluation in Taiwan." *Information Sciences* 271 (July): 196–212. https://doi.org/10.1016/j.ins.2014.02.118.

Dursun, Recep and Yenal Karakoç. 2019. "Field Hospital in Disasters." In *Afetlerde Acil Tıp Hizmetleri*, edited by DN Özüçelik, 60–3. Ankara: Turkiye Klinikleri. www.turkiyeklinikleri.com/article/en-afetlerde-sahra-hastanesi-84824.html.

"Governorship of Çanakkale." n.d. The Leaders Attending the 100th Anniversary Ceremonies Were Sent off by the Governor Ahmet Çınar. Accessed September 30, 2015. http://www.canakkale.gov.tr/hukumet-konagi.

Halpern, Pinchas, Boaz Rosen, Shemy Carasso, Patrick Sorkine, Yoram Wolf, Paul Benedek, and Giora Martinovich. 2003. "Intensive Care in a Field Hospital in an Urban Disaster Area: Lessons from the August 1999 Earthquake in Turkey*." *Critical Care Medicine* 31 (5): 1410–14. https://doi.org/10.1097/01.CCM.0000059439.07851.BD.

Han, Xueliang and Xi Chen. 2014. "A D-VIKOR Method for Medicine Provider Selection." In *2014 Seventh International Joint Conference on Computational Sciences and Optimization*, 419–23. IEEE. https://doi.org/10.1109/CSO.2014.87.

Kabak, Mehmet, and Yetkin Çınar. 2020. *Çok Kriterli Karar Verme Yöntemleri MS Excel Çözümlü Uygulamalar*. Nobel Akademik Yayıncılık.

Karakoç, Miraç Nevzat. 2020. "Developing Underground Hospital Models in Terms of Disaster Medicine: Bezmialem Vakif University Faculty of Medicine Underground Hospital Project." Bezmialem Vakif University. https://tez.yok.gov.tr/UlusalTezMerkezi/tezSorguSonucYeni.jsp.

Koçak, Hüseyin. 2018. "The Role of Disaster Medicine in Disaster Management and Preparedness." *Integrating Disaster Science and Management: Global Case Studies in Mitigation and Recovery*, January, 423–32. https://doi.org/10.1016/B978-0-12-812056-9.00025-7.

Koçak, Hüseyin, Cüneyt Çaliskan, Edip Kaya, Ömer Yavuz, and Kerim Hakan Altintas. 2015. "Determination of Individual Preparation Behaviors of Emergency Health Services Personnel towards Disasters." *Journal of Acute Disease* 4 (3): 180–5. https://doi.org/10.1016/j.joad.2015.04.004.

Koçak, Hüseyin, Cüneyt Çalışkan, Mehmet Şerafettin Sönmezler, Kenan Eliuz, and Fatih Küçükdurmaz. 2018. "Analysis of Medical Responses in Mass Gatherings: The Commemoration Ceremonies for the 100th Anniversary of the Battle of Gallipoli." *Prehospital and Disaster Medicine* 33 (3): 288–92. https://doi.org/10.1017/S1049023X18000353.

Koçak, Hüseyin, Kerem Kınık, Cüneyt Çalışkan, and Kurtuluş Açıksarı. 2021. "The Science of Disaster Medicine: From Response to Risk Reduction." *Medeniyet Medical Journal* 36 (4): 333–42. https://doi.org/10.4274/MMJ.galenos.2021.50375.

Kuru, Ayşegül, and Besim Akin. 2012. "Integrated Management Systems Approach to the Use of Multiple Criteria Decision Making Techniques and Applications." *Öneri Dergisi* 10 (38): 129–44. https://doi.org/10.14783/od.v10i38.1012000215.

Kutlu Gündoğdu, Fatma, and Cengiz Kahraman. 2019. "A Novel VIKOR Method Using Spherical Fuzzy Sets and its Application to Warehouse Site Selection." *Journal of Intelligent & Fuzzy Systems* 37 (1): 1197–211. https://doi.org/10.3233/JIFS-182651.

Louri, Nayef A., Jalal Abdulla Alkhan, Hayyan Hamad Isa, Yaser Asad, Abdulla Alsharooqi, Khalifa Ahmed Alomari, Nahed Kamal Hasan, et al. 2021. "Establishing a 130-Bed Field Intensive Care Unit to Prepare for COVID-19 in 7 Days in Bahrain Military Hospital." *Disaster Medicine and Public Health Preparedness* 15 (1): e34–43. https://doi.org/10.1017/dmp.2020.297.

Maharjan, Rajali, and Shinya Hanaoka. 2019. "Fuzzy Multi-Attribute Group Decision Making to Identify the Order of Establishing Temporary Logistics Hubs during Disaster Response." *Journal of Humanitarian Logistics and Supply Chain Management* 9 (1): 2–21. https://doi.org/10.1108/JHLSCM-02-2018-0013.

Miç, Pınar and Z. Figen Antmen. 2019. "A Healthcare Facility Location Selection Problem with Fuzzy TOPSIS Method for a Regional Hospital." *European Journal of Science and Technology* 16 (August): 750–7. https://doi.org/10.31590/ejosat.584217.

Moradian, Mohammad Javad, Ali Ardalan, Amir Nejati, Ali Darvishi Boloorani, Ali Akbarisari, and Behnaz Rastegarfar. 2017. "Risk Criteria in Hospital Site Selection: A Systematic Review." *PLoS Currents* 9 (May). http://currents.plos.org/disasters/article/risk-criteria-in-hospital-site-selection-a-systematic-review/.

Opricovic, S. 1998. Multicriteria Optimization of Civil Engineering Systems. Faculty of Civil Engineering, Belgrade.

Ortiz-Barrios, Miguel, Muhammet Gul, Pedro López-Meza, Melih Yucesan, and Eduardo Navarro-Jiménez. 2020. "Evaluation of Hospital Disaster Preparedness by a Multi-Criteria Decision Making Approach: The Case of Turkish Hospitals." *International Journal of Disaster Risk Reduction* 49 (October): 101748. https://doi.org/10.1016/j.ijdrr.2020.101748.

Öztürk, Ö.A. and İ. Güneş. 2019. "Sahra Tipi Hastanelerde Enerji İhtiyacının Güneş Enerjisi İle Karşılanması." Yalova: International Scientific Research Congress UBAK, 103–16. https://avesis.istanbulc.edu.tr/publication/details/001ba979-bee6-4971-96b0-bc188ccdece1/sahra-tipi-hastanelerde-enerji-ihtiyacinin-gunes-enerjisi-ile-karsilanmasi.

Paul, Jomon Aliyas and Leo MacDonald. 2016. "Location and Capacity Allocations Decisions to Mitigate the Impacts of Unexpected Disasters." *European Journal of Operational Research* 251 (1): 252–63. https://doi.org/10.1016/j.ejor.2015.10.028.

Rawls, Carmen G. and Mark A. Turnquist. 2010. "Pre-Positioning of Emergency Supplies for Disaster Response." *Transportation Research Part B: Methodological* 44 (4): 521–34. https://doi.org/10.1016/j.trb.2009.08.003.

Tekin, Erdal, Atif Bayramoglu, Mustafa Uzkeser, and Zeynep Cakir. 2017. "Evacuation of Hospitals during Disaster, Establishment of a Field Hospital, and Communication." *The Eurasian Journal of Medicine* 49 (2): 137–41. https://doi.org/10.5152/eurasianjmed.2017.16102.

Torkayesh, Ali Ebadi, Behnam Malmir, and Mehdi Rajabi Asadabadi. 2021. "Sustainable Waste Disposal Technology Selection: The Stratified Best-Worst Multi-Criteria Decision-Making Method." *Waste Management* 122 (March): 100–12. https://doi.org/10.1016/j.wasman.2020.12.040.

Uludağ, Ahmet Serhat, and Hatice Doğan. 2016. "A Service Quality Application Focusing on the Comparison of Multi-Criteria Decision Making Method." *Journal of The Faculty of Economics and Administrative Sciences* 6 (2): 17–47.

Vafaei, Nazanin and Basar Oztaysi. 2014. "Selecting the Field Hospital Place for Disasters: A Case Study in Istanbul." In: International Conference of the INFORMS GDN Section and the EURO Working Group on DSS, Special Focus: Group Decision Making and Web 3.03, 23–36. https://doi.org/10.13033/isahp.y2014.148.

Vahidnia, Mohammad H., Ali A. Alesheikh, and Abbas Alimohammadi. 2009. "Hospital Site Selection Using Fuzzy AHP and its Derivatives." *Journal of Environmental Management* 90 (10): 3048–56. https://doi.org/10.1016/j.jenvman.2009.04.010.

World Health Organization Pan American Health Organization. 2003. "WHO-PAHO Guidelines for the Use of Foreign Field Hospitals in the Aftermath of Natural Disasters." *International Meeting Hospital in Disasters – Handle with Care.* www.preventionweb.net/publication/guidelines-use-foreign-field-hospitals-aftermath-sudden-impact-disasters.

Wu, Cheng-Ru, Chin-Tsai Lin, and Huang-Chu Chen. 2007. "Optimal Selection of Location for Taiwanese Hospitals to Ensure a Competitive Advantage by Using the Analytic Hierarchy Process and Sensitivity Analysis." *Building and Environment* 42 (3): 1431–44. https://doi.org/10.1016/j.buildenv.2005.12.016.

Zhang, Fengyi, Li Luo, Huchang Liao, Ting Zhu, Yingkang Shi, and Wenwu Shen. 2016. "Inpatient Admission Assessment in West China Hospital Based on Hesitant Fuzzy Linguistic VIKOR Method." *Journal of Intelligent & Fuzzy Systems* 30 (6): 3143–54. https://doi.org/10.3233/IFS-152056.

11 Drone Evaluation and Selection for Training in Disaster Preparedness Phases with Interval MCDA Techniques

Zekiye Göktekin[1] *and Ahmet Bahadır Şimşek*[1]
[1]University of Gümüşhane

CONTENTS

11.1 Introduction 155
11.2 Related Studies 157
11.3 Methodology 158
11.3.1 Interval VIKOR 159
11.4 Implementation and Findings 161
11.5 Conclusion and Future Directions 165
References 165

11.1 INTRODUCTION

The successes achieved in unmanned aerial vehicle (UAV)-assisted disaster operations in recent years have made UAVs an essential component of disaster management (Erdelj & Natalizio, 2016; Rajan et al., 2021; Sandvik & Lohne, 2014). UAVs are commonly used in early reconnaissance (Pereira, 2021), logistics (Golabi et al., 2017), and communications (S. Zhang & Liu, 2018). They also have a critical role in guiding rescuers and survivors to the command center (Apvrille et al., 2014). The factors affecting the success of UAV operations can be reduced to the following two main factors: i) operational suitability of the UAV-equipment match, and ii) the ability of the rescue team to use the UAV.

The motivation for this chapter is grounded in a discussion of these two factors. Technically, UAVs are classified as fixed-wing and multi-rotor, which are in tradeoffs in various features (Iqbal et al., 2015). State-of-the-art technology offers both classes a wide range of specialized equipment such as thermal cameras and biological sensors. Therefore, the capabilities of the UAV, rather than the equipment, are decisive

DOI: 10.1201/9781003212904-11

in operational success. In terms of capabilities, UAV classes can be compared as follows. Fixed-wing ones can scan a large area in a short time and have a long operating time, while they require a take-off/landing area and are not suitable for low-altitude missions. Multi-rotor ones (hereafter "drones") are easy to use, suitable for low-altitude missions, especially observing hidden spots, but are slower than fixed-wings (but faster than humans), and have a short operating time. The differences between the UAV classes also differentiate the operations in which they are used effectively. In response-phase operations, drones provide remarkable advantages over fixed-wing classes thanks to their capability of hanging in the air and entering tight spaces (Ulukavak & Miman, 2021). In addition to choosing the right class of UAV, the mastery of the rescue team in drone technology is also effective in the success of the operation. To increase the mastery of the team, all members of the rescue team should know how to use the drone in which situations. Also, drone-oriented training should be conducted during the disaster preparedness phase so that the drone operator and rescue teams can act as a whole. The effect of training with the right equipment on the success of the training motivates the drone selection problem.

This chapter deals with the problem of evaluating and selecting available drones for rescue training. Keeping in mind that the selected drone will be used for the self-training of volunteers and rescue teams, the decision-making environment can be summarized as follows: i) the training drone should be ready to use, easy to handle, and readily supplied; ii) the training drone should be capable but economical; iii) the attributes of the training drone should be suitable for the economic and operating conditions of the trainees; and iv) the drone evaluation approach should seek a compromise on solutions and handle volatility in the drones' performance.

The first issue (i) describes the basic expectations from the training drone. It is possible to design special drones with superior capabilities at a high cost. However, custom-made drones can be complex to set up, use, and have a long lead-time. However, a drone available on the market exceeds basic expectations. The second issue (ii) concerns the drone evaluation criteria set. In addition to the benefit-oriented criteria such as control distance and flight time, it envisages the consideration of price, which is cost-oriented. Considering the ineptitude of novice trainees, it is clear that the economic life of drones will quickly run out. Therefore, high-cost drones will not be economical no matter how capable they are: it makes more sense to select the most capable one at an acceptable cost. The third issue (iii) is related to the importance weights assigned weights of importance attributed to the criteria. Drone capabilities, which have a key role in the success of operations, may differ in different geographies. For example, while the field of view is important in flat terrain, the control distance is critical in mountainous terrain. To maximize the benefit of the training, the characteristics of the geography in which the trainees will operate should be reflected in the decision. It also cannot be ignored in economic situations. Teams that do not have sufficient budgets give importance to price, as well as geographical features. The fourth issue (iv) describes the method used to evaluate drones. Selection requires consensus and approval of all criteria. For example, a drone has high speed but a low control distance. The control distance criterion vetoes the selection of this drone. The method should allow the decision-maker to manage the balance between acceptance and veto. In addition, the performances of the drones are expressed in specification sheets, with exact values measured in a controlled environment. However, in real-world use, performances may deviate from these values. For a realistic evaluation, the method should consider this volatility.

This decision environment was handled with multi-criteria decision making (MCDM) in this chapter. MCDM techniques allow for the evaluation of many alternatives according to their performance in benefit and cost criteria, and reflect the importance of the criteria to the decision. Thanks to the sensitivity analysis, critical managerial insights can be obtained by analyzing the effect of the change in parameters such as weight on the decision. In addition, the flexibility of the techniques provides effective results in uncertain data structures such as interval, fuzzy, and linguistic expressions. In this study, we used the interval extension of the VIKOR technique, which is selection-oriented, seeks a compromise solution, and handles uncertain data. Drones were evaluated based on their basic attributes such as weight, maximum speed, control distance, and price. Sensitivity analysis was performed with weight profiles representing different conditions.

The chapter is organized as follows. Section 11.2 discusses related studies. Section 11.3 gives detail on the methodology. Section 11.4 includes implementation and findings. Conclusions and future directions are provided in Section 11.5.

11.2 RELATED STUDIES

In this section, the background of the study, from motivation to evaluation, is presented in three stages. First, selection problems in disaster management are discussed. Then, studies involving drone selection problems are examined. Finally, common features of studies using interval numbers are highlighted.

Disaster management contains many selection-oriented problems, from supplier selection to the selection of temporary shelter locations. Olanrewaju et al. (2020) discussed integrating supplier selection into the preparedness phase to ensure timely distribution of evacuation materials and sought a solution with multi-stage stochastic programming. Chanta and Sangsawang (2012) determined the location of the shelters optimally with the capacity p-median model, aiming to cover the maximum number of people in flood disasters and to minimize the distance of the victims to the nearest shelter. Similarly, to fulfill the needs of victims and rescue teams, the problem of choosing a disaster logistics center has been solved by Turğut et al. (2011), who proposed a decision support system based on fuzzy analytical hierarchy process (AHP) methods. For more, the following studies can be examined: He et al. (2017), Kim et al. (2019), and Nappi and Souza (2015). The examples reveal that researchers seek solutions to selection problems that are strongly motivated by disaster management with operation research approaches.

The drone selection problem has attracted the attention of researchers from various fields, especially disaster management. Mehmood and Ahmed (2017) evaluated the potential of UAVs to be used in many emergencies such as forest fires. The best alternative was determined by scoring the usable UAVs on the cost–benefit axis using the simple weighted average method. Hristozov and Zlateva (2018) proposed an approach that takes into account geography, operation requirement, and UAV characteristics to determine the most suitable UAV in response to specific disaster conditions. Ulukavak and Miman (2021) evaluated the UAVs using the AHP method to transport light but important cargo to hard-to-reach areas in emergencies and chose the appropriate one. Studies have also focused on drone selection on different topics such as agriculture (Petkovics et al., 2017) and mining (Neroba, 2020). As with all decision problems, the drone selection problem requires a scientific approach suitable for the decision environment. As an exemplary study, Nur et al. (2020) adopted the fuzzy Technique for Order of Preference by Similarity to Ideal Solution (TOPSIS) approach to evaluate the interval linguistic expressions while evaluating the drones for use in the delivery to the last mile. The decision environment formed by the evaluation of realistic scenarios and criteria with linguistic expressions have been effective in the researchers' adoption of this solution approach.

In our decision environment, there was uncertainty whether the drones would be able to perform as stated in their specification sheets. Possible performance volatility should be considered for a more realistic evaluation. Interval numbers are a convenient way to handle imprecise and uncertain decision-making environments (Sayadi et al., 2009). This approach is adopted in selection problems on a variety of subjects (Hafezalkotob et al., 2016; Liang & Li, 2019; M. Zhang & Li, 2018). Apart from the selection problems, interval numbers are used to handle the reality within the scope of disaster management. Liu et al. (2019) used interval numbers to include uncertainty in reality on drought risk modeling. Li et al. (2015) dealt with the uncertainty in the amount of material with interval numbers on the optimal distribution of emergency materials under conditions of insufficient material. Huang et al. (2021) used interval numbers on the uncertainty of landslide susceptibility estimation. Tian et al. (2019) used interval numbers to better evaluate risk analysis in the oil and gas industry. These papers support conversion of the performance value in the specification sheets of drones to interval numbers for a more realistic evaluation.

11.3 METHODOLOGY

In this study, we adopted a flexible and replicable methodological approach. It can be explained step by step as follows.

Step 1. Identifying alternatives

It is a natural presupposition that all available drones are included in the alternative set. However, the presence of many alternatives increases the computational density. Moreover, many alternatives, especially weak ones, are dominated by the other, making it unnecessary to include them in the evaluation. Therefore, it makes sense to reduce the alternatives for a clean evaluation. We adopted a filtering approach to the reduction of alternatives. Three filters were applied to drones available from e-commerce sites: i) compliant with legal regulations; ii) suitable for expert or experienced users; and iii) image transfer.

Step 2. Determining the criteria

As criteria, drone specification sheets offer many distinctive features such as flight time, maximum speed, control distance, smart home return, obstacle avoidance, vibration suppression, etc. However, most of them can be ignored as they are not related to the problem. In addition, including too many criteria reduces the discriminating power of the method, as alternatives may dominate the others in at least one criterion. Therefore, the number of criteria should be sufficient for evaluation. In this context, the criteria were as follows: *maximum speed, weight, the field of view, control distance, flight time, and price.*

Step 3. Converting data to interval numbers

The performance of drones on criteria such as speed and control distance is volatile, and performance uncertainty can be handled with interval numbers. If the performance in the specification sheet is accepted as the upper value of the interval number, the lower value can be calculated by multiplying the upper value with a coefficient. The coefficient may vary for each criterion, depending on the perception of the decision-maker. We assume that the maximum speed, control distance, flight time performances, and prices of drones are volatile, and we converted all of these criteria into interval numbers with a coefficient of 90%. For example, if the maximum speed performance is 50 km/h in the specification sheet, it will be converted to the interval number as [45, 50].

Step 4. Evaluation

In this study, we adopted the interval VIKOR method to handle the decision environment mentioned earlier. The listed features of the technique were effective in its preference: i) is selection-oriented; ii) seeks a compromise; and iii) checks whether the selection satisfies the acceptability conditions.

Step 5. Sensitivity analysis

In MCDM, sensitivity analysis is often used to observe the effect of changes in weights and parameters on the result. In our problem, the differences in the economic and operating conditions of the trainees expected to affect the selection were represented by different weight profiles. Sensitivity analysis allows for determination of whether the difference between profiles affects the selection, and also which drone is suitable for which profile.

11.3.1 Interval VIKOR

The MCDA literature offers many extensions to handle interval data: interval AHP (Sugihara & Tanaka, 2001), interval ELECTRE (Vahdani et al., 2010), interval TOPSIS (Jahanshahloo et al., 2009), and interval VIKOR (Sayadi et al., 2009). In our study, we followed the interval extension of the VIKOR method proposed by Sayadi et al. (2009). Apart from the fact that VIKOR is a well-known and easy-to-apply method, it is preferable in that it seeks a compromise solution in the presence of conflicting criteria and is selection-oriented. In the interval VIKOR method, the decision matrix $[x]$ containing the interval numbers is expressed as follows.

In the presence of m alternatives $(A_1, A_2, \ldots, A_m)$ and n criteria $(C_1, C_2, \ldots, C_n)$, the decision matrix containing interval numbers is expressed as follows: $x = \left\{\left[f_{11}^L, f_{11}^U\right], \ldots, \left[f_{mn}^L, f_{mn}^U\right]\right\}$, $W = \{w_1, \ldots, w_n\}$. $\left[f_{ij}^L, f_{ij}^U\right]$ is an interval number that demonstrates the performance of A_i for C_j. $f_{ij} \in \left[f_{ij}^L, f_{ij}^U\right]$ can take any value between the lower limit f_{ij}^L and the upper limit f_{ij}^U. w_j is the weight of jth criterion. Interval VIKOR follows the steps below.

i) Determine the positive ideal solution (PIS) (A^*) and negative ideal solution (NIS) (A^-)

$$A^* = \left\{f_1^*, \ldots, f_n^*\right\} = \left\{\left(\max_i f_{ij}^U \mid j \in I\right) or \left(\min_i f_{ij}^L \mid j \in J\right)\right\} j = 1, 2, \ldots, n$$

$$A^- = \left\{f_1^-, \ldots, f_n^-\right\} = \left\{\left(\min_i f_{ij}^L \mid j \in I\right) \; or \left(\max_i f_{ij}^U \mid j \in J\right)\right\} \; j = 1, 2, \ldots, n$$

Sets I and J include the benefit and cost criteria, respectively.

ii) Determine $\left[S_i^L, S_i^U\right]$ and $\left[R_i^L, R_i^U\right]$

$$S_i^L = \sum_{j \in I} w_j \frac{f_j^* - f_{ij}^U}{f_j^* - f_{ij}^-} + \sum_{j \in J} w_j \frac{f_{ij}^L - f_j^*}{f_j^- - f_j^*} \quad i = 1, 2, \ldots, m$$

$$S_i^U = \sum_{j \in I} w_j \frac{f_j^* - f_{ij}^L}{f_j^* - f_{ij}^-} + \sum_{j \in J} w_j \frac{f_{ij}^U - f_j^*}{f_j^- - f_j^*} \quad i = 1, 2, \ldots, m$$

$$R_i^L = \max \left\{ w_j \frac{f_j^* - f_{ij}^U}{f_j^* - f_{ij}^-} \middle| j \in I, \; w_j \frac{f_{ij}^L - f_j^*}{f_j^- - f_j^*} \middle| j \in J \right\} \; i = 1, 2, \ldots, m$$

$$R_i^U = \max \left\{ w_j \frac{f_j^* - f_{ij}^L}{f_j^* - f_{ij}^-} \middle| j \in I, \; w_j \frac{f_{ij}^U - f_j^*}{f_j^- - f_j^*} \middle| j \in J \right\} \quad i = 1, 2, \ldots, m$$

The S and R indexes show group utility and individual regret of each alternative, respectively.

iii) Determine $\left[Q_i^L, Q_i^U\right]$

$$Q_i^L = v \frac{S_i^L - S^*}{S^- - S^*} + (1 - v) \frac{R_i^L - R^*}{R^- - R^*}$$

$$Q_i^U = v \frac{S_i^U - S^*}{S^- - S^*} + (1 - v) \frac{R_i^U - R^*}{R^- - R^*}$$

$S^* = \min\{S_i^L\}, \; S^- = \max\{S_i^U\}, \; R^* = \min\{R_i^L\}, \; R^- = \max\{R_i^U\} \;\; i = 1, 2, \ldots, m$

Q represents the consensus index of each alternative. $v$is the parameter representing the trade-off between maximum group utility and individual regret. It takes values in $[0,1]$, mostly 0.5.

iv) Determine the compromise solution

To determine the compromise solution, the S, R, and Q indices must be sorted by descending. To sort interval numbers, Sayadi et al. (2009) adopted an approach that compared interval numbers with each other to determine the minimum. However, we sorted the interval numbers by combining them into a single number using the midpoint approach, which is a simpler method, i.e. $Q_i = 0.5x\left(Q_i^L + Q_i^U\right)$ (Sengupta & Pal, 2000). The compromise alternative (a'), which is the first alternative of the sorted Q index, should satisfy the following two conditions:

Condition 1. Acceptable advantage

$Q(a'') - Q(a') \geq DQ$ where $DQ = \dfrac{1}{m-1}$ and a'' is the second alternative of sorted Q index.

Condition 2. Acceptable stability

a' must be first in the ranked S or/and R indexes.

If one of the conditions is not satisfied, the compromise solution set is formed according to the following conditions: i) a' and a'' if only condition 2 is not satisfied; or ii) $a', a'', \ldots, a^m$ if condition 1 is not satisfied. a^m is determined by the relation $Q(a^m) - Q(a') < DQ$. If both conditions are not satisfied, the compromise solution set is empty.

11.4 IMPLEMENTATION AND FINDINGS

In the implementation, we followed the steps in the methodology. In the first step, we identified 37 of 302 drones as alternatives by applying filters introduced in the methodology. In the second step, a total of six criteria were determined, namely benefit-oriented criteria (maximum speed, field of view, control distance, and flight time), and cost-oriented criteria (weight and price) (Nur et al., 2020). While weight and field of view criteria are considered to be exact values, maximum speed, price, control distance, and flight time are considered to be intervals. In the third step, the data were converted to interval numbers with a coefficient of 90%. Table 11.1 shows the decision matrix. A critical note about the alternatives is that some of them had the same performance in all criteria except price. These drones are technically the same, but the higher-priced ones (marked with and asterisk) have extra accessories.

Weight profiles and evaluation results are presented in Table 11.2. With the sensitivity analysis, the effect of 20 different weight sets representing different geographic conditions on the result was examined. In each of the first six profiles, one criterion was determined as dominant (50%), and the remainder were equally weighted (10%). In the seventh profile, all criteria were equally weighted (16.67%). In the remaining

TABLE 11.1
Decision Matrix

Drone Name	Data Type Orientation Drone Code	Interval Max Max Speed (km/h)	Interval Min Price (£)	Exact Min Weight (g)	Exact Max Field of View (°)	Interval Max Distance Control (m)	Interval Max Flight Time (mins)
Autel Robotics Evo II 8K	**D1***	[64.8–72]	[14977.8 –16642]	750	79	[8100–9000]	[36–40]
Autel Robotics Evo II 8K Rugged Bundle	**D1**	[64.8–72]	[13251.6–14724]	750	79	[8100–9000]	[36–40]
Xiaomi Fimi X8 SE 2020	**D2**	[58.5–65]	[3725.1–4139]	765	80	[7200–8000]	[31.5–35]
DJI Mavic Air 2	**D3**	[61.2–68]	[6824.7–7583]	570	84	[16650–18500]	[30.6–34]
DJI Mavic Air 2 Fly More Combo	**D3***	[61.2–68]	[9269.1–10299]	570	84	[16650–18500]	[30.6–34]
Yuneec Mantis G	**D4**	[64.8–72]	[6180.3–6867]	505	117	[1800–2000]	[29.7–33]
DJI Mavic 2 Pro Fly More Combo	**D5***	[64.8–72]	[16434–18260]	907	77	[4500–5000]	[27.9–31]
DJI Mavic 2 Pro	**D5**	[64.8–72]	[13681.8–15202]	907	77	[4500–5000]	[27.9–31]
DJI Mavic 2 Enterprise Dual Combo	**D6**	[64.8–72]	[35050.5–38945]	899	85	[4500–5000]	[27.9–31]
DJI Mavic 2 Zoom	**D7**	[64.8–72]	[10476–11640]	905	83	[4500–5000]	[27.9–31]
DJI Mavic 2 Enterprise Dual	**D8**	[64.8–72]	[32013.9–35571]	899	85	[4500–5000]	[27.9–31]
DJI Mavic 2 Zoom Fly More Combo	**D9**	[64.8–72]	[14097.6–15664]	905	83	[4500–5000]	[27.9–31]
DJI Mavic 2 Pro & DJI Goggles RE	**D10**	[64.8–72]	[17999.1–19999]	907	77	[4500–5000]	[27.9–31]
DJI Mavic 2 Zoom & DJI Goggles RE	**D11**	[64.8–72]	[17582.4–19536]	905	83	[4500–5000]	[27.9–31]
DJI Mavic 2 Enterprise	**D12**	[64.8–72]	[16649.1–18499]	905	82	[4500–5000]	[27.9–31]
DJI Phantom 4 Pro	**D13**	[64.8–72]	[16482.6–18314]	1388	84	[3150–3500]	[27–30]
Eachine EX5	**D14**	[31.5–35]	[2743.2–3048]	229	150	[900–1000]	[27–30]
DJI Mavic Pro Platinum	**D15**	[58.5–65]	[12259.8–13622]	734	79	[3600–4000]	[27–30]
DJI Phantom 4 Advanced+	**D16***	[64.8–72]	[15592.5–17325]	1368	84	[3150–3500]	[27–30]
DJI Phantom 4 Pro+ V2.0	**D17**	[64.8–72]	[16199.1–17999]	1375	84	[3150–3500]	[27–30]
DJI Phantom 4 Advanced	**D16**	[64.8–72]	[13504.5–15005]	1368	84	[3150–3500]	[27–30]
DJI Phantom 4 Pro+	**D18**	[64.8–72]	[17099.1–18999]	1388	84	[3150–3500]	[27–30]

DJI Mavic Pro Platinum Fly More Combo	**D19**	[58.5–65]	[14448.6–16054]	734	79	[3600–4000]	[27–30]
DJI Phantom 4 Pro V2.0	**D20**	[64.8–72]	[13791.6–15324]	1375	84	[3150–3500]	[27–30]
DJI Phantom 4	**D21**	[64.8–72]	[45179.1–50199]	1380	94	[2250–2500]	[25.2–28]
DJI Mavic Pro Fly More Combo	**D22**	[58.5–65]	[9396–10440]	734	79	[3600–4000]	[24.3–27]
Xiaomi Mi Drone 4K	**D23**	[58.5–65]	[7560–8400]	1390	94	[1800–2000]	[24.3–27]
DJI Mavic Pro Alpine White Combo	**D24**	[58.5–65]	[7586.1–8429]	734	79	[3600–4000]	[24.3–27]
DJI Inspire 2 Zenmuse X4S	**D25**	[84.6–94]	[38700–43000]	3440	84	[3150–3500]	[24.3–27]
Yuneec Typhoon H	**D26**	[63–70]	[19845–22050]	1695	98	[1440–1600]	[22.5–25]
Eachine EX4	**D27**	[32.4–36]	[3722.4–4136]	435	80	[2700–3000]	[22.5–25]
Yuneec Typhoon H Pro	**D28**	[63–70]	[7199.1–7999]	1695	98	[1440–1600]	[22.5–25]
LYZRC L109 Pro	**D29**	[48.6–54]	[3118.5–3465]	320	120	[1080–1200]	[22.5–25]
Hubsan Zino Pro H117P	**D30**	[32.4–36]	[3679.2–4088]	750	80	[3600–4000]	[20.7–23]
DJI Matrice 600 Pro	**D31**	[58.5–65]	[71941.5–79935]	9500	94	[3150–3500]	[14.4–16]
DJI Mavic Pro	**D32**	[58.5–65]	[10618.2–11798]	734	79	[3600–4000]	[24.3–27]
DJI Inspire 2 Zenmuse X5S	**D33**	[84.6–94]	[63835.2–70928]	3440	72	[3150–3500]	[24.3–27]

* The data were obtained from various e-commerce sites and drone company brochures.

TABLE 11.2
Weight Profiles and Evaluation Results

Weight Profiles	Max Speed	Price	Weight	Field of View	Distance Control	Flight Time	First Evaluation	Second Evaluation
1	50.00%	10.00%	10.00%	10.00%	10.00%	10.00%	D15	D15
2	10.00%	50.00%	10.00%	10.00%	10.00%	10.00%	D3,D3*	D3,D3*
3	10.00%	10.00%	50.00%	10.00%	10.00%	10.00%	D3,D3*	D3,D3*
4	10.00%	10.00%	10.00%	50.00%	10.00%	10.00%	D14	D14
5	10.00%	10.00%	10.00%	10.00%	50.00%	10.00%	D3,D3*	D3,D3*
6	10.00%	10.00%	10.00%	10.00%	10.00%	50.00%	D1,D1*	D1*,D1
7	16.67%	16.67%	16.67%	16.67%	16.67%	16.67%	D3,D3*	D3,D3*
8	5.00%	40.00%	5.00%	5.00%	5.00%	40.00%	D1,D1*	D1*,D1
9	15.00%	30.00%	5.00%	5.00%	5.00%	40.00%	D1,D1*	D1*,D1
10	15.00%	40.00%	5.00%	5.00%	5.00%	30.00%	D1,D1*	D1*,D1
11	20.00%	25.00%	5.00%	10.00%	5.00%	35.00%	D1,D1*	D1*,D1
12	15.00%	30.00%	5.00%	15.00%	5.00%	30.00%	D4	D4
13	20.00%	25.00%	5.00%	20.00%	5.00%	25.00%	D4	D4
14	5.00%	50.00%	5.00%	5.00%	5.00%	30.00%	D1,D2,D1*,D3	D1*,D1
15	5.00%	30.00%	5.00%	5.00%	5.00%	50.00%	D1,D1*	D1*,D1
16	10.00%	30.00%	5.00%	10.00%	10.00%	35.00%	D1,D1*	D1*,D1
17	20.00%	5.00%	5.00%	20.00%	20.00%	30.00%	D3,D3*	D3,D3*
18	15.00%	5.00%	5.00%	15.00%	10.00%	50.00%	D1,D1*	D1*,D1
19	15.00%	10.00%	5.00%	15.00%	15.00%	40.00%	D1,D1*,D3,D3*	D1*,D1,D3,D3*
20	15.00%	5.00%	5.00%	20.00%	5.00%	50.00%	D1,D1*	D1*,D1

profiles, the criteria were weighted at variable rates. We had two evaluation scenarios. The first evaluation covered six criteria, while the second covered five criteria, excluding price. The aim was to answer how the selection would change if price is no longer a criterion with the second evaluation. The criteria weights of the second evaluation were determined by distributing the weight of the price criterion to the other criteria in proportion to their weights.

A striking point in the results is that, in the selection, some profiles had a drone, while others had more than one drone. If we ignored drones that only differ in price, one suitable drone could be determined for all profiles except for profiles 14 and 19. In most profiles, cheap and expensive versions of the same drone appear in the solution set. Which one to choose depends on the economic situation of the decision-maker. For example, the solution set of profiles 2 and 17 included D3 and D3* drones. The D3 drone can be preferred since the second profile gives more weight to the price criterion, and the D3* drone can be preferred because the 17th profile gives a low weight to the price criterion. The fact that the selection was reduced to a single drone for most profiles demonstrates the success of the method in distinguishing drones. It also proves that the purpose of selecting a drone was achieved.

In profiles 14 and 19, there was more than one technically different drone in the solution set. Any drone in the solution set satisfied the profiles.

While the results showed that the D1 drone is the appropriate option for most profiles, the D3, D4, D14, and D15 drones are also suitable for different profiles. The dominant criterion and corresponding drone pairing are speed (D15), field of view (D14), price, weight, and control distance (D3), and flight time (D1). This finding proves that the importance attributed to the criteria is effective on the decision.

Another interesting finding was observed when comparing the scenarios. The first evaluation and second evaluation results were the same for all profiles except for profile 14, indicating that the price criterion did not affect the selection.

11.5 CONCLUSION AND FUTURE DIRECTIONS

Drones have an active role in disaster-management operations. The utility of drones in the field is limited to a team's ability to use drones. Rescue teams must be trained with the proper drones to get the maximum benefit from their drone-focused training. This chapter covers the evaluation and selection of drones that can be used for self-training of volunteers and rescuers.

As in many disaster-management problems, operations research techniques can be used effectively in drone selection. It is possible with MCDM techniques to clear the drone selection decision from decision-maker bias and to make the selection with a scientific approach. The decision environment of the problem is handled with the interval VIKOR method. The effect of different economic and operational conditions, as well as the price criterion, on the selection, was examined.

The results showed that drone selection changes when conditions change, and the selection should be made according to the conditions, to maximize benefit. This leads to the conclusion that drone selection is not an ordinary decision, and that a decision made under different conditions cannot be repeated for other conditions. In addition, the fact that the selection changes even in conditions that can be considered as similar to each other reveals that the decision-maker cannot determine the most suitable drone for the conditions without using a scientific methodology. In general, it is seen that the choices do not change when the price is removed from being a criterion. It shows that the power of price to influence the decision is weak, that similar drones are priced similarly, and that there is competition among them.

As a result, the methodology followed in the study ensures that the drone suitable for the operating conditions can be determined. Thanks to this systematic approach, the negativities of the wrong decision can be minimized, and maximum benefit can be obtained from the training.

The flexibility of the methodology allows the application to be customized to different decision environments. Thus, it is possible to expand its use for disaster-management professionals who do not have technical knowledge by turning it into a decision-support system form integrated into e-commerce sites in future studies.

REFERENCES

Apvrille, L., Tanzi, T., & Dugelay, J. L. (2014). Autonomous drones for assisting rescue services within the context of natural disasters. *2014 31th URSI General Assembly and Scientific Symposium, URSI GASS 2014*. https://doi.org/10.1109/URSIGASS.2014.6929384

Chanta, S. & Sangsawang, O. (2012). Shelter-site selection during flood disaster. *Lecture Notes in Management Science*, *4*, 282–288.

Erdelj, M. & Natalizio, E. (2016). UAV-assisted disaster management: applications and open issues. *2016 International Conference on Computing, Networking and Communications, ICNC 2016*. https://doi.org/10.1109/ICCNC.2016.7440563

Golabi, M., Shavarani, S. M., & Izbirak, G. (2017). An edge-based stochastic facility location problem in UAV-supported humanitarian relief logistics: a case study of Tehran earthquake. *Natural Hazards*, *87*(3), 1545–1565. https://doi.org/10.1007/s11069-017-2832-4

Hafezalkotob, A., Hafezalkotob, A., & Sayadi, M. K. (2016). Extension of MULTIMOORA method with interval numbers: an application in materials selection. *Applied Mathematical Modelling*, *40*(2), 1372–1386. https://doi.org/10.1016/J.APM.2015.07.019

He, J., Feng, C., Hu, D., & Liang, L. (2017). A decision model for emergency warehouse location based on a novel stochastic MCDA method: evidence from China. *Mathematical Problems in Engineering*, *2017*. https://doi.org/10.1155/2017/7804781

Hristozov, S. & Zlateva, P. (2018). Concept model for drone selection in specific disaster conditions. *The International Archives of the Photogrammetry, Remote Sensing and Spatial Information Sciences*, 245–249. https://doi.org/10.5194/isprs-archives-XLII-3-W4-245-2018

Huang, F., Ye, Z., Jiang, S. H., Huang, J., Chang, Z., & Chen, J. (2021). Uncertainty study of landslide susceptibility prediction considering the different attribute interval numbers of environmental factors and different data-based models. *CATENA*, *202*, 105250. https://doi.org/10.1016/J.CATENA.2021.105250

Iqbal, U., Irtiza, S., Shah, A., Jamil, M., Gillani, S. O., & Ayaz, Y. (2015). Selection of unmanned aerial system (UAS) for disaster relief operations: a comparison. *Science International (Lahore)*, *27*(4), 3199–3203.

Jahanshahloo, G. R., Hosseinzadeh Lotfi, F., & Davoodi, A. R. (2009). Extension of TOPSIS for decision-making problems with interval data: interval efficiency. *Mathematical and Computer Modelling*, *49*(5–6), 1137–1142. https://doi.org/10.1016/J.MCM.2008.07.009

Kim, S., Ramkumar, M., & Subramanian, N. (2019). Logistics service provider selection for disaster preparation: a socio-technical systems perspective. *Annals of Operations Research*, *283*(1–2), 1259–1282. https://doi.org/10.1007/s10479-018-03129-3

Li, Y., Zhou, Z., Li, B., Fan, Y., & Wang, Q. (2015). Optimization method for earthquake emergency material distribution based on possibility degree of interval numbers. *Journal of Basic Science and Engineering*, *23*(3), 637–644. https://doi.org/10.16058/J.ISSN.1005-0930.2015.03.022

Liang, D. & Li, Y. (2019). A decision making model base on the comparative advantage relation between three parameter interval numbers in railway location selection. *Proceedings of IEEE 14th International Conference on Intelligent Systems and Knowledge Engineering, ISKE 2019*, 209–213. https://doi.org/10.1109/ISKE47853.2019.9170291

Liu, X., Guo, P., Tan, Q., Xin, J., Li, Y., & Tang, Y. (2019). Drought risk evaluation model with interval number ranking and its application. *Science of the Total Environment*, *685*, 1042–1057. https://doi.org/10.1016/j.scitotenv.2019.06.260

Mehmood, S. & Ahmed, S. (2017). *Incorporation of Drones into Fire and Rescue Service of Esbjerg Municipality for a Robust Emergency Response*. Aalborg University.

Nappi, M. M. L. & Souza, J. C. (2015). Disaster management: hierarchical structuring criteria for selection and location of temporary shelters. *Natural Hazards*, *75*(3), 2421–2436. https://doi.org/10.1007/s11069-014-1437-4

Neroba, V. (2020). Development of methods for assessment and selection of unmanned aerial vehicle for mine reconnaissance. *ScienceRise*, *5*, 44–50. https://doi.org/10.21303/2313-8416.2020.001496

Nur, F., Alrahahleh, A., Burch, R., Babski-Reeves, K., & Marufuzzaman, M. (2020). Last mile delivery drone selection and evaluation using the interval-valued inferential fuzzy TOPSIS. *Journal of Computational Design and Engineering*, *7*(4), 397–411. https://doi.org/10.1093/jcde/qwaa033

Olanrewaju, O. G., Dong, Z. S., & Hu, S. (2020). Supplier selection decision making in disaster response. *Computers and Industrial Engineering*, *143*, 106412. https://doi.org/10.1016/j.cie.2020.106412

Pereira, F. L. (2021). Optimal control problems in drone operations for disaster search and rescue. *Procedia Computer Science*, *186*, 78–86. https://doi.org/10.1016/j.procs.2021.04.127

Petkovics, I., Simon, J., Petkovics, A., & Covic, Z. (2017). Selection of unmanned aerial vehicle for precision agriculture with multi-criteria decision making algorithm. *SISY 2017 – IEEE 15th International Symposium on Intelligent Systems and Informatics, Proceedings*, 151–155. https://doi.org/10.1109/SISY.2017.8080543

Rajan, J., Shriwastav, S., Kashyap, A., Ratnoo, A., & Ghose, D. (2021). Disaster management using unmanned aerial vehicles. In *Unmanned Aerial Systems* (pp. 129–155). Elsevier. https://doi.org/10.1016/b978-0-12-820276-0.00013-3

Sandvik, K. B. & Lohne, K. (2014). The rise of the humanitarian drone: giving content to an emerging concept. *Millennium: Journal of International Studies*, *43*(1), 145–164. https://doi.org/10.1177/0305829814529470

Sayadi, M. K., Heydari, M., & Shahanaghi, K. (2009). Extension of VIKOR method for decision making problem with interval numbers. *Applied Mathematical Modelling*, *33*(5), 2257–2262. https://doi.org/10.1016/J.APM.2008.06.002

Sengupta, A. & Pal, T. K. (2000). On comparing interval numbers. *European Journal of Operational Research*, *127*(1), 28–43. https://doi.org/10.1016/S0377-2217(99)00319-7

Sugihara, K. & Tanaka, H. (2001). Interval evaluations in the analytic hierarchy process by possibility analysis. *Computational Intelligence*, *17*(3), 567–579. https://doi.org/10.1111/0824-7935.00163

Tian, D., Zhao, C., Wang, B., & Zhou, M. (2019). A MEMCIF-IN method for safety risk assessment in oil and gas industry based on interval numbers and risk attitudes. *Engineering Applications of Artificial Intelligence*, *85*, 269–283. https://doi.org/10.1016/J.ENGAPPAI.2019.06.015

Turğut, B. T., Taş, G., Herekoğlu, A., Tozan, H., & Vayvay, O. (2011). A fuzzy AHP based decision support system for disaster center location selection and a case study for Istanbul. *Disaster Prevention and Management*, *20*(5), 499–520. https://doi.org/10.1108/09653561111178943

Ulukavak, M. & Miman, M. (2021). Selection of the most proper unmanned aerial vehicle for transportation in emergency operations by using analytic hierarchy process. *International Journal of Environment and Geoinformatics*, *8*(1), 78–91. https://doi.org/10.30897/ijegeo.760758

Vahdani, B., Jabbari, A. H. K., Roshanaei, V., & Zandieh, M. (2010). Extension of the ELECTRE method for decision-making problems with interval weights and data. *International Journal of Advanced Manufacturing Technology*, *50*(5–8), 793–800. https://doi.org/10.1007/s00170-010-2537-2

Zhang, M. & Li, G. (2018). Combining TOPSIS and GRA for supplier selection problem with interval numbers. *Journal of Central South University*, *25*(5), 1116–1128. https://doi.org/10.1007/S11771-018-3811-Y

Zhang, S. & Liu, J. (2018). Analysis and optimization of multiple unmanned aerial vehicle-assisted communications in post-disaster areas. *IEEE Transactions on Vehicular Technology*, *67*(12), 12049–12060. https://doi.org/10.1109/TVT.2018.2871614

12 Emergency Department Preparedness Assessment for Outbreaks by a Combined Bayesian BWM–RIM Model

Huai-Wei Lo,[1] Melike Erdogan,[2] Melih Yucesan,[3] and Muhammet Gul[4]

[1]Department Of Industrial Engineering And Management, Chaoyang University Of Technology, Taichung, Taiwan

[2]Department Of Industrial Engineering, Faculty Of Engineering, Düzce University, Düzce, Turkey

[3,*]Department Of Emergency And Disaster Management, Munzur University, Tunceli, Turkey

[4]Department Of Transportation And Logistics, Istanbul University, Istanbul, Turkey

CONTENTS

12.1 Introduction ... 170
12.2 Literature Review ... 171
12.2.1 Assessing Preparedness of Health Facilities in the Face of Outbreaks ... 171
12.2.2 An Analytical Tool for Outbreak Preparedness Assessment: MADM Concepts ... 172
12.2.3 Research Gaps and Contributions of the Study ... 174
12.3 Methodology ... 175
12.3.1 BWM and Bayesian BWM ... 175
12.3.2 RIM ... 178
12.3.3 Description of the Combined MADM Framework ... 179
12.4 Case Study ... 179
12.4.1 Description of the Observed EDS, the Questionnaire, and Decision-Making Group ... 180

DOI: 10.1201/9781003212904-12

12.4.2 ED Performance Metrics used in the Model 180
12.4.3 Implementation of the Combined Bayesian BWM–RIM Model to the Problem 181
12.4.4 Usability of the Proposed Combined Bayesian BWM–RIM Model Within the Scope of Managerial Insight 184
12.5 Conclusion 185
References 185

12.1 INTRODUCTION

The SARS-CoV-2 (COVID-19) virus, which the whole world has been struggling with, seems to cause trouble until social immunity for countries is achieved. In addition to lack of healthcare staff, medical supplies, and support services during epidemics, many problems arise, such as effective management of overcrowding (Seyedin et al. 2021). The fact that global disasters have increased almost three times compared to about forty years ago has made it key that hospitals and emergency services are prepared to overcome the crises (Al-Hajj et al. 2020). Hospitals are institutions where disaster preparedness is essential because of their unique missions, size, complexity, types of materials used, and the types of patients they encounter (Ginter et al. 2007). Confusion over roles and responsibilities, inadequate planning and training, and lack of integration of hospitals into community disaster planning can cause hospitals to be unprepared (Manley et al. 2006). Preparing hospitals by identifying their strengths and weaknesses is crucial in providing a more effective and coordinated response for future emergencies (Munasinghe et al. 2021). Because pandemics can affect the capacity of hospitals and the overall functioning of the health system, hospitals need to be prepared to cope with the challenge of an epidemic disaster before is occurs (Gul and Yucesan, 2021). Hospitals must anticipate how much clinical demand they may face when determining how best to put forward plans for pandemics and other disasters (Safavi et al. 2021). Evaluation of the disaster preparedness of hospitals is essential in determining and arranging the level of readiness of health facilities in the fight against disasters (Firdausi et al. 2021). In case of disasters, hospitals have to adapt other services such as intensive care units, but modifications to emergency departments (EDs) are among the most challenging (Jachetti et al. 2021). The COVID-19 pandemic has been an unthinkable challenge for EDs worldwide, making ED preparedness for the COVID-19 patient surge quite challenging while maintaining adequate capacity for ordinary emergencies (Barten et al. 2020). In disaster situations and normal conditions, early diagnosis and surveillance should be provided in EDs, and isolation measures should be implemented. There should be a triage system that predicts the correct orientation of the case according to the severity of the patient and/or isolation needs (Casalino et al. 2020). However, it should adopt a standard, multifaceted approach to manage patients presenting with COVID-19, especially recently, and be prepared to avoid risks to other patients in the ED or hospital (Pothiawala, 2020).

Many factors need to be considered to measure and determine the preparedness of emergency services during the COVID-19 pandemic. Some of these factors can be evaluated qualitatively, while others can be considered quantitatively. At this point, one of the most effective ways to consider all criteria and evaluate them successfully in the same analysis is the multi-attribute decision-making (MADM) approach. At this point, some approaches can used to handle uncertainties in the process in the best way and reflect them in Bayesian theory analysis, for example. This approach is one of most used for decision problems in uncertain environments.

Considering all these factors, in this study the problem of assessing ED preparedness in the COVID-19 pandemic was addressed. The preparedness levels of five EDs located in a densely populated area in a metropolitan city of Turkey were evaluated according to the determined criteria. For this purpose, criteria weights were calculated using the Bayesian best-worst method (BWM) and the preparedness levels of emergency services were compared with the reference ideal method (RIM) approach. Thus, we tried to determine the EDs' preparedness for pandemic disasters such as COVID-19. With real-life analysis carried out within the scope of this study, the preparedness levels of other EDs can be measured, and policies that can be adopted during the pandemic period can be revealed and strategies developed to ensure the safe use of EDs. The main motivation of this study was to evaluate emergency services in disaster situations by considering the uncertainty, and to utilize an approach that can achieve satisfactory results even in the presence of multiple conflicting factors such as multi-criteria decision making. The main contribution was to apply a Bayesian-based hybrid decision-making methodology for the first time to assess EDs under pandemic conditions. To the best of our knowledge, there have not been any papers that have considered COVID-19 restrictions to determine the preparedness level of EDs with the proposed Bayesian-based methodology. We assert that using the proposed methodology presented in this paper will be a source of motivation for researchers who want to work in this field.

Section 12.2 presents the literature review. Section 12.3 presents the adopted methodology in detail. Section 12.4 covers the real-life application, while Section 12.5 includes conclusions and suggestions for future studies.

12.2 LITERATURE REVIEW

The literature review related to the study is included and presented in the sub-sections.

12.2.1 Assessing Preparedness of Health Facilities in the Face of Outbreaks

The COVID-19 pandemic highlighted the importance for organizations to develop preventive measures, testing procedures, and emergency response plans (Hulland et al. 2019). Countries actively promoted health and safety plans to contain and slow the spread of the virus. Potentially infected persons were to self-isolate for 14 days, to ensure they became asymptomatic or complete recovered from symptoms. Patients who were diagnosed as positive were to arrange for treatment in medical institutions until they recovered (Wang et al. 2020). There is no doubt that health facilities are the institutions that bear the brunt of an epidemic. Because medical institutions provide many services (e.g., consultation, screening, and treatment, care), medical staff are placed in a high-risk environment for infection (Kramer et al. 2021).

To ensure that health facilities can maintain operations during an outbreak, epidemic prevention measures and preparations are urgent tasks. Seyedin et al. (2021) developed a hospital preparedness checklist to evaluate the ability to react to the COVID-19 pandemic. They interviewed 32 experts with senior experience in hospital management and epidemic prevention to construct several preparation projects. Epidemic prevention measures include pre-hospital measures (Marques et al. 2020); environmental monitoring (Marshall et al. 2020); prioritization and triage (Jöbges et al. 2020); internal and external integration of hospitals (Garcia-Lopez et al. 2021); information and communication systems (Singh et al. 2021); drug and medical equipment management (Garzotto et al. 2021); surge capacity (physical space, staff, supplies, and processes) (Carenzo et al. 2020); COVID-19 patient discharge and non-COVID-19 patient management (Gul and Yucesan, 2021); intensive care unit; medical staff management (Perinel et al. 2020); security, nutrition, and food health (Mardones et al. 2020); quarantine, exercises, and education (Hsu et al. 2021); laboratory (Mancini et al. 2020); personal protective equipment (PPE; Suleiman et al. 2020); prevention and control of hospital infection (Islam et al. 2020); and management of people who have died (www.who.int/publications/i/item/infection-prevention-and-control-for-the-safe-management-of-a-dead-body-in-the-context-of-covid-19-interim-guidance).

Focusing on medical and healthcare staff, Griffin et al. (2020) planned a hospital preparedness guide from a critical care perspective. The guide covers multiple areas, including intensive care unit surge capacity, and simultaneously focused on infection control, clinical operational challenges, ethics, staffing, and maintenance of staff wellness. Furthermore, Gul and Yucesan (2021) proposed a preparedness ranking model for Turkish hospitals in response to COVID-19, including a total of 99 criteria for the hospital's epidemic prevention work. In addition to the above-mentioned documents, many researchers have contributed to the epidemic prevention work of health facilities. Table 12.1 presents relevant references regarding the preparedness of health facilities.

12.2.2 An Analytical Tool for Outbreak Preparedness Assessment: MADM Concepts

The decision-making problems in today's world are complex and challenging. When encountering unstructured problems, there is no way to solve them by constructing mathematical models. MADM has an excellent performance in solving the above-mentioned issues. It can analyze the most appropriate choice for decision-makers (Huang et al. 2021). MADM can usually be divided into three procedures, including the establishment of evaluation criteria, the calculation of criteria weights, and the determination of the priority of alternatives (Chang et al. 2021). In the global COVID-19 pandemic, many epidemic prevention decisions are typical MADM issues.

TABLE 12.1
References of Preparedness of Health Facilities

Authors	Themes	Content and Results
Suleiman et al. (2020)	Preparedness of frontline doctors in Jordan healthcare facilities.	Their study revealed that the lack of PPE will cause doctors to feel fear and anxiety when dealing with COVID-19 patients.
Gupta and Federman (2020)	Hospital preparedness based on experience of Veterans Affairs Connecticut healthcare system.	They shared several rapid preparation and response work items to help other healthcare systems for epidemic prevention.
AlTakarli (2020)	A model for epidemic preparedness and management of China.	This paper pointed out that efforts to develop a resilient system against infectious diseases should be the most important concern for any country.
Boufkhed et al. (2021)	Preparedness and capacity of palliative care services in Middle-Eastern and North African countries.	The authors call on governments to integrate palliative care into preparedness projects.
Garzotto et al. (2020)	Assessing whether medical equipment is adequate to cope with the COVID-19 pandemic	The research integrates the opinions of scientists, technical experts, and medical experts to evaluate the management of medical equipment and the risks of its use.
Noh et al. (2020)	Strategies for safe hospital operations based on the experiences of Korea.	They proposed a Swiss cheese model to assess hospitals' anti-epidemic ability. The results indicate that the hospital needs to strengthen prevention in long-term care institutions.
Perinel et al. (2020)	Drug control for intensive care unit (ICU) patients with COVID-19.	The authors believe that inappropriate dosing regimens will cause the incidence of adverse events in patients. They proposed a prospective study to assess the pharmacokinetic properties of hydroxychloroquine (HCQ) in ICU COVID-19 patients.
Mukhtar (2021)	A proactive infection prevention measure approach for the outbreak control in Pakistan.	This study introduced comprehensive anti-epidemic measures in Pakistan and emphasized the importance of personal protection, waste disposal, and route planning.

In the past year, many studies have used the MADM-based concept to assess the preparedness for COVID-19 outbreaks (Gul and Yucesan, 2021). For example, Ortiz-Barrios et al. (2021) proposed a hybrid model to evaluate sales departments' preparedness level for COVID-19. In their study, analytic hierarchy process (AHP) was used to obtain the weights of criteria and sub-criteria. The technique for order preference by similarity to ideal solution (TOPSIS) was applied to calculate the preparedness level and rank the evaluated companies. The results could help the electrical industry to improve their prevention ability in the COVID-19 pandemic. In a similar AHP–TOPSIS model, Hezam et al. (2021) added neutrosophic set theory to optimize the adaptability of general AHP–TOPSIS in uncertain environments. The model was used to decide the priority of vaccine alternatives. Their study showed that medical staff, elderly people, high-risk groups, essential workers, and pregnant and lactating women are the top-priority groups for vaccination.

Gul and Yucesan (2021) developed a new approach, interval-valued spherical fuzzy sets (IVSFS) and TOPSIS (IVSFS–TOPSIS), to measure the hospital preparedness performance of Turkish tertiary healthcare services. IVSFSs improve many shortcomings of Pythagorean fuzzy sets and neutrosophic sets, so that the analysis results are more reliable in real situations. Moheimani et al. (2021) used the rough set theory to classify the level of hospital preparedness. They emphasized that hospitals need to have sufficient medical equipment and medicines to avoid frustration in providing the support and services that COVID-19 patients need. Belhadi et al. (2020) introduced an integration model to select an optimal infectious waste-management strategy, which used interval-valued fuzzy (IVF) AHP–VIKOR to measure the performance of alternative sustainable technologies. The above-mentioned articles have put forward many valuable and contribution criteria for the preparedness of various aspects. The use of MADM as an analysis tool for COVID-19 prevention/preparation issues is effective and practical.

12.2.3 Research Gaps and Contributions of the Study

When the literature is analyzed in depth, it is seen that many researchers and health-related decision-making institutions are working on the preparation of hospital emergency departments in case of epidemics such as the COVID-19. In some of these, the criteria used in the evaluation of the preparation were determined and MCDM methods were used to determine their importance levels. However, in the current study, Bayesian–BWM (BBWM) was preferred, because, for such a problem, BBWM is an appropriate and effective method for the reasons mentioned below. Similarly, the use of the RIM method, which is the second pillar of the integrated approach, is appropriate since the criteria determined for the evaluation of the pandemic preparedness of EDs consist of numerical data and include ideal reference values. In this case, such an integrated approach was considered to be unique to the problem.

In particular, the model contributions and advantages to the relevant literature can be listed as follows: (i) BBWM is a novel weighting method. Compared with AHP, BBWM dramatically reduces the number of pairwise comparisons of criteria and obtains more reliable and consistent results. In addition, BBWM integrates the opinions of multiple experts based on Bayesian statistical theory to train more reasonable final aggregated weights (Mohammadi and Rezaei 2020a; Yang et al. 2020). (ii) RIM is also a relatively new method of performance integration, proposed by Cables et al. (2016). The method overcomes the problem that TOPSIS and VIKOR are not applicable in the "nominal-the-type characteristic" criterion. Furthermore, it also sets a reference ideal for each criterion, so that the gap between the alternatives and the reference ideal can be more clearly defined (Sánchez-Lozano and Rodríguez 2020). (iii) This paper demonstrates the feasibility of combining the BBWM and RIM methods through case studies, and the management implications can support decision-makers to formulate the most appropriate epidemic-prevention strategies for EDs.

12.3 METHODOLOGY

In this section, the adopted methodology, which is a combination of the BWM and RIM approaches, is presented.

12.3.1 BWM and Bayesian BWM

BWM is one of the most important methods, initially proposed by Rezaei (2015). It requires fewer pairwise comparisons $(2n-3)$ than a full comparison $\left(n(n-1)\right)/2$. BWM has many significant advances, shown in Figure 12.1. Rezaei (2016) carried out an interval analysis of BWM. Hafezalkotob and Hafezalkotob (2017) adapted group and individual evaluations to fuzzy best–worst method (FBWM) for aggregating experts' judgments. Guo and Zhao (2017) aimed to express uncertainties by integrating BWM with the fuzzy set. Aboutorab et al. (2018) presented the Z-number extension of BWM. Safarzadeh et al. (2018) proposed a novel group-based decision-making method, according to BWM. Liao et al. (2019) proposed a hesitant fuzzy extension of BWM. Brunelli and Rezaei (2019) proposed a new metric for BWM called multiplicative. Liang et al. (2020) proposed a new consistency measurement for BWM. Mohammadi and Rezaei (2021a) have proposed Bayesian BWM by taking BWM from the probability perspective. Dong et al. (2021) suggested a fuzzy BWM to reveal some deficiencies/falsities of the approach of Guo and Zhao (2017).

The main problem with the original BWM is that it has to calculate the weight of each expert's evaluation separately, and finally use the arithmetic mean to aggregate the weights of multiple experts; however, this practice leads to potential information loss (Mohammadi and Rezaei 2020a;Rezaei 2015, 2016). To overcome the above issues, Mohammadi and Rezaei (2020a) proposed BBWM. The technique effectively aggregates the judgments of multiple experts to gain a set of optimal group criteria weights. All experts are requested to select the most important and least important criteria, and then they are compared pairwise with other criteria to constitute two information sets (best-to-others (BO) and others-to-worst (OW) vectors). The optimal group criteria weights are estimated by statistical distribution. The operational procedures of the BBWM approach are laid out in the following steps.

Step 1. Constructing evaluation criteria

The criteria $\{c_1, c_2, \ldots, c_n\}$ were determined through a literature review and multiple expert decision.

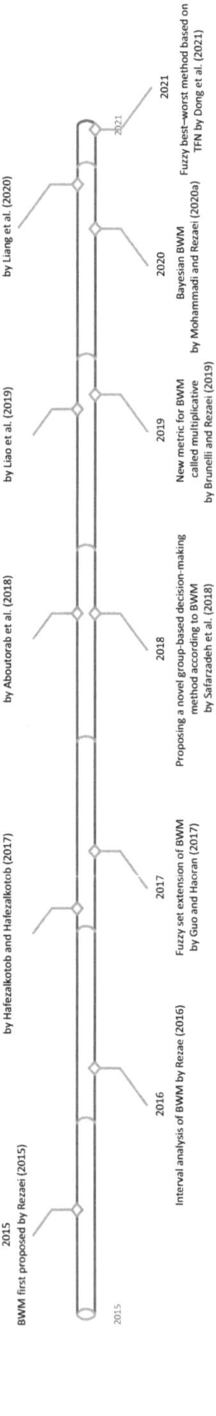

FIGURE 12.1 The historical development of BMW.

Step 2. Selecting the best and worst criteria
According to step 1, each expert selects what they think are the best and worst criteria.

Step 3. Comparing the best criteria with other criteria
Each of the experts assess the best criteria's relative importance to other criteria to produce the BO vector (Eq. 12.1). The rating scale is from 1 to 9.

$$A_{Bj} = \left(a_{B1}, a_{B2}, \ldots, a_{Bn}\right) \tag{12.1}$$

where a_{Bj} represents the importance of the significant multiple of best criterion B over criterion j.

Step 4. Comparing other criteria with the worst criterion
As in step 3, each expert assess the relative importance of the other criteria to the worst criterion to produce the OW vector (Eq. 12.2).

$$A_{jW} = \left(a_{1W}, a_{2W}, \ldots, a_{nW}\right)^T \tag{12.2}$$

where a_{jW} represents the importance of the significant multiple of criterion j over the worst criterion W.

Step 5. Gaining the group criteria weights
According to steps 1–4, the assessment data of all experts can be obtained. MATLAB software (by Mohammadi and Rezaei 2020a) can be used to gain the optimal group criteria weights w_j^{agg}.

Step 6. Checking ranking confidence
Whether the weight ranking is consistent must be checked, using the concept of credal ranking to check the confidence of c_i and c_j. The probability that c_i is better than c_j is

$$P\left(c_i > c_j\right) = \int I\left(w_i^{agg} > w_j^{agg}\right) P\left(w^{agg}\right) \tag{12.3}$$

where $P\left(w^{agg}\right)$is the posterior probability of w^{agg}, and I is the condition parameter, which can be computed when $\left(w_i^{agg} > w_j^{agg}\right)$ is true, otherwise it is 0. The Markov chain Monte Carlo (MCMC) approach is adopted to operate various simulations, and the sample size Q gained to use in computing its average confidence.

$$P\left(c_i > c_j\right) = \frac{1}{Q}\sum_{q=1}^{Q} I\left(w_i^{agg_q} > w_j^{agg_q}\right);$$

$$P\left(c_j > c_i\right) = \frac{1}{Q}\sum_{q=1}^{Q} I\left(w_j^{agg_q} > w_i^{agg_q}\right). \tag{12.4}$$

where w^{agg_q} represents q w^{agg}s from the MCMC sample. When $P\left(c_i > c_j\right) > 0.5$, it indicates that criterion i is more important than criterion j, and the probability presented is the confidence level. In addition, $P\left(c_i > c_j\right) + P\left(c_j > c_i\right) = 1$.

12.3.2 RIM

RIM is a MCDM method proposed by Cables et al. (2016). It has similar aspects to TOPSIS and VIKOR. However, it is novel in terms of the ideal solution concept and type of data (Cables et al. 2016; Sánchez-Lozano et al. 2019; Sofuoğlu, 2021). Moreover, the ranking reversal problem can be effectively avoided by RIM. The procedural steps of the RIM MCDM method are summarized below.

Step 1. Study context is determined.

The required execution parameters of RIM are established, and the following three elements are set for each criterion, as follows:

- The value range of the criterion: $t_j = [A_j, B_j]$.
- The reference ideal (RI) of the criterion: $s_j = [C_j, D_j]$.
- The weight of the criterion: w_j.
- It is noted that the RI includes two values as the maximum and minimum.

*Step 2. Valuation matrix **X** is determined*

Assuming that there are m alternatives $\left\{ED_1, ED_2, \ldots, ED_m\right\}$ and n criteria $\left\{c_1, c_2, \ldots, c_n\right\}$, each performance value of alternative ED_i under criteria c_j is investigated. The valuation matrix $\boldsymbol{X}$ is shown in Eq. (12.5).

$$X = \left[x_{ij}\right]_{m \times n}, i = 1,2,\ldots,m; j = 1,2,\ldots,n \tag{12.5}$$

*Step 3. Normalized valuation matrix **Y** is determined via RI*

An element of matrix $\boldsymbol{Y}$ is $f\ (x_{ij}, t_j, s_j)$ (Eq. 12.6). Computation of this element is explained in Eq. (12.7) as follows:

$$Y = \left[y_{ij}\right]_{m \times n},\ i = 1,2,\ldots,m; j = 1,2,\ldots,n. \tag{12.6}$$

$$y = f\left(x,[A,B],[C,D]\right) = \begin{cases} 1 & , x \in [C,D] \\ 1 - \dfrac{d\left(x,[C,D]\right)_{min}}{|A-C|} & , x \in [A,C] \wedge A \neq C \\ 1 - \dfrac{d\left(x,[C,D]\right)_{min}}{|D-B|} & , x \in [D,B] \wedge D \neq B \end{cases} \tag{12.7}$$

where $d\left(x,[C,D]\right)\min\left(|x-C|,|x-D|\right)_{min}$, x is a valuation of an approach, $[C,D]$is reference ideal, $[A,B]$is the range for universe of discourse,

$x \in [A,B],[C,D] \subset [A,B]$, function *f* allows finding a value that belongs to the unitary interval, *m* refers to the alternatives, and *n* refers to the criteria.

Step 4. Weighted normalized valuation matrix **Z** *is determined using multiply of matrix* ***Y*** *and* w^{agg}

The criterion weights w^{agg} gained from BBWM are multiplied with matrix ***Y*** to obtain the weighted normalized evaluation matrix **Z**, as shown in Eq. (12.8).

$$Z = \left[y_{ij} \cdot w_j^{agg} \right]_{m \times n}, \mathrm{i} = 1,2,\ldots,\mathrm{m}; \mathrm{j} = 1,2,\ldots,\mathrm{n}. \tag{12.8}$$

Step 5. The variation to the normalized RI for each alternative

The concept of Euclidean distance is used to determine the distance to the ideal and non-ideal alternatives, as shown in Eqs. (12.9) and (12.10).

$$I_i^+ = \sqrt{\sum_{j=1}^{n} \left(z_{ij} - w_j \right)^2}, \mathrm{i} = 1,2,\ldots,\mathrm{m}; \mathrm{j} = 1,2,\ldots,\mathrm{n}. \tag{12.9}$$

$$I_i^- = \sqrt{\sum_{j=1}^{n} \left(z_{ij} \right)^2}, \mathrm{i} = 1,2,\ldots,\mathrm{m}; \mathrm{j} = 1,2,\ldots,\mathrm{n}. \tag{12.10}$$

Step 6. Relative index for each alternative

The ranking index of the alternative can be calculated as Eq. (12.11). The value range of RI_i is from 0 to 1, and when RI_i is 1, which indicates that it is consistent with the RI.

$$RI_i = \frac{I_i^-}{I_i^+ + I_i^-}, \mathrm{i} = 1,2,\ldots,\mathrm{m} \tag{12.11}$$

12.3.3 Description of the Combined MADM Framework

The combined MADM framework structured for this study is demonstrated in Figure 12.2. The framework consists of three main processes. The first is on the definition of the problem and determination of the three main criteria and 17 sub-criteria. The first process was handled by a comprehensive literature review and expert opinions. The second process concerned determination of the importance weights of three main criteria and 17 sub-criteria regarding ED preparedness against outbreaks. The method used in this process was BBWM. The final process aimed to determine a preparedness level for the studied EDs. RIM was applied to find the index value and ranking order of the EDs in this final process.

12.4 CASE STUDY

The real-life application analysis conducted for the study is presented in this section.

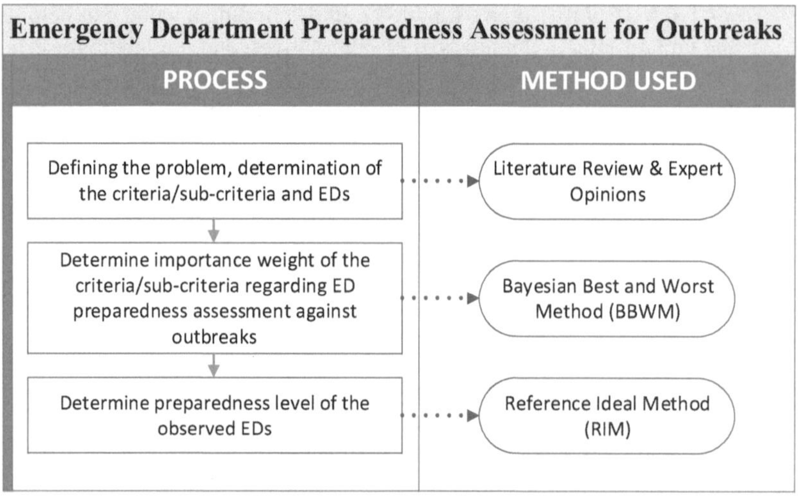

FIGURE 12.2 Developed combined framework for ed preparedness assessment of outbreaks.

12.4.1 Description of the Observed EDs, the Questionnaire, and Decision-Making Group

Five EDs located in a highly populated area in a metropolitan city of Turkey participated in the study. Three of the hospitals are major state hospitals and experienced in many medical branches. The other two are local state hospitals and have relatively lower daily patient arrivals. To perform BBWM computations in the second process of the framework, a special questionnaire was designed as a decision-support tool. This tool enabled the decision-making team to perform easily the BBWM pairwise comparisons between the main criteria and 17 sub-criteria. The decision-making group consisted of four experienced medical experts. Expert 1 was a healthcare officer with 25 years of experience. He had a Bachelor degree and was 44 years old. Experts 2 and 4 were both laboratory technicians; Expert 4 had a Bachelor degree with 20 years of experience and was 40 years old. Expert 2 had a two-year higher-education degree with five years' experience, and was 30 years old. Finally, Expert 3 was a registered nurse. She had a two-year higher education degree with 12 years' experience and was 37 years old.

12.4.2 ED Performance Metrics used in the Model

Three main criteria and 17 sub-criteria regarding ED readiness against outbreaks were determined with the aid of related literature: Pegoraro et al. (2020); Tabriz et al. (2020); Manolitzas et al. (2020); Akkan et al. (2020); Sørup et al. (2013); and Welch et al. (2011). These metrics and their corresponding sources are provided in Table 12.2. The two main criteria focus on general patient flow analysis of hospital EDs. The first main criterion, "ED time metric", consisted of criteria used to measure efficiency of time. The second main criterion was to measure the level of use of

TABLE 12.2
ED Performance Metrics Used in the Model

Code	*Definition*	*Source*
C1	**ED Time Metric**	Pegoraro et al. (2020); Pegoraro et al. (2020); Tabriz et al. (2020); Manolitzas et al. (2020); Akkan et al. (2020); Sørup et al. (2013); Welch et al. (2011)
C11	Doors to doctor time	
C12	Doors to triage time	
C13	Length of stay in the ED	
C14	Percentage of discharged from ED	
C15	Percentage of left without being seen	
C16	Boarding time	
C2	**Utilization of Medical Staff and Treatment Area**	
C21	Utilization of doctors	
C22	Utilization of nurses	
C23	Occupancy of ED beds	
C24	Utilization of medical devices (e.g., X-ray, tomography)	
C25	Sufficiency of medicine and medical equipment	
C3	**Patient Safety and ED Preparedness Level Against Outbreaks**	
C31	Mortality rate during ED care	
C32	Mortality rate during admission	
C33	Hospital infection rate	
C34	Response-ability to unexpected patient demand	
C35	Availability of PPEs of medical staff	
C36	Ability of ED to adopt to epidemic conditions (e.g., disinfection, social distance, isolation)	

Note: *Bold font in code column refers to the main criteria; others refer to the sub-criteria regarding ED preparedness for outbreaks.*

doctors, nurses, equipment, and physical spaces. The last main criterion, "Patient safety and ED preparedness level against outbreaks", was directly related to the special conditions of EDs in times an outbreak disaster. It included six sub-criteria from infection rate to the ability of am ED to adopt to epidemic conditions, as given in Table 12.2.

12.4.3 Implementation of the Combined Bayesian BWM–RIM Model to the Problem

The criteria/sub-criteria weights regarding ED preparedness assessment were identified by using BBWM. Within this framework, the problem was handled by following the steps of BBWM introduced in Section 12.3.2. The detailed process of this decision-making problem is explained below.

The C1 ("ED time metric") criterion included six sub-criteria. C2 ("Utilization of medical staff and treatment area") consists of five sub-criteria; C3 ("Patient safety and ED preparedness level against outbreaks") had six sub-criteria. Initially, the best and worst criteria/sub-criteria were defined by four experts, and then "best-to-others" and "others-to-worst" evaluation matrices were formed. Meanwhile, a user-friendly decision support system in MS Excel was designed to gather the decision-makers' assessments simply. The evaluations were aggregated through the proposed decision-support system. The 1–9 scale in Rezaei (2015) was adopted for the evaluations. Comparing the criteria/sub-criteria with themselves was considered equally important in the evaluation of both criteria and sub-criteria. In this study, four evaluations were included and two tables were fulfilled (best-to-others and others-to-worst) for each evaluation. The evaluations obtained from the proposed decision-support system were then moved to the MATLAB platform for Bayesian BWM computations. The procedure of Mohammadi and Rezai (2020a) was followed. In carrying out the computations, the local weight of the criteria and sub-criteria were determined. Credal rankings were determined, and special graphics related to creedal rankings were also shown. The criteria/sub-criteria weights calculated BBWM are presented in Table 12.3. The credal ranking graph of the criteria/sub-criteria is shown in Figure 12.3. The "C3" criterion was found to be the most important criterion (0.654), followed by "C2" (0.216) and "C1" (0.130). When the sub-criteria were considered from the perspective of the global weight, there was a big difference between two sub-criteria (C32 and C36) and the remaining ones. These two

TABLE 12.3
Weights of Criteria/Sub-Criteria Obtained by Bayesian BWM

Criteria/Sub-criteria	Local weight		Global weight
C1		**0.130**	
C11	0.116		0.015
C12	0.100		0.013
C13	0.230		0.030
C14	0.184		0.024
C15	0.228		0.030
C16	0.143		0.019
C2		**0.216**	
C21	0.385		0.083
C22	0.189		0.041
C23	0.125		0.027
C24	0.136		0.029
C25	0.165		0.036
C3		**0.654**	
C31	0.092		0.060
C32	0.347		0.227
C33	0.105		0.069
C34	0.112		0.073
C35	0.140		0.092
C36	0.204		0.134

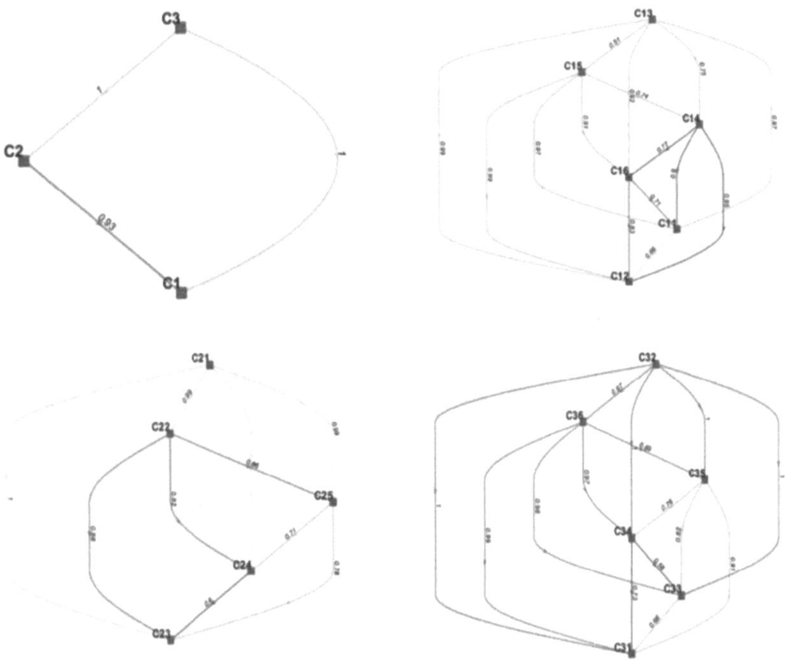

FIGURE 12.3 Credal ranking results for criteria/sub-criteria.

sub-criteria have higher importance weight values than the remaining sub-criteria. These sub-criteria concerned mortality rate during admission and the ability of ED to adopt to epidemic conditions.

The purpose of the credit ranking by Mohammadi and Rezaei (2020a, 2020b) is to show how superior one criterion is to another criterion. The values written above the arrows in Figure 12.3 show that criterion A is more important than criterion B in $(A \xrightarrow{D} B)$ D reliability. In this regard, the graphs of C32 and C36 have higher D values over the arrows.

In order to determine ED disaster preparedness levels with the RIM MCDM approach, a score was obtained by utilizing the criteria weights calculated with BBWM and the valuation matrix data obtained for five EDs. The data for the EDs are given in Table 12.4. In this case study, the sub-criteria excluding C13, C21, C22, and C23 had range values defined by linguistic terms. These linguistic terms are based on a 5-point scale. Each linguistic term had a crisp integer value from 5 to 1 (high to low). Following the procedures in Section 12.3.3, indexes for each ED were determined as presented in Table 12.5.

In accordance with the outcomes of RIM index values, ED-1 was the hospital ED most prepared for disasters under the specified criteria (RI = 0.690), followed by ED-3 (RI = 0.673) and ED-2 (RI = 0.608), and ED-4 and ED-5 (RI = 0.511 for both).

TABLE 12.4
Description of the Data Used in RIM Computations

Code	Weight	Range	Reference Ideal
C11	0.015	High, normal high, normal, normal low, low	Low
C12	0.013	High, normal high, normal, normal low, low	Low
C13	0.030	[0,320]	Lower than 240
C14	0.024	High, normal high, normal, normal low, low	High
C15	0.030	High, normal high, normal, normal low, low	Low
C16	0.019	High, normal high, normal, normal low, low	Low
C21	0.083	[0.20,0.70]	0.40–0.55
C22	0.041	[0.60–1.00]	0.70–0.85
C23	0.027	[0.50–1.00]	Lower than 0.6
C24	0.029	High, normal high, normal, normal low, low	High
C25	0.036	High, normal high, normal, normal low, low	High
C31	0.060	High, normal high, normal, normal low, low	Low
C32	0.227	High, normal high, normal, normal low, low	Low
C33	0.069	High, normal high, normal, normal low, low	Low
C34	0.073	High, normal high, normal, normal low, low	High
C35	0.092	High, normal high, normal, normal low, low	High
C36	0.134	High, normal high, normal, normal low, low	High

TABLE 12.5
Index Results for the Studied EDs

#ED	I_i^+	I_i^-	RI_i
ED-1	0.104	0.231	0.690
ED-2	0.134	0.208	0.608
ED-3	0.116	0.239	0.673
ED-4	0.173	0.181	0.511
ED-5	0.173	0.181	0.511

12.4.4 Usability of the Proposed Combined Bayesian BWM–RIM Model Within the Scope of Managerial Insight

Our proposed combined BWM–RIM method will guide decision-makers in many ways. A vast pool of criteria was used to evaluate ED performance. Thanks to the BWM, which was the first stage of the study, the decision-makers had the opportunity to determine which essential criteria affect ED performance. Thanks to the credal ranking, a feature of BWM, ranking information was available, and the information at which level the criteria are superior to each other presented. Thus, hospital managers will be able to identify the strengths of and threats to their hospitals. An action plan can be created in light of this information. In the second phase of the proposed model, by ranking the hospitals in terms of ED performance, decision-makers had the opportunity to compare their hospitals with competitors.

12.5 CONCLUSION

The responsibility of health institutions in the fight against COVID-19 is enormous. EDs are among the units that are on the alert in case of such pandemic disasters. EDs should be prepared for all disasters, including the current pandemic. EDs should reorganize their operations according to changing needs and increasing patient arrivals, and should be prepared for all disasters. So, in this study, we addressed the problem of EDs preparedness for epidemics and aimed to measure the preparedness of emergency services by proposing a combined decision-making methodology. In order to evaluate the EDs, we determined 17 criteria by using expert opinions and related literature, and we used the BBWM approach to determine the weights of these criteria in the decision process. The five EDs identified were evaluated with the RIM method, and their preparedness level examined. As a result, ED-1 was the hospital ED found to be most prepared for disasters under the evaluated criteria, followed by ED-3 and ED-2. The factor that should be considered and improved in the first place in increasing the preparedness level of EDs was determined to be “C3: Patient safety and ED preparedness level against outbreaks”, and the sub-criterion was “C32: Mortality rate during admission”.

With the proposed method, the preparedness level of emergency services in disaster situations can be measured and policies and practices can be put forward so that they can be better prepared.

For future studies, a comparative analysis could be conducted to analyze the validity and the reliability of the application using different MCDM approaches.

REFERENCES

Aboutorab, Hamed, Morteza Saberi, Mehdi Rajabi Asadabadi, Omar Hussain, and Elizabeth Chang. 2018. “ZBWM: The Z-Number Extension of Best Worst Method and Its Application for Supplier Development.” *Expert Systems with Applications* 107: 115–25. doi:10.1016/j.eswa.2018.04.015.

Akkan, Can, Melis Almula Karadayi, Yeliz Ekinci, Füsun Ülengin, Nimet Uray, and Elif Karaosmanoğlu. 2020. “Efficiency Analysis of Emergency Departments in Metropolitan Areas.” *Socio-Economic Planning Sciences* 69 (100679): 100679. doi:10.1016/j.seps.2019.01.001.

Al-Hajj, Samar, Hadi Abou-El-Hassan, Lana Khalil, Haytham M. A. Kaafarani, and Mazen El Sayed. 2020. “Hospital Disaster and Emergency Preparedness (HDEP) in Lebanon: A National Comprehensive Assessment.” *International Journal of Disaster Risk Reduction* 51 (101889): 101889. doi:10.1016/j.ijdrr.2020.101889.

AlTakarli, Nourah S. 2020. “China’s Response to the COVID-19 Outbreak: A Model for Epidemic Preparedness and Management.” *Dubai Medical Journal* 3 (2): 44–49. doi:10.1159/000508448.

Barten, Dennis G., Renske W. J. Kusters, and Nathalie A. L. R. Peters. 2020. “A Swift and Dynamic Strategy to Expand Emergency Department Capacity for COVID-19.” *Disaster Medicine and Public Health Preparedness* 16, 1190–1193. doi:10.1017/dmp.2020.430.

Belhadi, Amine, Sachin S. Kamble, Syed Abdul Rehman Khan, Fatima Ezahra Touriki, and Dileep Kumar M. 2020. "Infectious Waste Management Strategy during COVID-19 Pandemic in Africa: An Integrated Decision-Making Framework for Selecting Sustainable Technologies." *Environmental Management* 66 (6): 1085–1104. doi:10.1007/s00267-020-01375-5.

Boufkhed, Sabah, Richard Harding, Tezer Kutluk, Abdullatif Husseini, Nasim Pourghazian, and Omar Shamieh. 2021. "What Is the Preparedness and Capacity of Palliative Care Services in Middle-Eastern and North African Countries to Respond to COVID-19? A Rapid Survey." *Journal of Pain and Symptom Management* 61 (2): e13–50. doi:10.1016/j.jpainsymman.2020.10.025.

Brunelli, Matteo, and Jafar Rezaei. 2019. "A Multiplicative Best–Worst Method for Multi-Criteria Decision Making." *Operations Research Letters* 47 (1): 12–15. doi:10.1016/j.orl.2018.11.008.

Cables, E., M. T. Lamata, and J. L. Verdegay. 2016. "RIM-Reference Ideal Method in Multicriteria Decision Making." *Information Sciences* 337–338: 1–10. doi:10.1016/j.ins.2015.12.011.

Carenzo, L., E. Costantini, M. Greco, F. L. Barra, V. Rendiniello, M. Mainetti, R. Bui, et al. 2020. "Hospital Surge Capacity in a Tertiary Emergency Referral Centre During the COVID-19 Outbreak in Italy." *Anaesthesia* 75 (7): 928–34. doi:10.1111/anae.15072.

Casalino, Enrique, Donia Bouzid, Arsalene Ben Hammouda, Mathias Wargon, Sonja Curac, Romain Hellmann, Christophe Choquet, and Daniel Aiham Ghazali. 2020. "COVID-19 Preparedness among Emergency Departments: A Cross-Sectional Study in France." *Disaster Medicine and Public Health Preparedness* 16: 245–253. doi:10.1017/dmp.2020.331.

Chang, Tai-Wu, Chun-Jui Pai, Huai-Wei Lo, and Shu-Kung Hu. 2021. "A Hybrid Decision-Making Model for Sustainable Supplier Evaluation in Electronics Manufacturing." *Computers & Industrial Engineering* 156 (107283): 107283. doi:10.1016/j.cie.2021.107283.

Dong, Jiuying, Shuping Wan, and Shyi-Ming Chen. 2021. "Fuzzy Best-Worst Method Based on Triangular Fuzzy Numbers for Multi-Criteria Decision-Making." *Information Sciences* 547: 1080–1104. doi:10.1016/j.ins.2020.09.014.

Firdausi, Nobella, Fatma Lestari, and Avinia Ismiyati. 2021. "Disaster Preparedness Analysis of Public Health Centers in DKI Jakarta Province in 2020." *International Journal of Safety and Security Engineering* 11 (1): 91–99. doi:10.18280/ijsse.110110.

Garcia-Lopez, Joan, Joaquin Delgadillo, Anna Vilarrodona, Sergi Querol, Joan Ovejo, Ruth Coll, Anna Millan, et al. 2021. "SARS-CoV-2/COVID-19 Pandemic: First Wave, Impact, Response and Lessons Learnt in a Fully Integrated Regional Blood and Tissue Bank. A Narrative Report." *Trasfusione Del Sangue [Blood Transfusion]* 19 (2): 158–67. doi:10.2450/2021.0259-20.

Garzotto, Francesco, Erica Ceresola, Sofia Panagiotakopoulou, Giovanni Spina, Francesca Menotto, Marco Benozzi, Maurizio Casarotto, et al. 2020. "COVID-19: Ensuring Our Medical Equipment Can Meet the Challenge." *Expert Review of Medical Devices* 17 (6): 483–89. doi:10.1080/17434440.2020.1772757.

Ginter, Peter M., W. Jack Duncan, and Maziar Abdolrasulnia. 2007. "Hospital Strategic Preparedness Planning: The New Imperative." *Prehospital and Disaster Medicine* 22 (6): 529–36. doi:10.1017/s1049023x00005380.

Griffin, Kelly M., Maria G. Karas, Natalia S. Ivascu, and Lindsay Lief. 2020. "Hospital Preparedness for COVID-19: A Practical Guide from a Critical Care Perspective." *American Journal of Respiratory and Critical Care Medicine* 201 (11): 1337–44. doi:10.1164/rccm.202004-1037CP.

Gul, Muhammet, and Melih Yucesan. 2021. "Hospital Preparedness Assessment against COVID-19 Pandemic: A Case Study in Turkish Tertiary Healthcare Services." *Mathematical Problems in Engineering* 2021: 1–18. doi:10.1155/2021/2931219.

Guo, Sen, and Haoran Zhao. 2017. "Fuzzy Best-Worst Multi-Criteria Decision-Making Method and Its Applications." *Knowledge-Based Systems* 121: 23–31. doi:10.1016/j.knosys.2017.01.010.

Gupta, Shaili, and Daniel G. Federman. 2020. "Hospital Preparedness for COVID-19 Pandemic: Experience from Department of Medicine at Veterans Affairs Connecticut Healthcare System." *Postgraduate Medicine* 132 (6): 489–94. doi:10.1080/00325481.2020.1761668.

Hafezalkotob, Ashkan and Arian Hafezalkotob. 2017. "A Novel Approach for Combination of Individual and Group Decisions Based on Fuzzy Best-Worst Method." *Applied Soft Computing* 59: 316–25. doi:10.1016/j.asoc.2017.05.036.

Hezam, Ibrahim M., Moddassir Khan Nayeem, Abdelaziz Foul, and Adel Fahad Alrasheedi. 2021. "COVID-19 Vaccine: A Neutrosophic MCDM Approach for Determining the Priority Groups." *Results in Physics* 20 (103654): 103654. doi:10.1016/j.rinp.2020.103654.

Hsu, Wan-Chi Jackie, Huai-Wei Lo, and Chin-Cheng Yang. 2021. "The Formulation of Epidemic Prevention Work of COVID-19 for Colleges and Universities: Priorities and Recommendations." *Sustainability* 13 (4): 2081. doi:10.3390/su13042081.

Huang, Chun-Nen, James J. H. Liou, Huai-Wei Lo, and Fu-Jung Chang. 2021. "Building an Assessment Model for Measuring Airport Resilience." *Journal of Air Transport Management* 95 (102101): 102101. doi:10.1016/j.jairtraman.2021.102101.

Hulland, E. N., K. E. Wiens, S. Shirude, J. D. Morgan, A. Bertozzi-Villa, T. H. Farag, N. Fullman, et al. 2019. "Travel Time to Health Facilities in Areas of Outbreak Potential: Maps for Guiding Local Preparedness and Response." *BMC Medicine* 17 (1): 232. doi:10.1186/s12916-019-1459-6.

Islam, M. Saiful, Kazi M. Rahman, Yanni Sun, Mohammed O. Qureshi, Ikram Abdi, Abrar A. Chughtai, and Holly Seale. 2020. "Current Knowledge of COVID-19 and Infection Prevention and Control Strategies in Healthcare Settings: A Global Analysis." *Infection Control and Hospital Epidemiology* 41 (10): 1196–1206. doi:10.1017/ice.2020.237.

Jachetti, A., G. Colombo, B. Brignolo-Ottolini, J. Franchi, M. Solbiati, M. Pecorino Meli, P. Bosco, and G. Costantino. 2021. "Emergency Department Reorganisation to Cope with COVID-19 Outbreak in Milan University Hospital: A Time-Sensitive Challenge." *BMC Emergency Medicine* 21 (1): 74. doi:10.1186/s12873-021-00464-w.

Jöbges, Susanne, Rasita Vinay, Valerie A. Luyckx, and Nikola Biller-Andorno. 2020. "Recommendations on COVID-19 Triage: International Comparison and Ethical Analysis." *Bioethics* 34 (9): 948–59. doi:10.1111/bioe.12805.

Kramer, Victoria, Irina Papazova, Andreas Thoma, Miriam Kunz, Peter Falkai, Thomas Schneider-Axmann, Anke Hierundar, Elias Wagner, and Alkomiet Hasan. 2021. "Subjective Burden and Perspectives of German Healthcare Workers during the COVID-19 Pandemic." *European Archives of Psychiatry and Clinical Neuroscience* 271 (2): 271–81. doi:10.1007/s00406-020-01183-2.

Liang, Fuqi, Matteo Brunelli, and Jafar Rezaei. 2020. "Consistency Issues in the Best Worst Method: Measurements and Thresholds." *Omega* 96 (102175): 102175. doi:10.1016/j.omega.2019.102175.

Liao, Huchang, Xiaomei Mi, Qin Yu, and Li Luo. 2019. "Hospital Performance Evaluation by a Hesitant Fuzzy Linguistic Best Worst Method with Inconsistency Repairing." *Journal of Cleaner Production* 232: 657–71. doi:10.1016/j.jclepro.2019.05.308.

Mancini, Fabiola, Fabrizio Barbanti, Maria Scaturro, Giulia Errico, Angelo Iacobino, Antonino Bella, Flavia Riccardo, et al. 2020. "Laboratory Management for SARS-CoV-2

Detection: A User-Friendly Combination of the Heat Treatment Approach and RT-Real-Time PCR Testing." *Emerging Microbes & Infections* 9 (1): 1393–96. doi:10.1080/22221751.2020.1775500.

Manley, William G., Paul M. Furbee, Jeffery H. Coben, Sharon K. Smyth, Daniel E. Summers, Ronald C. Althouse, Rodney L. Kimble, Aaron T. Kocsis, and James C. Helmkamp. 2006. "Realities of Disaster Preparedness in Rural Hospitals." *Disaster Management & Response* 4 (3): 80–87. doi:10.1016/j.dmr.2006.05.001.

Manolitzas, Panagiotis, Evangelos Grigoroudis, Jason Christodoulou, and Nikolaos Matsatsinis. 2020. "S-MEDUTA: Combining Balanced Scorecard with Simulation and MCDA Techniques for the Evaluation of the Strategic Performance of an Emergency Department." *Advances in Experimental Medicine and Biology* 1194: 1–22. doi:10.1007/978-3-030-32622-7_1.

Mardones, Fernando O., Karl M. Rich, Lisa A. Boden, Andrea I. Moreno-Switt, Marisa L. Caipo, Natalia Zimin-Veselkoff, Abdulaziz M. Alateeqi, and Isabelle Baltenweck. 2020. "The COVID-19 Pandemic and Global Food Security." *Frontiers in Veterinary Science* 7: 578508. doi:10.3389/fvets.2020.578508.

Marques, Lorraine Cichowicz, Danieley Cristini Lucca, Everson Orlandini Alves, Gisele Cristina Manfrini Fernandes, and Keyla Cristiane do Nascimento. 2020. "Covid-19: Nursing Care for Safety in the Mobile Pre-Hospital Service." *Texto & Contexto Enfermagem* 29. doi:10.1590/1980-265x-tce-2020-0119.

Marshall, Douglas L., Frederic Bois, Soren K. S. Jensen, Svend A. Linde, Richard Higby, Yvoine Rémy-McCort, Sean Murray, Bryan Dieckelman, Fitri Sudradjat, and Gilles G. Martin. 2020. "Sentinel Coronavirus Environmental Monitoring Can Contribute to Detecting Asymptomatic SARS-CoV-2 Virus Spreaders and Can Verify Effectiveness of Workplace COVID-19 Controls." *Microbial Risk Analysis* 16: 100137. doi:10.1016/j.mran.2020.100137.

Mohammadi, Majid and Jafar Rezaei. 2020a. "Bayesian Best-Worst Method: A Probabilistic Group Decision Making Model." *Omega* 96 (102075): 102075. doi:10.1016/j.omega.2019.06.001.

———. 2020b. "Evaluating and Comparing Ontology Alignment Systems: An MCDM Approach." *Web Semantics* 64 (100592): 100592. doi:10.1016/j.websem.2020.100592.

Moheimani, Arash, Reza Sheikh, Seyed Mohammad Hassan Hosseini, and Shib Sankar Sana. 2021. "Assessing the Preparedness of Hospitals Facing Disasters Using the Rough Set Theory: Guidelines for More Preparedness to Cope with the COVID-19." *International Journal of Systems Science Operations & Logistics*, 1–16. doi:10.1080/23302674.2021.1904301.

Mukhtar, Sonia. 2021. "Preparedness and Proactive Infection Control Measures of Pakistan during COVID-19 Pandemic Outbreak." *Research in Social & Administrative Pharmacy* 17 (1): 2052. doi:10.1016/j.sapharm.2020.04.011.

Munasinghe, Nimali Lakmini, Gerard O'Reilly, and Peter Cameron. 2021. "Examining the Experience and Lessons Learnt for Disaster-Preparedness in Sri Lankan Hospitals: A Scoping Review." *International Journal of Disaster Risk Reduction* 64 (102494): 102494. doi:10.1016/j.ijdrr.2021.102494.

Noh, Ji Yun, Joon Young Song, Jin Gu Yoon, Hye Seong, Hee Jin Cheong, and Woo Joo Kim. 2020. "Safe Hospital Preparedness in the Era of COVID-19: The Swiss Cheese Model." *International Journal of Infectious Diseases* 98: 294–96. doi:10.1016/j.ijid.2020.06.094.

Ortiz-Barrios, Miguel, Arlen Alaine Borrego-Areyanes, Iván Darío Gómez-Villar, Fabio De Felice, Antonella Petrillo, Muhammet Gul, and Melih Yucesan. 2021. "A Multiple Criteria Decision-Making Approach for Increasing the Preparedness Level of Sales Departments against COVID-19 and Future Pandemics: A Real-World Case." *International Journal of Disaster Risk Reduction* 62 (102411): 102411. doi:10.1016/j.ijdrr.2021.102411.

Pegoraro, Fábio, Eduardo Alves Portela Santos, and Eduardo de Freitas Rocha Loures. 2020. "A Support Framework for Decision Making in Emergency Department Management." *Computers & Industrial Engineering* 146 (106477): 106477. doi:10.1016/j.cie.2020.106477.

Perinel, Sophie, Manon Launay, Élisabeth Botelho-Nevers, Éric Diconne, Aurore Louf-Durier, Raphaël Lachand, Martin Murgier, et al. 2020. "Towards Optimization of Hydroxychloroquine Dosing in Intensive Care Unit COVID-19 Patients." *Clinical Infectious Diseases* 71 (16): 2227–29. doi:10.1093/cid/ciaa394.

Pothiawala, Sohil. 2020. "SARS-CoV-2 and Its Dynamic Impact on Emergency Department Preparedness and Management." *Journal of Emergency Practice and Trauma* 6 (2): 102–8. doi:10.34172/jept.2020.25.

Rezaei, Jafar. 2015. "Best-Worst Multi-Criteria Decision-Making Method." *Omega* 53: 49–57. doi:10.1016/j.omega.2014.11.009.

———. 2016. "Best-Worst Multi-Criteria Decision-Making Method: Some Properties and a Linear Model." *Omega* 64: 126–30. doi:10.1016/j.omega.2015.12.001.

Safarzadeh, Soroush, Saba Khansefid, and Morteza Rasti-Barzoki. 2018. "A Group Multi-Criteria Decision-Making Based on Best-Worst Method." *Computers & Industrial Engineering* 126: 111–21. doi:10.1016/j.cie.2018.09.011.

Safavi, Kyan C., Ann L. Prestipino, Ana Cecilia Zenteno Langle, Martin Copenhaver, Michael Hu, Bethany Daily, Allison Koehler, Paul D. Biddinger, and Peter F. Dunn. 2021. "The Power of Modeling in Emergency Preparedness for COVID-19: A Moonshot Moment for Hospitals." *Disaster Medicine and Public Health Preparedness*, 1–3. doi:10.1017/dmp.2021.51.

Sánchez-Lozano, J. M., M. Fernández-Martínez, and M. T. Lamata. 2019. "Near-Earth Asteroid Impact Dates: A Reference Ideal Method (RIM) Approach." *Engineering Applications of Artificial Intelligence* 81: 157–68. doi:10.1016/j.engappai.2019.02.010.

Sánchez-Lozano, J. M., and O. Naranjo Rodríguez. 2020. "Application of Fuzzy Reference Ideal Method (FRIM) to the Military Advanced Training Aircraft Selection." *Applied Soft Computing* 106061: 106061. doi:10.1016/j.asoc.2020.106061.

Seyedin, Hesam, Shandiz Moslehi, Fazeleh Sakhaei, and Mohsen Dowlati. 2021. "Developing a Hospital Preparedness Checklist to Assess the Ability to Respond to the COVID-19 Pandemic." *La Revue de Sante de La Mediterranee Orientale [Eastern Mediterranean Health Journal]* 27 (2): 131–41. doi:10.26719/2021.27.2.131.

Singh, Pranav Kumar, Sukumar Nandi, Kayhan Zrar Ghafoor, Uttam Ghosh, and Danda B. Rawat. 2021. "Preventing COVID-19 Spread Using Information and Communication Technology." *IEEE Consumer Electronics Magazine* 10 (4): 18–27. doi:10.1109/mce.2020.3047608.

Sofuoğlu, Mehmet Alper. 2021. "A New Biomaterial Selection Approach Using Reference Ideal Method." *Sadhana* 46 (1). doi:10.1007/s12046-021-01559-7.

Sørup, C. M., P. Jacobsen, and J. L. Forberg. 2013. "Evaluation of Emergency Department Performance-a Systematic Review on Recommended Performance and Quality-in-Care Measures." *Scandinavian Journal of Trauma, Resuscitation and Emergency Medicine* 21: 18–27.

Suleiman, Aiman, Isam Bsisu, Hasan Guzu, Abeer Santarisi, Murad Alsatari, Ala' Abbad, Ahmad Jaber, et al. 2020. "Preparedness of Frontline Doctors in Jordan Healthcare Facilities to COVID-19 Outbreak." *International Journal of Environmental Research and Public Health* 17 (9): 3181. doi:10.3390/ijerph17093181

Tabriz, A. A., J. G. Trogdon, and B.J. Fried. 2020. "Association between adopting emergency department crowding interventions and emergency departments' core performance measures." *The American Journal of Emergency Medicine* 38 (2): 258–265.

Wang, C. Jason, Chun Y. Ng, and Robert H. Brook. 2020. "Response to COVID-19 in Taiwan: Big Data Analytics, New Technology, and Proactive Testing: Big Data Analytics, New Technology, and Proactive Testing." *JAMA* 323 (14): 1341–42. doi:10.1001/jama.2020.3151.

Welch, Shari J., Brent R. Asplin, Suzanne Stone-Griffith, Steven J. Davidson, James Augustine, Jeremiah Schuur, and Emergency Department Benchmarking Alliance. 2011. "Emergency Department Operational Metrics, Measures and Definitions: Results of the Second Performance Measures and Benchmarking Summit." *Annals of Emergency Medicine* 58 (1): 33–40. doi:10.1016/j.annemergmed.2010.08.040.

Yang, Jen-Jen, Huai-Wei Lo, Chen-Shen Chao, Chih-Chien Shen, and Chin-Cheng Yang. 2020. "Establishing a Sustainable Sports Tourism Evaluation Framework with a Hybrid Multi-Criteria Decision-Making Model to Explore Potential Sports Tourism Attractions in Taiwan." *Sustainability* 12 (4): 1673. doi:10.3390/su1204167

13 Ranking of Districts of Istanbul According to Earthquake Loss Estimates with Entropy and CoCoSo Integration

Pelin Toktaş
Başkent University

CONTENTS

13.1 Introduction 191
13.2 Literature Review 192
13.3 Proposed Risk Assessment Approach: Entropy–CoCoSo Integration 196
 13.3.1 Stage 1: Obtaining Criteria Weights Using the Entropy Method 196
 13.3.2 Stage 2: Obtaining Final Rankings of The Alternatives Using the CoCoSo Method 197
13.4 Application of the Proposed Approach: Ranking of Districts of Istanbul According to Earthquake Loss Estimates 198
 13.4.1 Stage 1: Obtaining Criteria Weights Using the Entropy Method 198
 13.4.2 Stage 2: Obtaining Final Rankings of the Alternatives Using the CoCoSo Method 202
13.5 Conclusion 204
References 206

13.1 INTRODUCTION

Istanbul is one of the most popular cities in the world with its location and historical buildings. However, it is one of the cities with the highest earthquake risk due to its proximity to fault lines, crowded population, and building density. There have been many devastating earthquakes in the Marmara region in the past, and future earthquakes are likely to cause great loss of life and property (Erdik et al 2003, Kundak and Türkoğlu 2010). To be prepared for an earthquake, it is important to understand and reduce the risk posed by an earthquake. Pre-disaster studies can provide the opportunity to keep bad results at a minimum level.

DOI: 10.1201/9781003212904-13

In this context, within the scope of the "Istanbul Province, Probable Earthquake Loss Estimates Update Project (2019)", done in cooperation with Istanbul Metropolitan Municipality and Boğaziçi University Kandilli Observatory Earthquake Research Institute, building damage, possible deaths and injuries, infrastructure damage, and temporary accommodation needs in districts and neighborhoods in a possible earthquake were analyzed and reported in booklets produced for each district (İBB 2020). These booklets were prepared by carrying out subjective analyses and mapping for each district to present earthquake hazard analyses and a risk analysis of urban superstructure and infrastructure elements. In the study, a deterministic approach was taken as a basis and the results were achieved by using an earthquake scenario with a magnitude of Mw 7.5. In the integrated disaster-management system, studies of pre-disaster protection such as loss and damage reduction, preparedness, forecasting, and early warning, disaster impact analysis are called risk management. Studies of post-disaster recovery such as intervention, recovery, and restructuring are called crisis management (Kadıoğlu et al 2005).

If the districts of Istanbul where destruction and damage will be high are determined before an earthquake occurs and measures are taken within the scope of damage reduction, the vicious cycle of restructuring after an earthquake can be eliminated. In this context, this study aimed to rank the 39 districts of Istanbul from the least risky to the riskiest in the Mw 7.5 earthquake scenario considered within the scope of the project. In the district risk-assessment process, it is appropriate to apply multi-criteria decision-making (MCDM) approaches, as it includes ranking of the districts, considering the 12 criteria determined after a magnitude Mw 7.5 earthquake scenario, including possible deaths and injuries, infrastructure damages, and temporary accommodation needs. The entropy method, which is an objective and easy-to-apply method, was used to calculate the criterion weights. In the entropy method, data in the initial decision matrix are used to calculate the weights of the criteria. It was suggested that the combined compromise solution (CoCoSo) method, a new-generation method, be used to rank the districts. The criterion weights obtained with the entropy method were integrated into the CoCoSo method.

The rest of the chapter is organized as follows. In Section 13.2, a detailed literature review of the CoCoSo method and its applications is presented. Section 13.3 consists of the steps of the proposed risk assessment approach: the integrated entropy-CoCoSo method. The application of the proposed risk assessment approach is given in Section 13.4. Conclusions, limitations, and future directions are given in Section 13.5.

13.2 LITERATURE REVIEW

In this study, ranking of the districts was done using integrated entropy and CoCoSo methods. Calculation of criterion weights is an important issue in a decision problem that includes more than one criterion. In the literature, different MCDM methods have been proposed for ranking criteria. It is suggested that entropy is used, which is an easy-to-apply method, without the need for decision makers' evaluations in weighting criteria. The rest of the literature review concentrated on the CoCoSo method and its applications. The CoCoSo method is an MCDM method that was first introduced in 2019 by Yazdani et al. (2019a), when the logistics provider selection problem was considered. Since then, it has been widely used in the literature.

Some selected papers from 2019 onwards are summarized in the following. Ecer et al. (2019) evaluated 12 OPEC members considering 41 sustainable development indicators of 10 dimensions using the CoCoSo method. They found that the most sustainable OPEC country was the United Arab Emirates (Ecer et al 2019). In the storage system, Erceg et al. (2019) suggested a new method combining ABC analysis, the full consistency method (FUCOM), and CoCoSo for stock management (Erceg et al 2019). A drug cold chain logistics supplier was selected by Wen et al. (2019a) using a combined model based on integration of the stepwise weight assessment ratio analysis (SWARA) and CoCoSo (Wen et al 2019a). An extended version of the CoCoSo method was implemented using a fuzzy hesitant fuzzy linguistic environment to solve the multi-expert MCDM problem of third-party logistics service supplier selection (Wen et al 2019b). Decision-making trial and evaluation laboratory (DEMATEL) and best–worst method (BWM) were proposed to weigh criteria and application to a supplier selection problem (Yazdani et al 2019b, Zolfani et al 2019).

An experimental design study on optimization of hardness for three different strengths of concrete was conducted to appraise the mechanical properties of concrete using combinative distance-based assessment (CODAS) and CoCoSo (Bagal et al 2020). A similar multi-response experimental study was done by Barua et al. (2020) for a composite. Optimization of parameters was carried out with CoCoSo and the results verified in confirmatory experiments (Barua et al 2019). Integrated criteria importance through inter criteria correlation (CRITIC) and CoCoSo method was proposed to rank vehicles considering different technical and operational criteria (Biswas et al 2020). Healthcare organizations in India were compared considering supply chain performance using the PIvot Pairwise RElative Criteria Importance Assessment (PIPRECIA) method for criteria weights and multi-attributive border approximation area comparison (MABAC), CoCoSo, and measurement of alternatives and ranking according to compromise solution (MARCOS) to rank organizations (Biswas 2020). In another study, the CoCoSo method was used with a combination of the normalized weighted method for supplier selection (Ecer and Pamucar 2020). Kharwar et al. proposed a hybrid approach, including artificial neural network modeling and the development of an integrated principal component analysis–CoCoSo, for the optimization of different outputs during the drilling of multiwall carbon nanotube nanocomposites (Kharwar et al 2020). A cloud service provider was selected by Lai et al. (2020) using the integration of the CoCoSo method with the maximum variance optimization model selection problem. They also justified the accuracy of the model with sensitivity and comparative analyses (Lai et al 2020).

The CoCoSo method was extended to the Pythagorean fuzzy environment and applied to logistics distribution center selection (Liao et al 2020). Mi and Liao suggested a hybrid procedure of stochastic multicriteria acceptability analysis (SMAA) and the CoCoSo method (Mi and Liao 2020). A Pythagorean fuzzy decision-making algorithm-based CoCoSo method was proposed for 5G industry evaluation by Peng et al. in 2020. Criteria weights were calculated using the criteria importance through inter-criteria correlation (CRITIC) method (Peng et al 2020). Peng and Huang used

CRITIC–CoCoSo integration for financial risk evaluation (Peng and Huang 2020). Peng and Smarandache advanced a neutrosophic soft decision-making method based on CoCoSo with the CRITIC weighting procedure for the issue of China's rare earth industry security evaluation (Peng and Smarandache 2020). Stanujkic et al. calculated criteria weights using the entropy method and then used them in the CoCoSo method to determine the position of European Union countries relative to the Sustainable Development Goals (Stanujkic et al 2020). Ulutaş et al. carried out a study to select a location for a logistic center based on fuzzy SWARA and the CoCoSo method (Ulutaş et al 2020).

A model was developed to select the location of a logistics center. In the first stage of the study, data envelopment analysis was employed to determine alternatives and then the rough full consistency (R-FUCOM) and CoCoSo methods were used to evaluate the performance of the efficient communities (Yazdani et al 2020). An integrated BWM and CoCoSo method based on interval rough boundaries was utilized to achieve supplier selection (Zhang et al 2020). Using CRITIC-CoCoSo integration, Zolfani et al. applied a gray-based decision support framework to select the location of a temporary hospital during the COVID-19 pandemic (Zolfani et al 2020). Alrasheedi et al. combined CoCoSo and an interval-valued intuitionistic fuzzy set (IVIFS) to rank and evaluate the indicators of green growth in sustainable manufacturing (Alrasheedi et al 2021). The CoCoSo method was used for 2020 Global Knowledge Index (GKI) components to measure the knowledge performance of the G7 countries (Altıntaş 2021). Cui et al. proposed a hybrid SWARA–CoCoSo method based on Pythagorean fuzzy sets (PFSs) to rank various organizations in the manufacturing sector under identified barriers (Cui et al 2021). Deveci et al. carried out a study on the fuzzy power heronian function-based CoCoSo method for prioritization of autonomous vehicles in real-time traffic management (Deveci et al 2021). Circular economy practices were evaluated using CoCoSo and prioritized based on their importance (Khan and Haleem 2021). Kieu et al. utilized spherical fuzzy analytic hierarchy process (SF-AHP) and CoCoSo in an integrated manner to support the distribution location selection problem of perishable agricultural products (Kieu et al 2021). Lahane and Kant developed a hybrid framework of Pythagorean fuzzy analytic hierarchy process (PF-AHP) and Pythagorean fuzzy CoCoSo (PF-CoCoSo) to rank the performance outcomes of a circular supply chain (Lahane and Kant 2021). Luo et al. suggested a model based on probabilistic linguistic term sets and the integrated determination of objective criteria weights (IDOCRIW)–CoCoSo to establish an online evaluation ranking method for tourism attractions (Luo et al 2021). Peng et al. advanced a method based on a hesitant fuzzy soft CoCoSo method to evaluate the Internet of Everything (IoE) companies (Peng et al 2021a). In the same year, Peng et al. combined objective criteria weighting using CRITIC and subjective criteria weighting. Then, they suggested an interval-valued fuzzy soft decision-making algorithm-based CoCoSo for intelligent healthcare management evaluation (Peng et al 2021b). Peng and Garg performed an intuitionistic fuzzy soft decision-making method based on CoCoSo and CRITIC for CCN cache placement strategy selection (Peng and Garg 2021). Peng and Luo investigated improving the CoCoSo method by using the Renyi entropy method combining subjective and objective preferences with fuzzy picture information (Peng and Luo 2021).

A comparison study of MCDM methods including TOPSIS, VIKOR, MOOSRA, WASPAS, and CoCoSo was conducted by Štirbanović et al (2021). The rankings of different MCDM methods were aggregated for the final ranking of acid mine drainage treatment methods (Štirbanović et al 2021). Topal carried out a financial performance analysis of electricity-generation companies with the entropy-based CoCoSo method (Topal 2021).

Torkayesh et al. assessed the performance of the G7 countries using real data from an Organisation for Economic Co-operation and Development (OECD) data set, where 14 social sustainability indicators were selected based on CRITIC and Shannon's entropy methods, and the CoCoSo method (Torkayesh et al 2021a). Fuzzy PIvot Pairwise RElative Criteria Importance Assessment (FPIPRECIA), fuzzy Parameter Space Investigation (FPSI), and the CoCoSo methods were employed to solve transportation company selection problem (Ulutaş et al 2021a). In another study by Ulutaş et al., they selected insulation materials with the PSI-CRITIC-based CoCoSo method (Ulutaş et al 2021b). A novel fuzzy group MCDM approach through integrating triangular fuzzy hesitant sets (TFHS), failure mode and effect analysis (FMEA), and CoCoSo for outsourcing risk was suggested by Yazdani et al. As a result of this study, the authors suggested that risk evaluation of outsourcing providers must consider multi-experts, multi-criteria, multi-uncertainties, and measurability (Yazdani et al 2021). Another FUCOM–FMEA–CoCoSo integration study was conducted by Yousefi et al. to assess and prioritize failures in a production process (Yousefi et al 2021). For the evaluation of healthcare sectors in Eastern European countries, Torkayesh et al. proposed a novel hybrid weight determination model using BWM and level-based weight assessment (LBWA) to determine the weights of healthcare indicators. Then, the CoCoSo method was applied to evaluate healthcare performance (Torkayesh et al 2021b).

As a result of the literature search, only one earthquake disaster study was found in which the CoCoSo method was employed. The AHP, TOPSIS, and AHP-TOPSIS criterion weighting procedures, and the CoCoSo method, were used to prioritize the demand of relief regions during an earthquake. Five regions (south, east, west, north, and center) were ranked according to six criteria: Region strength, access to roads, population density, number of medical centers, post-accident hazards, and readiness (Shahkam and Cheraghchi, 2021).

This study is the first to rank the districts of Istanbul according to earthquake loss estimates by integrating entropy and CoCoSo methods. Thirty-nine districts of İstanbul were prioritized considering 12 different criteria. Issues such as determining the locations of aid centers in the event of an earthquake, raising the awareness of earthquakes in people living in this region, strengthening the structures, or putting urban transformation plans into practice may arise in the riskiest districts determined at the end of the study.

13.3 PROPOSED RISK ASSESSMENT APPROACH: ENTROPY–COCOSO INTEGRATION

13.3.1 Stage 1: Obtaining Criteria Weights Using the Entropy Method

Step 1.1 Obtain the initial decision matrix.

In the first step of the proposed approach, alternatives $(A_i, i = 1,2,\dots,m)$ and criteria $(C_j, j = 1,2,\dots,n)$ are obtained; then, the initial decision matrix is obtained as in Eq. 13.1.

$$B = \begin{bmatrix} x_{(1,1)} & x_{(1,2)} & \dots & x_{(1,J)} \\ x_{(2,1)} & x_{(2,2)} & & x_{(2,J)} \\ \vdots & & \ddots & \vdots \\ x_{(I,1)} & x_{(I,2)} & \cdots & x_{(I,J)} \end{bmatrix} \tag{13.1}$$

In Eq. 13.1, $x_{(i,j)}, i = 1,2,\dots,I; j = 1,2,\dots,J$ shows the ith alternative value of the jth criterion.

Step 1.2 Normalize the initial decision matrix.

In the decision problem considered, the value of the alternatives for each criterion may be in different units and structures. With the normalization process performed at this stage, the values of all the elements in the initial decision matrix are moved to the range [0,1] using Eq. 13.2.

$$n_{(i,j)} = \frac{x_{(i,j)}}{\sum_{i=1}^{I} x_{(i,j)}}, \quad i = 1,2,\dots,I; \; j = 1,2,\dots,J \tag{13.2}$$

Step 1.3 Calculate the entropy value for each criterion.

The entropy value of each criterion is calculated as in Eq. 13.3

$$e_j = -k \sum_{i=1}^{I} n_{(i,j)} \ln\left(n_{(i,j)}\right), \quad j = 1,2,\dots,J \tag{13.3}$$

In Eq. 13.3, the constant k depends on the number of alternatives and is calculated with the formula $k = 1/(ln(I))$. Entropy values take values in the range of $0 \le e_j \le 1$.

Step 1.4 Find the degrees of differentiation for each criterion.

By using the entropy values calculated for each criterion, the degrees of differentiation are found as in Eq. 13.4.

$$d_j = 1 - e_j, \qquad j = 1,2,\ldots,J \tag{13.4}$$

Step 1.5 Calculate the entropy weights for each criterion.

In the last step of the first stage, the entropy criterion weights are calculated as in Eq. 13.5.

$$w_j = \frac{d_j}{\sum_{j=1}^{J} d_j}, \qquad j = 1,2,\ldots,J \tag{13.5}$$

13.3.2 Stage 2: Obtaining Final Rankings of The Alternatives Using the CoCoSo Method

Step 2.1 Normalize the initial decision matrix.

The normalization procedure of the CoCoSo method is different from entropy. Min-max normalization is employed as in Eq. 13.6.

$$n^*_{(i,j)} = \begin{cases} \dfrac{x_{(i,j)} - \min\limits_{\forall i} x_{(i,j)}}{\max\limits_{\forall i} x_{(i,j)} - \min\limits_{\forall i} x_{(i,j)}}, & \textit{for the benefit type criterion} \\[2ex] \dfrac{\max\limits_{\forall i} x_{(i,j)} - x_{(i,j)}}{\max\limits_{\forall i} x_{(i,j)} - \min\limits_{\forall i} x_{(i,j)}}, & \textit{for the cost type criterion} \end{cases} \tag{13.6}$$

Step 2.2. Calculate the score (S_i) and the relative importance (P_i) of each alternative.

In this step, the score of each alternative (S_i) is calculated using the weighted sum method (WSM) as in Eq. 13.7.

$$S_i = \sum_{j=1}^{J} n^*_{(i,j)} w_j, \quad i = 1,2,\ldots,I \tag{13.7}$$

Thus, the relative importance of each alternative (P_i) is obtained by the weighted product method (WPM) based on weighted aggregated sum product assessment (WASPAS) multiplicative manner using Eq. 13.8.

$$P_i = \sum_{j=1}^{J} \left[n^*_{(i,j)} \right]^{w_j}, \quad i = 1,2,\ldots,I \tag{13.8}$$

The proposed approach uses the entropy weights of criteria obtained in stage 1 in Eqs. 13.7 and 13.8.

Step 2.3. Compute the relative weights for the alternatives.

To obtain the relative weights of the alternatives, three different scoring strategies given in Eqs. 13.9–13.11 are used.

$$(l_i)_1 = \frac{S_i + P_i}{\sum_{i=1}^{I}(S_i + P_i)}, \quad i = 1,2,\ldots,I \tag{13.9}$$

$$(l_i)_2 = \frac{S_i}{\min\limits_{\forall i} S_i} + \frac{P_i}{\min\limits_{\forall i} P_i}, \quad i = 1,2,\ldots,I \tag{13.10}$$

$$(l_i)_3 = \frac{\alpha S_i + (1-\alpha)P_i}{\alpha \max\limits_{\forall i} S_i + (1-\alpha)\max\limits_{\forall i} P_i}, \quad 0 \le \alpha \le 1, \quad i = 1,2,\ldots,I \tag{13.11}$$

Equation 13.9 shows the score of each alternative based on the arithmetic mean for the sum of WSM and WPM. Equation 13.10 calculates the sum of relative scores of WSM and WPM compared to the best. The CoCoSo method suggests using a score based on the balanced compromise of WSM and WPM given in Eq. 13.11.

Step 2.4. Obtain the final ranking of the alternatives.

Final scores of alternatives for ranking, based on three different scoring strategies given in the previous step, are calculated as in Eq. 13.12.

$$l_i = \left[(l_i)_1 (l_i)_2 (l_i)_3\right]^{1/3} + \frac{1}{3}\left[(l_i)_1 + (l_i)_2 + (l_i)_3\right] \tag{13.12}$$

In Eq. 13.12, l_i values are ordered from largest to the smallest. Then, the alternative with the largest l_i value is chosen as the best alternative.

13.4 APPLICATION OF THE PROPOSED APPROACH: RANKING OF DISTRICTS OF ISTANBUL ACCORDING TO EARTHQUAKE LOSS ESTIMATES

13.4.1 Stage 1: Obtaining Criteria Weights Using the Entropy Method

Step 1.1 Obtain the initial decision matrix.

In this study, the aim was to rank the 39 districts of Istanbul given in Table 13.1 from the least risky to the riskiest in a Mw 7.5 earthquake scenario considered within the scope of the project.

The 12 different criteria and definitions determined to make the risk assessment of the districts were as follows:

- The number of extremely damaged buildings (C_1): Estimation of the number of extremely damaged buildings in the district for the deterministic (Mw 7.5) scenario earthquake.

TABLE 13.1
Districts of Istanbul (Alternatives)

A_i	District	A_i	District	A_i	District
A_1	Adalar	A_{14}	Büyükçekmece	A_{27}	Maltepe
A_2	Arnavutköy	A_{15}	Çatalca	A_{28}	Pendik
A_3	Ataşehir	A_{16}	Çekmeköy	A_{29}	Sancaktepe
A_4	Avcılar	A_{17}	Esenler	A_{30}	Sarıyer
A_5	Bağcılar	A_{18}	Esenyurt	A_{31}	Şile
A_6	Bahçelievler	A_{19}	Eyüp	A_{32}	Silivri
A_7	Bakırköy	A_{20}	Fatih	A_{33}	Şişli
A_8	Başakşehir	A_{21}	Gaziosmanpaşa	A_{34}	Sultanbeyli
A_9	Bayrampaşa	A_{22}	Güngören	A_{35}	Sultangazi
A_{10}	Beşiktaş	A_{23}	Kadıköy	A_{36}	Tuzla
A_{11}	Beykoz	A_{24}	Kağıthane	A_{37}	Ümraniye
A_{12}	Beylikdüzü	A_{25}	Kartal	A_{38}	Üsküdar
A_{13}	Beyoğlu	A_{26}	Küçükçekmece	A_{39}	Zeytinburnu

- The number of heavily damaged buildings (C_2): Estimation of the number of heavily damaged buildings in the district for the deterministic (Mw 7.5) scenario earthquake.
- The number of moderately damaged buildings (C_3): Estimation of the number of moderately damaged buildings in the district for the deterministic (Mw 7.5) scenario earthquake.
- The number of slightly damaged buildings (C_4): Estimation of the number of slightly damaged buildings in the district for the deterministic (Mw 7.5) scenario earthquake.
- The number of deaths (C_5): Estimation of the number of deaths in the district for the deterministic (Mw 7.5) night scenario earthquake.
- The number of seriously injured people (C_6): Estimation of the number of seriously injured people in the district for the deterministic (Mw 7.5) night scenario earthquake.
- The number of people in need of hospital treatment (C_7): Estimation of the number of people who need to be treated in hospital in the district for the deterministic Mw 7.5 night scenario earthquake.
- The number of slightly injured people (C_8): Estimation of the number of slightly injured people in the district for the deterministic (Mw 7.5) night scenario earthquake.
- The number of damaged points in the natural gas pipeline (C_9): The estimation of how many points of the natural gas line will be damaged for the deterministic (Mw 7.5) scenario earthquake.
- The number of damaged points in the drinking water pipeline (C_{10}): The estimation of how many points of the drinking water pipeline will be damaged for the deterministic (Mw 7.5) earthquake.

- The number of damaged points in the wastewater pipeline (C_{11}): The estimation of how many points of the wastewater pipeline will be damaged for the deterministic (Mw 7.5) scenario earthquake.
- The number of temporary accommodations needs (C_{12}): The number of temporary accommodations needs of the district for (Mw 7.5) scenario earthquake.

An obtained initial decision matrix is given in Table 13.2.

Step 1.2 Normalize the initial decision matrix.

The normalized initial decision matrix was calculated using Eq. 13.2 and depicted in Table 13.3.

In Table 13.3, the normalized value of Adalar, the first district given Table 13.1, according to the first criterion (the number of extremely damaged buildings) was calculated as in Eq. 13.13 as an example:

$$n_{(1,1)} = \frac{x_{(1,1)}}{\sum_{i=1}^{39} x_{(i,1)}} = \frac{413}{13497} = 0.031 \quad (13.13)$$

Step 1.3 Calculate the entropy value for each criterion.

Equation 13.3 was used to calculate the entropy value of each criterion. For example, the entropy value of the second criterion (the number of heavily damaged buildings) was obtained in Eq. 13.14 where $k = 1/ln(I) = 1/\ln(39) = 0.273$.

$$\begin{aligned} e_2 = -k\sum_{i=1}^{7} n_{(i,2)} \ln\left(n_{(i,2)}\right) = \\ -0.273\left[0.022\ln(0.022) + 0.007\ln(0.007) + \cdots \right. \\ \left. + 0.030\ln(0.030)\right] = 0.925 \end{aligned} \quad (13.14)$$

The entropy values of all criteria are given in Table 13.4.

Step 1.4 Find the degrees of differentiation for each criterion.

By subtracting the entropy values given in Table 13.4 from 1, the degrees of differentiation for each criterion given in Table 13.5 were found. For example, the entropy value of the fifth criterion (the number of deaths) was calculated as $d_5 = 1 - e_5 = 1 - 0.815 = 0.185$.

Step 1.5 Calculate the entropy weights for each criterion.

The entropy criterion weights obtained using Eq. 13.5 are given in Table 13.6.

The entropy weight of the sixth criterion (the number of seriously injured people) was calculated as in Eq. 13.15 as an example.

$$w_6 = \frac{d_6}{\sum_{j=1}^{12} d_j} = \frac{0.191}{0.126 + 0.075 + \cdots + 0.085} = \frac{0.191}{1.22} = 0.156 \quad (13.15)$$

TABLE 13.2
Initial Decision Matrix

	C_1	C_2	C_3	C_4	C_5	C_6	C_7	C_8	C_9	C_{10}	C_{11}	C_{12}
A_1	413	743	1894	1842	76	61	217	378	3	5	8	3115
A_2	41	243	1704	4950	0	0	34	178	7	9	17	5815
A_3	123	471	2756	6833	89	47	330	755	6	7	18	14702
A_4	233	1261	5545	9285	465	239	1385	2799	19	22	54	34941
A_5	796	1825	8001	15397	1179	652	3148	5869	16	20	52	59513
A_6	796	1490	5668	9686	1633	879	4261	7599	17	21	52	61199
A_7	782	1306	3394	3939	1046	531	2701	4610	15	17	48	28910
A_8	115	575	2977	6243	71	45	287	670	6	9	20	13397
A_9	614	1182	4059	7415	520	340	1384	2529	5	8	19	23064
A_{10}	64	187	1053	3241	26	14	99	240	2	2	6	4096
A_{11}	113	443	2756	7356	25	16	112	284	6	6	13	5937
A_{12}	321	950	3004	4331	527	276	1456	2655	21	32	72	26670
A_{13}	410	841	3132	7158	217	150	609	1141	3	5	12	10295
A_{14}	439	1635	6622	9367	288	154	885	1754	26	33	54	24298
A_{15}	63	301	1656	4053	4	2	25	67	4	7	13	1947
A_{16}	16	113	998	3392	1	0	18	109	3	4	7	3978
A_{17}	391	813	3739	8302	638	352	1687	3133	6	7	20	31271
A_{18}	563	1768	7911	14672	1003	543	2789	5441	24	28	63	67410
A_{19}	300	841	3794	8237	168	110	525	1080	6	9	20	16066
A_{20}	2083	3496	9946	15758	1484	985	3897	6767	11	18	43	46794
A_{21}	129	403	2276	6969	140	83	418	898	3	4	10	13963
A_{22}	342	611	2407	4597	754	415	1965	3496	7	9	24	27478
A_{23}	209	505	2426	6717	190	93	592	1135	6	7	19	13585
A_{24}	94	275	1727	5816	84	44	275	601	3	4	10	9908
A_{25}	199	593	3282	8173	176	87	553	1148	7	8	22	18189
A_{26}	1240	2616	8978	13799	1515	925	4079	7413	21	25	63	72774
A_{27}	237	690	3388	8156	234	130	706	1438	8	10	22	20102
A_{28}	274	1041	5625	13449	195	101	707	1566	13	16	35	28281
A_{29}	84	416	2539	6733	48	24	216	574	6	8	19	12967
A_{30}	102	367	2486	6977	33	21	141	341	5	6	11	6616
A_{31}	22	106	774	2402	0	0	0	1	3	5	6	876
A_{32}	444	1707	7155	11997	58	27	246	566	21	29	36	13481
A_{33}	68	207	1008	4333	55	27	172	368	2	2	5	4845
A_{34}	109	648	3949	8225	73	38	300	756	7	9	22	16635
A_{35}	61	245	1804	6884	57	27	217	556	3	4	11	10524
A_{36}	446	1370	5310	8239	268	169	808	1565	14	19	48	21417
A_{37}	105	534	3626	10656	42	17	250	704	7	10	23	16345
A_{38}	146	493	3341	9610	95	42	365	849	6	6	15	15087
A_{39}	510	1025	3799	6335	668	374	1767	3206	8	11	30	31030

TABLE 13.3
Normalized Initial Decision Matrix

	C_1	C_2	...	C_{12}
A_1	0.031	0.022	...	0.004
A_2	0.003	0.007	...	0.007
⋮	⋮	⋮	...	⋮
A_{39}	0.038	0.030	...	0.037

TABLE 13.4
Entropy Values of Criteria

e_1	e_2	e_3	e_4	e_5	e_6	e_7	e_8	e_9	e_{10}	e_{11}	e_{12}
0.874	0.925	0.954	0.973	0.815	0.809	0.840	0.861	0.934	0.936	0.939	0.915

TABLE 13.5
Degrees of Difference for Criteria

d_1	d_2	d_3	d_4	d_5	d_6	d_7	d_8	d_9	d_{10}	d_{11}	d_{12}
0.126	0.075	0.046	0.027	0.185	0.191	0.160	0.139	0.066	0.064	0.061	0.085

TABLE 13.6
Entropy Criterion Weights

w_1	w_2	w_3	w_4	w_5	w_6	w_7	w_8	w_9	w_{10}	w_{11}	w_{12}
0.103	0.061	0.037	0.022	0.151	0.156	0.130	0.114	0.054	0.052	0.050	0.069

13.4.2 Stage 2: Obtaining Final Rankings of the Alternatives Using the CoCoSo Method

Step 2.1 Normalize the initial decision matrix.

Normalized values based on min-max normalization were obtained by Eq. 13.6 and are depicted in Table 13.7.

Step 2.2. Calculate the score (S_i) and the relative importance (P_i) of each alternative.

In this step, the score and the relative importance of each alternative were provided by Eq. 13.7 and 13.8. Table 13.8 shows the S_i and P_i values.

TABLE 13.7
Normalized Initial Decision Matrix for CoCoSo Method

$n^*_{(i,j)}$	C_1	C_2	...	C_{12}
A_1	0.808	0.812	...	0.969
A_2	0.988	0.960	...	0.931
⋮	⋮	⋮	...	⋮
A_{39}	0.761	0.729	...	0.581

TABLE 13.8
S_i and P_i Values

	S_i	P_i
A_1	0.924	11.919
A_2	0.947	11.942
⋮	⋮	⋮
A_{39}	0.761	0.729

Sample score calculation for the second alternative, Arnavutköy, using entropy weights was given as per Eq. 13.16.

$$S_2 = \sum_{j=1}^{12} n^*_{(2,j)} w_j = 0.988(0.103) + 0.812(0.061) + \dots + 0.969(0.069) = 0.947 \tag{13.16}$$

Similarly, the relative importance for the second alternative, Arnavutköy, was as given in Eq. 13.17.

$$P_2 = \sum_{j=1}^{12} \left[n^*_{(2,j)}\right]^{w_j} = (0.988)^{0.103} + (0.812)^{0.061} + \dots + (0.969)^{0.069} = 11.942 \tag{13.17}$$

Step 2.3. Compute the relative weights for the alternatives.

Three different scoring strategies using Eqs. 13.9–13.11 for the relative weights of the alternatives were obtained in Table 13.9.

Sample calculations of relative weights for alternative 39, Zeytinburnu, were given as per Eqs. 13.18–13.20:

$$(l_{39})_1 = \frac{S_{39} + P_{39}}{\sum_{i=1}^{39}(S_i + P_i)} = \frac{0.761 + 0.729}{472.541} = 0.026 \tag{13.18}$$

TABLE 13.9
Relative Weights of Alternatives for Different Scoring Strategies ($\alpha = 0.5$)

	$(I_i)_1$	$(I_i)_2$	$(I_i)_3$
A_1	0.027	9.361	0.989
A_2	0.027	9.553	0.993
⋮	⋮	⋮	⋮
A_{39}	0.026	7.034	0.940

$$(I_{39})_2 = \frac{S_{39}}{\min\limits_{\forall i} S_i} + \frac{P_{39}}{\min\limits_{\forall i} P_i} = \frac{0.761}{0.125} + \frac{0.729}{6.028} = 7.034 \tag{13.19}$$

$$\begin{aligned} \text{For } \alpha = 0.5,\ (I_{39})_3 &= \frac{\alpha S_{39} + (1-\alpha) P_{39}}{\alpha \max\limits_{\forall i} S_i + (1-\alpha) \max\limits_{\forall i} P_i} \\ &= \frac{0.5(0.761) + (1-0.5)0.729}{0.5(0.951) + (1-0.5)11.990} = 0.940 \end{aligned} \tag{13.20}$$

Step 2.4. Obtain the final ranking of the alternatives.

Final alternative scores for ranking were calculated using Eq. 13.12 and are depicted in Table 13.10. I_i scores were sorted from the largest to smallest to obtain the final ranking of the alternatives.

Equation 13.21 shows the sample calculation for the final score of the first alternative, Adalar.

$$\begin{aligned} I_1 &= \left[(I_1)_1 (I_1)_2 (I_1)_3\right]^{\frac{1}{3}} + \frac{1}{3}\left[(I_1)_1 + (I_1)_2 + (I_1)_3\right] \\ &= \left[0.027(9.361)(0.989)\right]^{\frac{1}{3}} \\ &\quad + \frac{1}{3}\left[0.027 + 9.361 + 0.989\right] = 4.090 \end{aligned} \tag{13.21}$$

According to Table 13.10, alternative 31, Şile, is the least risky district. The twentieth district, Fatih, is the riskiest.

13.5 CONCLUSION

The exact timing and magnitude of earthquakes cannot be predicted, which necessitates measures to prevent or reduce risks being taken before an earthquake occurs. Improvements should be made by strategic planning, action plans, and urban transformation projects in the risky districts of Istanbul. In this context, this study aimed to find the riskiest district of İstanbul in the case of an earthquake.

TABLE 13.10
Final Alternative Scores and Rankings

A_i	District	I_i	Rank	A_i	District	I_i	Rank
A_1	Adalar	4.090	11	A_{21}	Gaziosmanpaşa	4.021	13
A_2	Arnavutköy	4.162	6	A_{22}	Güngören	3.261	29
A_3	Ataşehir	4.002	14	A_{23}	Kadıköy	3.901	19
A_4	Avcılar	3.187	31	A_{24}	Kağıthane	4.118	10
A_5	Bağcılar	2.224	36	A_{25}	Kartal	3.857	21
A_6	Bahçelievler	1.576	37	A_{26}	Küçükçekmece	1.407	38
A_7	Bakırköy	2.718	34	A_{27}	Maltepe	3.758	23
A_8	Başakşehir	4.001	15	A_{28}	Pendik	3.575	24
A_9	Bayrampaşa	3.401	27	A_{29}	Sancaktepe	4.053	12
A_{10}	Beşiktaş	4.255	3	A_{30}	Sarıyer	4.130	8
A_{11}	Beykoz	4.119	9	A_{31}	Şile	4.294	1
A_{12}	Beylikdüzü	3.011	33	A_{32}	Silivri	3.574	25
A_{13}	Beyoğlu	3.861	20	A_{33}	Şişli	4.221	4
A_{14}	Büyükçekmece	3.019	32	A_{34}	Sultanbeyli	3.953	18
A_{15}	Çatalca	4.215	5	A_{35}	Sultangazi	4.138	7
A_{16}	Çekmeköy	4.274	2	A_{36}	Tuzla	3.466	26
A_{17}	Esenler	3.330	28	A_{37}	Ümraniye	3.970	17
A_{18}	Esenyurt	2.272	35	A_{38}	Üsküdar	3.981	16
A_{19}	Eyüp	3.831	22	A_{39}	Zeytinburnu	3.221	30
A_{20}	Fatih	1.115	39				

A risk-assessment study was conducted in which districts of Istanbul were listed based on entropy–CoCoSo integration, according to magnitude Mw 7.5 earthquake scenario estimates. When the results obtained with the application of the entropy method were examined, it was revealed that the number of seriously injured people, defined as the sixth criterion, is the most important criterion in terms of risk. This criterion was followed by the number of deaths and the number of people in need of hospital treatment.

According to the results of the CoCoSo method, while Şile was the least risky district, Fatih was the riskiest. When the initial decision matrix of the study is examined where all the selected criteria are of a cost type, it is seen that all the estimation values of criteria for Şile are very low, while they are very high for Fatih. In terms of the riskiest districts, Küçükçekmece and Bahçelievler follow Fatih. In the districts with the highest risk:

- Research should be conducted to recover from the effects of the earthquake or to minimize their effects on society.
- Various plans should be developed, and these should be put into practice.
- It is necessary to raise the awareness earthquakes in people living in these districts, and to initiate disaster-management studies that will allow them to be affected less or not at all in the case of an earthquake.

The limitations of this study are in the selection of criteria. For example, road closure rate, power outage time estimation after an earthquake, etc., could have been added as criteria. Estimations of the criteria values for each district require analytical studies in the earthquake engineering literature or empirical models based on earthquake experiences. In future studies, these criteria can be obtained and added to the proposed method. Entropy–CoCoSo integration can be used for risk assessment in different areas. Different criteria weighting procedures can be developed and used in CoCoSo. The proposed approach can also be tried for different decision problems.

REFERENCES

Alrasheedi, M., Mardani, A., Mishra, A. R., Streimikiene, D., Liao, H., & Al-nefaie, A. H. (2021). Evaluating the green growth indicators to achieve sustainable development: A novel extended interval-valued intuitionistic fuzzy-combined compromise solution approach. *Sustainable Development*, 29(1), 120–142.

Altıntaş, F. F. (2021). G7 ülkelerinin bilgi performanslarının analizi: CoCoSo yöntemi ile bir uygulama. *Journal of Life Economics*, 8(3), 337–347.

Bagal, D. K., Naik, B., Parida, B., Barua, A., Jeet, S., Singh, S. K., & Pattanaik, A. K. (2020). Comparative mechanical characterization of M30 concrete grade by fractional replacement of Portland Pozzolana cement with industrial waste using CoCoSo and CODAS methods. *IOP Conference Series: Materials Science and Engineering*, 970, 012015.

Barua, A., Jeet, S., Bagal, D. K., Satapathy, P., & Agrawal, P. K. (2019). Evaluation of mechanical behavior of hybrid natural fiber-reinforced nano sic particles composite using hybrid Taguchi-CoCoSo method. *International Journal of Innovative Technology and Exploring Engineering*, 8(10), 3341–3345.

Biswas, T., Chatterjee, P., & Choudhuri, B. (2020). Selection of commercially available alternative passenger vehicle in automotive environment. *Operational Research in Engineering Sciences: Theory and Applications*, 3(1), 16–27.

Biswas, S. (2020). Measuring performance of healthcare supply chains in India: A comparative analysis of multi-criteria decision-making methods. *Decision Making: Applications in Management and Engineering*, 3(2), 162–189.

Cui, Y., Liu, W., Rani, P., & Alrasheedi, M. (2021). Internet of Things (IoT) adoption barriers for the circular economy using Pythagorean fuzzy SWARA-CoCoSo decision-making approach in the manufacturing sector. *Technological Forecasting and Social Change*, 171, 120951.

Deveci, M., Pamucar, D., & Gokasar, I. (2021). Fuzzy Power Heronian function-based CoCoSo method for the advantage prioritization of autonomous vehicles in real-time traffic management. *Sustainable Cities and Society*, 69, 102846.

Ecer, F., Pamucar, D., Zolfani, S. H., & Eshkalag, M. K. (2019). Sustainability assessment of OPEC countries: Application of a multiple attribute decision-making tool. *Journal of Cleaner Production*, 241, 118324.

Ecer, F. & Pamucar, D. (2020). Sustainable supplier selection: A novel integrated fuzzy best-worst method (F-BWM) and fuzzy CoCoSo with Bonferroni (CoCoSo'B) multi-criteria model. *Journal of Cleaner Production*, 266, 121981.

Erceg, Ž., Starčević, V., Pamučar, D., Mitrović, G., Stević, Ž., & Žikić, S. (2019). A new model for stock management in order to rationalize costs: ABC-FUCOM-interval rough CoCoSo model. *Symmetry*, 11(12), 1527.

Erdik, M., Fahjan, Y., Ozel, O., Alcik, H., Mert, A., & Gul, M. (2003). Istanbul earthquake rapid response and the early warning system. *Bulletin of Earthquake Engineering*, 1(1), 157–163.

İstanbul Büyükşehir Belediyesi Deprem Risk Yönetimi ve Kentsel İyileştirme Daire Başkanlığı Deprem ve Zemin İnceleme Müdürlüğü. 2020. İstanbul'un 39 ilçesi için "*İlçe Olası Deprem Kayıp Tahmini Kitapçıkları*" https://depremzemin.ibb.istanbul/guncelcalismalarimiz/ (accessed October 28, 2021).

Kadıoğlu, M., İ. Helvacıoğlu, N. Okay, A. Tezer, L. Trabzon, H. Türkoğlu, Y.S. Ünal, & R. Yiğiter, 2005. *Okullar İçin Afet Yönetimi ve Acil Yardım Planı Kılavuzu*, Mayıs-2003, İTÜ Afet Yönetim Merkezi Yayınları, İTÜ Press.

Khan, S. & Haleem, A. (2021). Investigation of circular economy practices in the context of emerging economies: a CoCoSo approach. *International Journal of Sustainable Engineering*, 14(3), 357–367.

Kharwar, P. K., Verma, R. K., & Singh, A. (2020). Neural network modeling and combined compromise solution (CoCoSo) method for optimization of drilling performances in polymer nanocomposites. *Journal of Thermoplastic Composite Materials*, doi: 10.1177/0892705720939165.

Kieu, P. T., Nguyen, V. T., Nguyen, V. T., & Ho, T. P. (2021). A spherical fuzzy analytic hierarchy process (SF-AHP) and combined compromise solution (CoCoSo) algorithm in distribution center location selection: A case study in agricultural supply chain. *Axioms*, 10(2), 53.

Kundak, S. & Türkoğlu, H. (2010). İstanbul'da deprem riski analizi. İTÜ DERGİSİ/a, 6(2).

Lahane, S. & Kant, R. (2021). A hybrid Pythagorean fuzzy AHP–CoCoSo framework to rank the performance outcomes of circular supply chain due to adoption of its enablers. *Waste Management*, 130, 48–60.

Lai, H., Liao, H., Wen, Z., Zavadskas, E. K., & Al-Barakati, A. (2020). An improved CoCoSo method with a maximum variance optimization model for cloud service provider selection. *Engineering Economics*, 31(4), 411–424.

Liao, H., Qin, R., Wu, D., Yazdani, M., & Zavadskas, E. K. (2020). Pythagorean fuzzy combined compromise solution method integrating the cumulative prospect theory and combined weights for cold chain logistics distribution center selection. *International Journal of Intelligent Systems*, 35(12), 2009–2031.

Luo, Y., Zhang, X., Qin, Y., Yang, Z., & Liang, Y. (2021). Tourism attraction selection with sentiment analysis of online reviews based on probabilistic linguistic term sets and the IDOCRIW-CoCoSo Model. *International Journal of Fuzzy Systems*, 23(1), 295–308.

Mi, X. & Liao, H. (2020). Renewable energy investments by a combined compromise solution method with stochastic information. *Journal of Cleaner Production*, 276, 123351.

Peng, X., Zhang, X., & Luo, Z. (2020). Pythagorean fuzzy MCDM method based on CoCoSo and CRITIC with score function for 5G industry evaluation. *Artificial Intelligence Review*, 53(5), 3813–3847.

Peng, X. & Huang, H. (2020). Fuzzy decision-making method based on CoCoSo with critic for financial risk evaluation. *Technological and Economic Development of Economy*, 26(4), 695–724.

Peng, X. & Smarandache, F. (2020). A decision-making framework for China's rare earth industry security evaluation by neutrosophic soft CoCoSo method. *Journal of Intelligent & Fuzzy Systems*, 39(5), 7571–7585.

Peng, X., Garg, H., & Luo, Z. (2021a). Hesitant fuzzy soft combined compromise solution method for IoE companies' Evaluation. *International Journal of Fuzzy Systems*, 24(1), 457–473.

Peng, X., Krishankumar, R., & Ravichandran, K. S. (2021b). A novel interval-valued fuzzy soft decision-making method based on CoCoSo and CRITIC for intelligent healthcare management evaluation. *Soft Computing*, 25(6), 4213–4241.

Peng, X. & Garg, H. (2021). Intuitionistic fuzzy soft decision-making method based on CoCoSo and CRITIC for CCN cache placement strategy selection. *Artificial Intelligence Review*, 55(2), 1576–1604.

Peng, X. & Luo, Z. (2021). Decision-making model for China's stock market bubble warning: the CoCoSo with picture fuzzy information. *Artificial Intelligence Review*, 54(8), 5675–5697.

Stanujkic, D., Popovic, G., Zavadskas, E. K., Karabasevic, D., & Binkyte-Veliene, A. (2020). Assessment of progress towards achieving Sustainable Development Goals of the "Agenda 2030" by using the CoCoSo and the Shannon entropy methods: The case of the EU countries. *Sustainability*, 12(14), 5717.

Shadkam, E. & Cheraghchi, M. (2021). Prioritization of earthquake relief using a hybrid two-phase approach. *Journal of Applied Research on Industrial Engineering*.

Štirbanović, Z., Gardić, V., Stanujkić, D., Marković, R., Sokolović, J., & Stevanović, Z. (2021). Comparative MCDM analysis for AMD treatment method selection. *Water Resources Management*, 35(11), 3737–3753.

Topal, A. (2021). Çok kriterli karar verme analizi ile elektrik üretim şirketlerinin finansal performans analizi: Entropi tabanlı Cocoso yöntemi. *Business & Management Studies: An International Journal*, 9(2), 532–546.

Torkayesh, A. E., Ecer, F., Pamucar, D., & Karamaşa, Ç. (2021a). Comparative assessment of social sustainability performance: Integrated data-driven weighting system and CoCoSo model. *Sustainable Cities and Society*, 71, 102975.

Torkayesh, A. E., Pamucar, D., Ecer, F., & Chatterjee, P. (2021b). An integrated BWM-LBWA-CoCoSo framework for evaluation of healthcare sectors in Eastern Europe. *Socio-Economic Planning Sciences*, 101052.

Ulutaş, A., Popovic, G., Radanov, P., Stanujkic, D., & Karabasevic, D. (2021a). A new hybrid fuzzy PSI-PIPRECIA-CoCoSo MCDM based approach to solving the transportation company selection problem. *Technological and Economic Development of Economy*, 27(5), 1227–1249.

Ulutaş, A., Balo, F., Sua, L., Karabasevic, D., Stanujkic, D., & Popovic, G. (2021b). Selection of insulation materials with PSI-CRITIC based CoCoSo method. *Revista de la Construcción*, 20(2), 382–392.

Ulutaş, A., Karakuş, C. B., & Topal, A. (2020). Location selection for logistics center with fuzzy SWARA and CoCoSo methods. *Journal of Intelligent & Fuzzy Systems*, 38(4), 4693–4709.

Wen, Z., Liao, H., Ren, R., Bai, C., Zavadskas, E. K., Antucheviciene, J., & Al-Barakati, A. (2019a). Cold chain logistics management of medicine with an integrated multi-criteria decision-making method. *International Journal of Environmental Research and Public Health*, 16(23), 4843.

Wen, Z., Liao, H., Kazimieras Zavadskas, E., & Al-Barakati, A. (2019b). Selection of third-party logistics service providers in supply chain finance by a hesitant fuzzy linguistic combined compromise solution method. *Economic Research*, 32(1), 4033–4058.

Yazdani, M., Zarate, P., Zavadskas, E. K., & Turskis, Z. (2019a). A Combined Compromise Solution (CoCoSo) method for multi-criteria decision-making problems. *Management* Decision, 57(9), 2501–2519.

Yazdani, M., Wen, Z., Liao, H., Banaitis, A., & Turskis, Z. (2019b). A grey combined compromise solution (CoCoSo-G) method for supplier selection in construction management. *Journal of Civil Engineering and Management*, 25(8), 858–874.

Yazdani, M., Chatterjee, P., Pamucar, D., & Chakraborty, S. (2020). Development of an integrated decision-making model for location selection of logistics centers in the Spanish autonomous communities. *Expert Systems with Applications*, 148, 113208.

Yazdani, M., Mohammed, A., Bai, C., & Labib, A. (2021). A novel hesitant-fuzzy-based group decision approach for outsourcing risk. *Expert Systems with Applications*, 184, 115517.
Yousefi, S., Valipour, M., & Gul, M. (2021). Systems failure analysis using Z-number theory-based combined compromise solution and full consistency method. *Applied Soft Computing*, 113, 107902.
Zhang, Z., Liao, H., Al-Barakati, A., Zavadskas, E. K., & Antuchevičienė, J. (2020). Supplier selection for housing development by an integrated method with interval rough boundaries. *International Journal of Strategic Property Management*, 24(4), 269–284.
Zolfani, S. H., Chatterjee, P., & Yazdani, M. (2019). A structured framework for sustainable supplier selection using a combined BWM-CoCoSo model. In *International Scientific Conference in Business, Management and Economics Engineering*. Vilnius, Lithuania (pp. 797–804).

14 Evaluation of Criteria Affecting the Preparedness Levels of Hospitals in Pandemic Periods by the Entropy Weighting Method

A Case Study for the COVID-19 Pandemic

Elif Kılıç Delice[1] *and Gülin Feryal Can*[2]
[1]University of Atatürk
[2]University of Başkent

CONTENTS

14.1 Introduction Background and Driving Forces 211
14.2 Literature Review on Preparedness Levels of Hospitals 213
14.3 EWM Implementation Steps 216
14.4 Application 218
14.5 Results and Discussion 221
References 221

14.1 INTRODUCTION: BACKGROUND AND DRIVING FORCES

The COVID-19 virus emerged in late 2019 and has spread rapidly to many parts of the world. Governments worldwide have implemented social distancing measures, companies have closed, and people have worked from home. Additionally, the social and economic effects of the COVID-19 virus still persist. Countries are trying to normalize daily life in terms of economic and social aspects.

The COVID-19 era has been one of the biggest economic and social shocks, deeply affecting health systems in all countries and hospitals, which were suddenly exposed

DOI: 10.1201/9781003212904-14

to this virus. This unexpected situation also caused the collapse of the health system in many countries. In order to cope with the pandemic, healthcare workers (HCWs) had to work for long time periods. They could not go home for months due to the danger of transmitting the virus to their families, and many HCWs even lost their lives while treating patients. This troublesome process was a result of many patients suddenly arriving at health units at the same time and health units were unprepared for this.

The pandemic continues due to the inability to eradicated the COVID-19 virus and incomplete vaccination programs. During this period, the need for other health services has increased gradually. Therefore, the implementation of other health services was inevitable. This situation reveals the necessity of providing services to patients with and without COVID-19 in the same system and increases the workload of HCWs.

The COVID-19 pandemic has created unexpected needs in healthcare practices around the world, such as funding, personnel, crisis management, equipment, ethics, and operational plans. However, the fight against the pandemic has allowed healthcare organizations to see how prepared they are for such situations. Along with the pandemic, many hospitals have started studies to determine in which aspects they need to improve in order to provide adequate diagnosis and treatment services. Researchers are also performing many studies on this subject.

Hospitals, which are a critical component of a healthcare system, play an important role in providing essential health services to people. Healthcare systems worldwide have faced some problems in providing services to patients, especially in pandemics. Recently, the world as a whole has been combating the COVID-19 pandemic. COVID-19 causes an increase in the number of patient demands, which may affect hospital capacity in terms of medical staff, healthcare processes, medical supplies, and equipment. To deal with the COVID-19 challenges, hospitals must have fulfilled their preparations before these events occur. In this context, this chapter aims to help hospitals determine the criteria that need to be strengthened in order to improve their diagnosis and treatment systems. Comments on the importance of criteria according to weight scores are provided and there is discussion of the preparedness level of the hospitals. To meet these aims, Entropy Weighting Method (EWM), which is one of the Multi Criteria Decision Making (MCDM) approaches, is proposed to help hospitals determine which criteria should be improved for their own diagnosis and treatment systems during a pandemic process, in order to increase their preparedness levels. In this context, five public hospitals in Turkey were considered to evaluate their preparedness levels for the COVID-19 pandemic. EWM was performed to determine the criteria that have the most impact on the preparedness level for a pandemic by obtaining criteria importance weights. EWM considers variability in preparedness criteria values for different hospitals ("performance values of hospitals"). There are various weighting approaches in the literature, including CRiteria Importance Through Inter-criteria Correlation (CRITIC), standard deviation, and mean weight. However, EWM is different. While EWM determines the importance weights of criteria, it also considers the uncertainty brought about by the change between the values of these criteria according to the alternatives. According to EWM, if the values of a criterion relative to the alternatives are close to each other, this criterion should have a lower importance level than the alternatives because, considering such criteria, it is difficult to choose between alternatives. The values of such criteria compared to alternatives contain high uncertainty. However, if the values of a criterion vary significantly with respect to the alternatives, the importance of this criterion should be high and more consideration should be given to the selection of alternatives because the uncertainty of the values of this criterion compared to the alternatives is low. The entropy value calculated for the criteria indicates the level of uncertainty. Criteria with high entropy values will have lower importance weights. In terms of hospital preparedness level, EWM was used to determine the importance weights of criteria that affect the readiness levels of hospitals by considering the uncertainty brought about by the changing values of these criteria according to the hospitals.

As a result of the study, within the scope of the criteria discussed, it was determined which hospital was most ready to fight the pandemic, and suggestions were made for the criteria that hospitals should improve themselves.

The organization of the remaining sections is as follows. Section 14.2 presents a literature review of the preparedness levels of hospitals. Section 14.3 consists of the EWM implementation steps and Section 14.4 covers EWM applications for the preparedness levels of hospitals. Section 14.5 includes the result and discussion.

14.2 LITERATURE REVIEW ON PREPAREDNESS LEVELS OF HOSPITALS

Hospitals have complex processes that mostly require external support and supply lines. In normal times, all hospitals provide intensive care to patients. Under pandemic conditions, with an increasing surge demand, hospitals may experience challenges in continuing necessary services, and their capacity may no longer be able to meet this increasing demand (Haghani et al., 2020). An effective hospital management policy is required to reduce these difficulties. In the literature, with the emergence of COVID-19, it is seen that researchers have conducted some studies to form an effective hospital management policy in terms of the preparedness of hospitals for pandemics. Most of these studies focused on determining the level of preparedness of hospitals against different disasters. However, the theme of hospital readiness against COVID-19 has not been widely investigated in the literature. Researchers have dealt with this topic with a conceptual/questionnaire-based structure. These studies are discussed below in brief.

For example, Blavin and Arnos (2020) sent the American Hospital Association Annual Survey to 6500 U.S. non-federal, general, and surgical community hospitals to review hospital organization structure, facilities, and services. As a result of the study, it was determined that staff capacity, and the use of hospital structure and physical space should be maximized. Additionally, creating internal rapid-response groups should be formed and equipment should be transferred to essential units; patient caseloads should be categorized and prioritized, and elective surgeries should be canceled. Ogoina et al. (2021) aimed to evaluate the preparedness of hospitals in Nigeria to respond to the COVID-19 outbreak. They received data from the updated World Health Organization (WHO) COVID-19 hospital preparedness checklist. The study concluded that most hospitals in Nigeria are not ready to respond to the COVID-19 pandemic and must prioritize the challenges related to critical care for COVID-19 patients and staff protection. El-Ghitany (2020) developed a short-term forecast scenario to predict the hospital preparedness for COVID-19 by using secondary data from daily COVID-19 reports and the report issued on 8 April 2020 by the Egyptian Ministry of Health and Population. An exponential growth rate model was performed to analyze the complete pandemic timeline. He proposed providing adequate intensive care unit (ICU) beds and ventilators for the estimated number of cases needed by the end of April 2020 at the latest. Brown et al. (2020) performed a study that considered rural healthcare center preparation and preparedness response to the threat of COVID-19 in Robert Packer Hospital, Sayre, PA, USA. At the end of the study, they stated that elective procedures should be canceled, a shortage of personnel protective equipment (PPE) and healthcare providers should be overcome, high-speed internet should be provided to meet HCWs' audiovisual needs, and uncomfortable changes across the nation's healthcare system should be avoided. Chobdar and Rahdar (2020) determined the preparedness of hospitals in Sistan and Baluchestan province, Iran, during the COVID-19 crisis. They considered 24 hospitals and received their data by implementing a Comprehensive Hospital Preparedness Checklist for COVID-19, which was translated and validated. They found that the low number of healthcare services and surge capacity were some of the reasons for lower hospital preparedness levels. Hamal et al. (2020) performed non-probability sampling to collect responses from 56 doctors working in different hospitals in Nepal. An online questionnaire was performed, based on WHO and Centers for Disease Control and Prevention guidelines. It was found that the majority of medical doctors working in the government hospitals of Nepal thought that the provision of face masks, airborne isolation rooms, critical care preparedness, and hands-on staff training were inadequate. Sanders et al. (2021) developed a qualitative hospital services questionnaire modeled on the basis of the WHO COVID-19 assessment tool, and suggested that the availability of pulse oximetry, oxygen capacity, and high-flow nasal cannula (HFNC) devices should be increased. Gül and Yücesan (2021) revised a WHO checklist to confirm Turkey's management of the COVID-19 pandemic. Ninety-nine sub-components in the ten-component checklist were weighted with the interval-valued global weighted arithmetic mean (IVSWAM) operator. They then ranked the alternatives using IVSWAM weights using the Technique for Order Preference by Similarity to Ideal Solution (TOPSIS) method. Hosseini et al. (2021), assessed the preparedness of hospitals in response to the COVID-19 pandemic in Mazandaran province, Iran. They performed the standard checklist of the Pan American Health Organization (PAHO). They considered 39 hospitals in Mazandaran, including university hospitals, social security hospitals, and private hospitals. They showed that the preparedness of the hospitals was at a good level in Mazandaran. However, due to the uncertainty related to a pandemic, monitoring the key components is important to maintain a high level of preparation. Lailani et al. (2021) analyzed hospitals' preparedness in dealing with COVID-19, and evaluated the strategies and follow-up carried out by the hospitals. They used the WHO hospital preparedness checklist for three time periods. It was determined that Pupuk Kaltim Hospital's level of preparedness for COVID-19 increased as it followed the checklist. Labib et al. (2020) analyzed the preparedness of ICUs in Cairo University Hospitals to deal efficiently and effectively with COVID-19 breakout. An exploratory cross-sectional study was performed at Cairo University ICUs: six pediatric ICUs and two adult ICUs. The WHO checklist for hospital preparedness was implemented. As a result of the study, was determined that adult ICUs are more prepared than pediatric ICUs in areas such as diagnosis, isolation, and case management. In addition, it was recommended that continuous training of health workers is provided and that active communication is ensured. Kaito et al. (2021) reviewed the literature regarding hospital preparedness, the initial management of COVID-19, and surveillance of HCWs. At the end of the study, they emphasized that each hospital should develop an incident-management system and establish a designated multidisciplinary medical team to deal with COVID-19. Saravi et al. (2021) implemented a standard PAHO checklist at the Mazandaran University of Medical Sciences, and the preparedness of the Center for Disaster Management and Medical Emergencies in the face of the COVID-19 pandemic was found to be at a good level. However, it was revealed that the supply chain and personnel need to be strengthened for sustainable preparedness. Shrivastava and Shrivastava (2020) performed a study to improve hospital preparedness to meet rising health-related needs during the COVID-19 pandemic. They stated that to advance the preparedness of hospitals, they need to be well equipped with the desired resources to respond to patient requirements. Additionally, they emphasized that the detailed checklist will help hospital authorities to design an effective response without becoming overwhelmed or compromising their emergency services. Kazzaz et al. (2020) provided information on key aspects of pediatric preparation at the hospital level, such as workforce, equipment, procurement, capacity planning, and infection-prevention strategies. They also addressed the management of pediatric COVID-19 patients in high-risk areas such as critical care areas, emergency departments, and operating rooms. Shrestha et al. (2020) presented the level of preparedness achieved with limited resources and lessons learned while isolating patients in Patan Hospital. They stated that during the COVID-19 period it is necessary that all hospitals have a central coordination mechanism to manage their resources. In the light of the studies mentioned above, many contributions were made to hospital disaster preparedness. Most of the papers have contributed to the literature by proposing cross-sectional questionnaire-based frameworks. However, conceptual-based models suggesting new attributes in hospital disaster readiness and review papers that provide a comprehensive overview of the topic are also widespread. The literature review performed in this study presents the gap in hospital disaster preparedness assessment from the viewpoint of MCDM. There have been limited studies regarding the applications of MCDM methods in hospital disaster preparedness; the paper by Gul and Yucesan (2021) was one study on this topic. Moving from this point, the aim was to develop an MCDM model supported by EWM for hospital preparedness assessment against the COVID-19 pandemic. In this study, which was different to other studies in the literature, hospitals were evaluated according to their readiness levels for the pandemic, considering criteria with real values. EWM was used to determine criteria importance weights that affect the readiness levels of hospitals by considering the uncertainty brought about by the changing values of these criteria according to the hospitals. No other study in the literature has compared different criteria related to the readiness levels of hospitals, has used EWM in this comparison, or has considered the differentiation in criteria values when determining these importance levels.

14.3 EWM IMPLEMENTATION STEPS

EWM is an objective evaluation methods that uses the quantitative data of criteria and the interactions between these data in a decision matrix (Mukhametzyanov, 2021). EWM is an effective method that can be used in different decision-making problems and to measure the amount of useful information in a data set. It is highly reliable for information measurement and provides high accuracy in weight-determination problems. It is based on the opinion that the amount or quality of the information obtained in the decision-making environment affects the accuracy and reliability of the solution (Wu et al., 2011). The EWM steps are explained below (Mukhametzyanov, 2021).

Step 1. Determine criteria and alternatives.
Criteria are denoted as $C_j; j = 1,\ldots,n$, alternatives are indicated as $A_i, i = 1,\ldots,m$.

Step 2. Form the initial decision matrix.
Initial decision matrix presented as [*X*], shows the performance values of alternatives for the criteria and the element of [*X*] is indicated as x_{i_j}. and resents the *ith* alternative's performance value for the *jth* criterion for $(x_{i_j}, i = 1,\ldots,m; j = 1,\ldots,n.)$

Step 3. Normalize the initial decision matrix.
[*X*] is normalized by using Equations (14.1) and (14.2) given below considering benefit and cost type criteria separately. In this way, x_{i_j} values are transformed into [0,1] and the normalized initial decision matrix indicated as [*N*] is structured. The element of [*N*] is presented as x_{i_j}. It is the normalized performance value of the *ith* alternative for the *jth* criterion $(x_{i_j}^*; i = 1,\ldots,m; j = 1,\ldots,n)$.

For benefit-type criteria,

$$x_{i_j}^{*} = \frac{x_{i_j} - x_j^{\min}}{x_j^{\max} - x_j^{\min}} \tag{14.1}$$

For cost-type criteria,

$$x_{i_j}^{*} = \frac{x_j^{\max} - x_{i_j}}{x_j^{\max} - x_j^{\min}} \tag{14.2}$$

where $x_j^{\max} = \max_i(x_{i_j})$ and $x_j^{\min} = \min_i(x_{i_j})$

Step 4. Determine entropy values for criteria.
In this step, the intensity values of criteria for alternatives shown as $p_{i_j}, i = 1,\ldots,m; j = 1,\ldots,n$ is computed first by Equation (14.3).

$$p_{i_j} = \frac{x_{i_j}^{*}}{\sum_{i=1}^{m} x_{i_j}^{*}}; \sum_{i=1}^{m} p_{i_j} = 1 \tag{14.3}$$

Second, entropy values for each criterion denoted as e_j are calculated as in Equation (14.4).

$$e_j = -\frac{1}{\ln m}\sum_{i=1}^{m} p_{i_j} \ln p_{i_j} \tag{14.4}$$

If $p_{i_j} = 0 \Rightarrow p_{i_j} \ln p_{i_j} = 0$. e_j values show the uncertainty level for each criterion. If the values of the criteria vary according to the alternatives, the entropy values will also be low because the level of uncertainty will be low. According to the EWM, it is understood that the criteria in question have a distinctive feature in the selection of alternatives. If the values taken by the criteria according to the alternatives do not vary or are close to each other, the uncertainty in the values of the related criteria is high. Therefore, the entropy values will also be high. In terms of method, the evaluation is that such criteria have not got a distinctive feature in the selection of alternatives.

Third, the value away from entropy is named as the key indicator indicated as q_j and computed as in Equation (14.5).

$$q_j = 1 - e_j \tag{14.5}$$

Step 5. Calculate criteria weights.

To compute the criteria weights indicated as w_j, q_j values are considered as in Equation (14.6).

$$w_j = \frac{q_j}{\sum_{j=1}^{n} q_j} \tag{14.6}$$

where w_j is weight of *jth* criterion. According to EWM, if a criterion has a lower e_j value, this criterion should have more importance than the other criteria.

14.4 APPLICATION

During the COVID-19 pandemic, hospitals faced many challenges in patient care and staff management, and the level of preparedness of hospitals became an important issue in the process. For this reason, measuring preparedness levels and defining aspects that should be improved are crucial to hospitals in overcoming the pandemic. In this study, EWM was performed to determine the most important preparedness criterion. According to this criterion, five public hospitals were evaluated in terms of their weaknesses. The implementation steps of EWM for the readiness assessment are given below.

Step 1. Determine criteria and alternatives.

It was determined as a result of the literature review that the capacity of human resources such as doctors and nurses (Gul and Yucesan, 2021; Hamal et al., 2020; Kaito et al., 2021; Kazzaz et al., 2020); number of beds; number of ICU beds; number of HFNC devices and number of ventilators (Blavin and Arnos, 2020; El-Ghitany, 2020; Brown et al., 2020; Hamal et al., 2020; Kaito et al., 2021; Kazzaz et al., 2020); and number of laboratories and oxygen capacity (Gul and Yucesan, 2021; Kazzaz et al., 2020; Ogoina et al., 2021; Sanders et al., 2021) are important in determining the preparedness level of hospitals for COVID-19. In this application, six related criteria C_j, $j = 1,\ldots,7$ were considered by reviewing the literature and consulting health experts. The number of doctors, nurses, ICU beds, service beds, and laboratories, and oxygen capacity and number of HFNC devices were denoted as C_1, C_2, C_3, $C_{4,}$ $C_{5,}$ C_6, and C_7 respectively. After the criteria were determined, the five largest public hospitals in different cities $\left(A_j, i = 1,\ldots,5\right)$ were evaluated according to these criteria to form alternatives. Information on the human resources, medical equipment, and critical unit capacities of these hospitals are given in Table 14.1.

Step 2. Form the initial decision matrix.

Firstly, $[X]$ consisting of criteria values for five hospitals was created as shown Table 14.1.

In the evaluation of hospital preparedness level, all criteria considered were of a benefit type. If the hospital preparedness level for COVID-19 is to increase, criteria considered in the analysis should have higher values. For example, if the number of beds of any hospital is high, more COVID-19 patients can have treatment.

Step 3. Normalize the initial decision matrix.
Since all criteria were benefit type criteria, [X] was normalized as in Equation (14.1) by using the $x_j^{\min}$ and $x_j^{\max}$ values depicted in Table 14.1. In this way, [N] was obtained as shown in Table 14.2.

Step 4. Determine entropy values for criteria.
In this step, the p_{ij} values presented in Table 14.3 were first calculated with Equation (14.3) with the help of $\sum_{i=1}^{m} x_{i_j}$ values in Table 14.2.

Then, the e_j value of each criterion was determined using Equation (14.4) considering p_{ij} values; additionally, q_j values were computed using the determined entropy values as in Equation (14.5). Entropy and key indicator values for each criterion are shown in Table 14.4.

TABLE 14.1
Initial Decision Matrix

	C_1	C_2	C_3	C_4	C_5	C_6	C_7
A_1	203	1102	145	870	2	423	27
A_2	301	1203	207	1242	3	580	30
A_3	153	850	122	854	1	351	20
A_4	351	1403	220	1320	4	708	35
A_5	170	952	156	936	2	309	25
$x_j^{\min}$	153	850	122	854	1	309	20
$x_j^{\max}$	351	1403	220	1320	4	708	35

TABLE 14.2
Normalized Initial Decision Matrix

	C_1	C_2	C_3	C_4	C_5	C_6	C_7
A_1	0.253	0.456	0.235	0.034	0.333	0.286	0.467
A_2	0.747	0.638	0.867	0.833	0.667	0.679	0.667
A_3	0.000	0.000	0.000	0.000	0.000	0.105	0.000
A_4	1.000	1.000	1.000	1.000	1.000	1.000	1.000
A_5	0.086	0.184	0.347	0.176	0.333	0.000	0.333
$\sum_{i=1}^{m} x_{i_j}$	2.086	2.278	2.449	2.043	2.333	2.070	2.467

TABLE 14.3
Intensity Values of Criteria

	C_1	C_2	C_3	C_4	C_5	C_6	C_7
A_1	0.121	0.200	0.096	0.017	0.143	0.138	0.189
A_2	0.358	0.280	0.354	0.408	0.286	0.328	0.270
A_3	0.000	0.000	0.000	0.000	0.000	0.051	0.000
A_4	0.479	0.439	0.408	0.489	0.429	0.483	0.405
A_5	0.041	0.081	0.142	0.086	0.143	0.000	0.135

TABLE 14.4
Entropy and Key Indicator Values of Criteria

	C_1	C_2	C_3	C_4	C_5	C_6	C_7	
e_j	0.668	0.773	0.767	0.618	0.793	0.710	0.811	
q_j	0.312	0.227	0.233	0.382	0.207	0.290	0.189	$\sum_{j-1}^{n} q_j = 1.840$

TABLE 14.5
Criteria Weights

	C_1	C_2	C_3	C_4	C_5	C_6	C_7
w_j	0.170	0.124	0.126	0.207	0.112	0.158	0.103
Ranking	2	5	4	1	6	3	7

Step 5. Calculate criteria weights.

To calculate the w_j values given in Table 14.5, Equation (14.6) was implemented using the $\sum_{j=1}^{n} q_j$ value. These weights were only determined for hospitals participating in the study. Different weights could be obtained according to the different hospitals. This study was a pilot study on how to determine the importance weights of the criteria affecting the hospital readiness level. The study can be expanded by increasing the number of hospitals included.

The most important criterion affecting the level of preparedness of hospitals during the COVID-19 period was determined to be the number of service beds (C_4), as seen in Table 14.5. The second and third most important criteria were identified as the number of doctors (C_1) and capacity of oxygen (C_6). Additionally, the least important criterion was the number of HFNC devices (C_7). The performance values of the hospitals differ in regard to the number of service beds, the number of doctors, and capacity of oxygen, and therefore their importance weights should be higher than the other criteria. These criteria should be considered in ranking the hospitals because such criteria facilitate the evaluation of hospitals. Since the performance values for the number of HFNC devices are similar to each other, the importance weight determined by EWM was lower than the other criteria. For this reason, this criterion does not have a distinctive feature in the ranking of hospitals' preparedness for the pandemic, and it can be said that the performance values of the hospitals for this criterion include uncertainty. For an effective fight against the COVID-19 pandemic, it would be appropriate for five hospitals to evaluate their level of preparedness primarily according to number of service beds, the number of doctors, and oxygen capacity. Five hospitals considered in this study will be able to determine aspects that need to be improved in terms of raising the level of preparedness, according to the calculated importance weights of the criteria, considering patient demand.

14.5 RESULTS AND DISCUSSION

The COVID-19 virus has caused serious discomfort to most people, and therefore the number of people treated in hospitals during the pandemic has increased day by day. However, many people could not be treated in hospitals because there was a lack of bed capacity. In addition, an insufficient number of doctors increased the workload and shift hours of HCWs. Patients suffering from shortness of breath due to insufficient oxygen capacity in hospitals had a hard time and therefore hospitals tried to increase their oxygen capacity. Similarly, it was determined that the hospitals analyzed in this pilot study must first provide adequate service beds, doctors, and oxygen capacity so that patients can receive the necessary treatment and not lose their lives.

These results should also be considered in increasing the preparedness level for ananother pandemic. This study could be improved by considering more criteria or using different MCDM methods. In addition, the preparedness levels of other hospitals in the public and private sectors during pandemics can be compared.

REFERENCES

Blavin, F. and Arnos, D. 2020. Hospital preparedness for COVID-19. Robert Wood Johnson Foundation. www.rwjf.org/en/library/research/2020/03/hospital-readiness-for-COVID 19-analysis-of-bed-capacity-and-how-it-varies-across-the-country.html (accessed October 27,2021).

Brown, J., Guru, S., Williams, K. et al. 2020. Rural healthcare center preparation and preparedness response to threat of COVID-19. *Journal of the American College of Surgeons* 230(6):1105–1110.

Chobdar, K. M. and Rahdar, M. A. 2020. Investigating the preparedness of hospitals in Sistan and Baluchestan province in crisis of COVID-19. *Journal of Military Medicine* 22(6):553–561.

El-Ghitany, E. 2020. A short-term forecast scenario for COVID-19 epidemic and allocated hospital preparedness in Egypt. Preprints (doi: 10.20944/preprints202004.0473.v1).

Gul, M. and Yucesan, M. 2021. Hospital preparedness assessment against COVID-19 pandemic: A case study in Turkish tertiary healthcare services. *Mathematical Problems in Engineering* https://doi.org/10.1155/2021/2931219.

Hosseini, S. H., Saleh Tabari, Y., Assadi, T. et al. 2021. Hospitals preparedness in response to COVID-19 pandemic in Mazandaran province, Iran 2020. *Journal of Mazandaran University of Medical Sciences* 31(196):71–81.

Hamal, P. K., Pokhrel, N., Pandey, D. et al. 2020. Perspective of doctors for COVID-19 pandemic preparedness in Government Hospitals of Nepal. *Journal of the Nepal Health Research Council* 18:166–71.

Haghani, M., Bliemer, M. C., Goerlandt, F., and Li, J. 2020. The scientific literature on Coronaviruses, COVID-19 and its associated safety-related research dimensions: a scientometric analysis and scoping review. *Safety Science*, 129. doi: 10.1016/j.ssci.2020.104806

Kaito, D., Matsumura, K., and Yamamoto, R. 2021. Hospital preparedness for COVID-19: the known and the known unknown. *The Keio Journal of Medicine* 70(2):25–34.

Kazzaz, Y. M., Alkhalaf, H., Alharbi, M. et al. 2020. Hospital preparedness and management of pediatric population during COVID-19 outbreak. *Annals of Thoracic Medicine* 15(3):107–117.

Lailani, D., Rivai, F., Pasinringi, S. A. et al. 2021. The preparedness of the Pupuk Kaltim Bontang hospital in facing the COVID-19 pandemic. *Turkish Journal of Physiotherapy and Rehabilitation* 15: 8108–8113.

Labib, J. R., Kamal, S., Salem, M. R. et al. 2020. Hospital preparedness for critical care during COVID-19 pandemic: exploratory cross-sectional study. *Open Access Macedonian Journal of Medical Sciences* 8(T1):429–432.

Mukhametzyanov I.Z. 2021. Specific character of objective methods for determining weights of criteria in MCDM problems: Entropy, CRITIC, SD. *Decision Making: Applications in Management and Engineering* 4(2):76–105.

Ogoina, D., Mahmood, D., Oyeyemi, A. S. et al. (2021). A national survey of hospital preparedness during the COVID-19 pandemic in Nigeria. *PLoS One* 16(9), e0257567.

Saravi, R.H., Saleh Tabari, Y., Hosseini, S. H. et al. 2021. Evaluation of operational preparedness of pre-hospital emergency department of Mazandaran University of Medical Sciences in the face of COVID-19 epidemic in 2020. *Health in Emergencies and Disasters* 6:217–223.

Shrivastava, S. R. and Shrivastava, P. S. 2020. Coronavirus disease 2019 pandemic: Improving hospital preparedness to meet the rising health-related needs. *International Journal of Academic Medicine* 6(2):54–55.

Shrestha, A., Rajbhandari, P., and Bajracharya, S. 2020. Hospital preparedness for outbreak at Patan hospital: lesson learnt from COVID-19. *Journal of Nepal Health Research Council* 18(1):142–143.

Sanders, J. E., Chakare, T., Mapota-Masoabi, L. et al. 2021. National hospital preparedness for COVID-19 in Lesotho: evidence for oxygen ecosystem strengthening. www.medrxiv.org/content/10.1101/2021.04.27.21256199v1 (accessed October 27, 2021).

Wu, J., Sun, J., Liang, L., and Zha, Y. 2011. Determination of weights for ultimate cross efficiency using Shannon entropy. *Expert Systems with Applications* 38(5):5162–5165.

15 Prioritizing Post-Disaster Debris for Recycling

Özer Eroglu[1] *and Cevriye Gencer*[2]

[1]Turkish Land Forces, Erzincan, Türkiye

[2]Gazi University,Ankara, Türkiye

CONTENTS

15.1 Introduction 223
15.2 A Brief Literature Review 224
15.3 The SWARA Method 225
15.4 The CoCoSo Method 226
15.5 Case Study 228
15.6 Conclusions 232
References 233

15.1 INTRODUCTION

Disasters are serious events occurring over short or long periods, and directly affect human life and economies. After every disaster, many deaths, epidemics, and economic depressions occur. Naturally, humans have gained experience after every disaster and have tried to develop disaster-management plans. Humans, who are vulnerable to natural phenomena, have developed methods to forecast disasters. In addition, we have also tried to develop action plans for during and after disasters. In order to develop and implement these plans, many counties have made very high-cost investments. Depending on the size of the disaster and the environment, infrastructure may be damaged, the amount and type of product demand may change, production may stop, and epidemics may occur. After search, rescue, and temporary accommodation activities, restructuring and normalization of life emerge as a big problem area. The biggest burden of this phase on countries' budgets is the removal and disposal of debris, and the construction of new buildings. For example, the removal, transportation, recycling, and disposal of approximately 900 000 cubic yards of vegetal debris/waste that occurred after Hurricane Isabel costs \$8 million alone, and the removal of the wreckage that occurred during Hurricane Katrina resulted in a cost of approximately \$4.4 million (Fetter and Rakes, 2012). The direct and indirect burden of the İstanbul Earthquake in 1999 on the Turkish Economy is around \$21 million (Avdar, 2017).

In the reconstruction phase, some studies have shown that disaster debris can be reused by recycling, in order to reduce costs. It is known that 80% of 23 million tons of debris was recycled after the Japan Earthquake in 2011 and regained in the

DOI: 10.1201/9781003212904-15

economy. It was stated that household wastes (white goods, electronic wastes, etc.), hand tools, vehicle scraps, materials obtained from brick and building wreckage, and wooden materials were recycled and reused (Brown and Milke, 2016).

Although the disposal of post-disaster debris/waste and their contribution to the economy are within the scope of waste management, post-disaster debris has been evaluated within the scope of disaster management due to the amount of debris/waste generated after the disaster, the need for quick debris removal, and the need for life to return to normal within a very short time. Therefore, the recycling of some raw materials needed in the post-disaster normalization process and its contribution to the economy with regard to long-term gains is important. Considering debris as a resource will be beneficial both in the restructuring process and disposal of debris/waste. Studies and plans for the disposal of debris/waste before and after the disaster (waste estimation, disposal method, bringing it into the economy, etc.) should be evaluated within the scope of disaster-management studies.

In this study, the recycling of debris after an earthquake will be discussed. The study was conducted within the framework of expert evaluations in İstanbul, Turkey. Alternatives are determined as debris classes and criteria are determined as factors affecting the utilization priorities of debris in recycling. Implementation was carried out by determining the importance of the criteria by an expert group. Due to the fact that the decision process consisted of decision-makers, criteria, and alternatives, this problem is considered to be a multi-criteria decision-making problem (MCDM).

In this study, the stepwise weight assessment ratio analysis (SWARA) method, which can be used in studies where experts from various disciplines are together and there is a low mathematical processing load, was preferred to determine the criteria weight. The weight values obtained from SWARA were integrated into the combined compromise solution (CoCoSo) method, which is a compendium of compromise solutions. Using the CoCoSo method, utility values can be determined by different addition and aggregation operators. The aggregation function is used to provide a compromise solution and aggregate the utility values of each decision alternative. It is a combination of the simple additive weighting (SAW) and exponentially weighted product (EWP) methods. Therefore, it is presented to users as an easy-to-use sorting method consisting of simple mathematical operations.

This study is organized as follows: a brief literature review about the debris and the processes performed after the disasters in Section 15.2; an explanation of the methods in Sections 15.3 and 15.4; a case study in Section 15.5; and the conclusions in Section 15.6.

15.2 A BRIEF LITERATURE REVIEW

The Emergency Events Database (EM-DAT) was created with the initial support of the World Health Organization (WHO) and the Belgian Government, and classifies disasters into two main groups. The first group includes natural disasters: geophysical (earthquake, mass movement, volcanic activity), meteorological (extreme temperature, fog, storm), hydrological (flood, landslide, wave action), climatological (drought, glacial lake outburst, wildfire), biological (epidemic, insect infestation, animal accident), and extraterrestrial (impact, space weather). The second group includes technological disasters, that is, industrial transport and miscellaneous accidents (EM-DAT, 2021). There are many studies of post-disaster recycling advices and post-disaster waste-management benefits. These benefits are determined as landfill space reduction, raw material demand reduction, waste-management cost reduction, and job creation (Gulledge, 1995; Kobayashi, 1995; Jones, 1996; Baycan and Peterson, 2002; Baycan, 2004; Basnayake et al., 2005; Brown et al. 2011a; Brown et al. 2011b, Brown and Milke, 2016). In some studies, the waste generated during disasters have generally been examined in three stages. These stages include waste that requires urgent intervention, recyclable waste, and the rebuilding phase. These three stages are not independent of each other. However, the duration of these stages is different in each disaster. Generally, in waste management, in terms of public health and safety, the emergency phase is defined as a short-term action plan and includes eliminating threats. It usually takes a few days to about two weeks (Baycan and Peterson, 2002; Güler, 2018; Kuramoto, 1995). Brown (2011b) identified post-disaster recycling barriers as time constraints, limited resource availability, the mixed nature of waste, hazards in the waste matrix, displaced population, post-disaster market challenges (capacity, availability, disruption, storage space limitations), contractual arrangements, and the availability and feasibility of alternative waste-management options. Total direct damage caused by hurricanes Florence, Michael, Harvey, Irma, and Maria cost US$300 billion (Howell and Elliott, 2018); the total cost of debris-management activities following Hurricane Katrina (100 million cubic yards) exceeded US$4 billion. Debris-management operations are determined in three phases: pre-disaster (debris event and forecasts, debris procurement planning, debris management sites planning); response (debris estimation, resource requirements assessment, debris clearance); and recovery (debris collection, temporary facility installment, debris disposal, debris recycling) (Lorca et al., 2015). When the literature is examined studies are concentrated in three main groups: precautions taken before and after disasters; the activities carried out before and after disasters; and the impacts on the economy due to disasters. Since the aim of the study was to evaluate the debris generated after the disasters, the literature review was kept short. Although there are many studies on disaster and post-disaster management, there are few on the contribution of post-disaster debris to the restructuring process and are mostly in report form (UNDP, 2015; UNDP, 2017; EPA, 2019). There is no study in the literature regarding determination of the order and prioritization of debris generated after the disaster, which was the main aim of this study. In this study, for the first time, the prioritization of post-disaster debris for reuse was discussed and evaluated as an MCDM problem due to the criteria, decision-makers, and alternatives available.

In the next section, the MCDM methods implemented in the case study are introduced.

15.3 THE SWARA METHOD

The SWARA method, outlined step by step below, has simple mathematical operations, which is preferred when experts cannot agree on the importance of criteria (Keršuliene et al., 2010; Eroğlu and Gencer, 2021).

Step 1 Determining the general list of criteria and expert survey.

Experts define the criteria (c_j) and list the determined criteria from most important to least important. Thus, according to their own experience, lists in order of importance are obtained from experts (n).

Step 2 Calculation of relative scores for all criteria.

Starting from the second criterion, experts compare each criterion (j) with the next criterion ($j+1$) and assign a value (s_j) to the criteria between 0 and 1. A value is not given to the most important criterion.

Step 3 Determination of coefficient (k_j)

By Equation 15.1, a coefficient value can be obtained. The value of "1" is given to the most important criterion, which is at top of the list.

$$k_j = \begin{cases} 1, & j = 1 \\ s_j + 1, & j > 1 \end{cases} \tag{15.1}$$

Step 4 Determination of recalculated weight (v_j).

Recalculated weights (new significance vector) are calculated by using Equation 15.2.

$$v_j = \begin{cases} 1, & j = 1 \\ \dfrac{v_{j-1}}{k_j} & j > 1 \end{cases} \tag{15.2}$$

Step 5 Determining criteria weights (w_j).

Criteria weights are calculated by Equation 15.3.

$$w_j = \frac{v_j}{\sum_{j=1}^{n} v_j} \tag{15.3}$$

15.4 THE COCOSO METHOD

CoCoSo, the steps of which are given below, is a method based on an integrated simple additive weighting and exponentially weighted product model (Yazdani et al., 2019).

Step 1 Obtaining the initial decision matrix.

This matrix, which is shown in Table 15.1, includes the data (x_{ij}) and weight information (w_j) about the criteria (C_j) and the alternatives (A_i).

TABLE 15.1
Initial Decision Matrix

Weights	W_1	W_2	.	.	W_j
Optimal Value	Max	Max	.	.	Min
Criterion	C1	C2	.	.	C_j
A1	X_{11}	X_{12}	.	.	X_{1j}
A2	X_{21}	X_{22}	.	.	X_{2j}
.	.	.	.	.	.
.	.	.	.	.	.
A_i	X_{i1}	X_{i2}			X_{ij}

Step 2 Normalizing decision matrix and obtaining comparability sequence measures (r_{ij}).

Comparability sequence measures can be obtained by using Equation 15.4 for benefit (max.) criteria and Equation 15.5 for cost (min.) criteria.

$$r_{ij} = \frac{x_{ij} - \min x_{ij}}{\max x_{ij} - \min x_{ij}} \tag{15.4}$$

$$r_{ij} = \frac{\max x_{ij} - x_{ij}}{\max x_{ij} - \min x_{ij}} \tag{15.5}$$

Step 3 Obtaining S_i and P_i values

After obtaining comparability sequence measures by using Equations 15.6 and 15.7, the S_i (sum of the weighted comparability sequence for each alternative) and P_i (sum of the exponentially weighted comparability sequence for each alternative) values can be determined respectively.

$$S_i = \sum_{j=1}^{n} \left(w_j r_{ij}\right) \tag{15.6}$$

$$P_i = \sum_{j=1}^{n} (r_{ij})^{w_j} \tag{15.7}$$

Step 4 Determining relative weights.

To determine the relative weights of alternatives, three appraisal score strategies, the mathematical notations of which are given below, can be used. In Equation 15.10, λ is chosen by experts (usually 0.5).

$$k_{ia} = \frac{P_i + S_i}{\sum_{0=1}^{n} \left(P_i + S_i\right)} \tag{15.8}$$

$$k_{ib} = \frac{S_i}{Min\ S_i} + \frac{P_i}{Min\ P_i} \tag{15.9}$$

$$k_{ic} = \frac{\lambda S_i + (1-\lambda) P_i}{\lambda MaxS_i + \lambda MaxP_i}, \quad 0 \le \lambda \le 1 \tag{15.10}$$

Step 5 Final ranks.

The final ranks are calculated using Equation 15.11.

$$k_i = \sqrt[3]{k_{ia} k_{ib} k_{ic}} + \frac{1}{3}\left(k_{ia} + k_{ib} + k_{ic}\right) \tag{15.11}$$

15.5 CASE STUDY

The data used in this case study were obtained from the "Probable Earthquake Loss Estimates of Istanbul Province Updating Project" report, which was prepared in partnership with Istanbul Metropolitan Municipality and Boğaziçi University. This report includes possible building damage, loss of life, and estimates of casualties and infrastructure damage as a result of an Mw 7.5 earthquake in 15 different scenarios (İBB and Deprem ve Zemin Şube Müdürlüğü, 2021).

Istanbul is a metropolitan city with a population of 15 062 677 and 1 166 300 buildings/facilities (masonry structures: 177 904; timber structures: 7 715; prefabricated structures: 2 812; steel structures: 1 846; reinforced concreted structures: 976 023) located on the North Anatolian Fault. According to the Probable Earthquake Loss Estimates of Istanbul Province Updating Project, in the case of an earthquake of 7.5 magnitude, 16.7% of buildings/facilities will be totally damaged and cannot be used to shelter (very badly damaged: 13 495; badly damaged: 34 345; moderately damaged: 146 552), 25.9% will be able to be used as they will only be slightly damaged, and 57.5% will be undamaged. According to the dollar exchange rate in the year the project was carried out in, there will be a loss of $12.4 million due to structural damage; in addition, approximately $21.9 million will be lost due to non-structural damage such as production stoppages or reductions. It was evaluated that there would be 25 million tons of debris (debris can be transported in 1 million trips with a truck capacity of 25 tons). It was estimated that the content of this wastes will comprise approximately 14.4 million tons of reinforced concrete and steel, and 10.5 million tons of brick and wood. Another study also provided information on the volumes of concrete, steel, and bricks suitable for recycling/reuse that could be obtained from 1 215 200 masonry and 484 800 reinforced concrete-type structures (each of which is 100 m^2) and is presented in Table 15.2 (Aktaş, 2015).

In addition, there will be electronic devices, household appliances, and hazardous wastes in debris. Except for hazardous materials, with good separation, these materials can be recycled.

This study was carried out with the opinions of a decision-maker (DM) group of five civil engineers. Three engineers from the decision-making group continue

TABLE 15.2
Initial Decision Matrix

Type	Masonry (Million Tons)	Reinforced Concrete (Million Tons)	Total (Million Tons)
Concrete	72.9	44.2	117.1
Steel	2.7	1.7	4.4
Bricks	21.9	6.5	28.4

to work in the academic field and the other two work in the sector. The competence level of the DM group is good according to their experience. One of the decision makers, who is in the academic field, has many publications on disaster management; one has worked as a volunteer (10 years) in post-disaster assembly zones and their construction, and also was a member of a municipality in İstanbul; and one of the decision-makers had worked as an engineer in Disaster and Emergency Management Presidency (12 years). The two other engineers were previously members of the Ministry of Environment and Urbanization (18–20 years).

First of all, the decision-makers classified the debris into groups (concrete, bricks, woods, plastics, glass, electronic materials, household appliances, and steel and other metals) for reconstruction, recycling, and regaining. These materials constituted the alternatives of the study. Concrete and brick debris can be used for new concrete production, new brick production, pavement and asphalt construction, and as filling materials. Wood, glass, and plastic debris can be used in the production of timber, reprocessing (furniture, etc.), decorative filling materials, insulation materials, and the production of windows/doors and new plastic devices/materials. Household appliances and electronic devices can be reused with selective separation as plastic, metal components, and electronic components. Steel and other metal goods such as steel used in building foundations and structures, water pipes, metals in electrical lines (copper cable, etc.) can be reused. The production of steel from scrap requires 90% less energy than producing it from ore. The decision-makers assumed that an earthquake with a magnitude of 7.5 had occurred and 16.7% (194 392 buildings) of the structures had become completely unusable, as per the "Probable Earthquake Loss Estimates of Istanbul Province Updating Project" report. Based on the assumption that 194 392 buildings collapsed after the possible earthquake, had to be demolished due to severe damage, or needed extensive repairs, it was found that 75% of all waste (25 million tons) could be usable and recycled. In this context, the decision-makers estimated that 7.1 million tons of concrete (Alternative-1), 3.5 million tons of steel (Alternative-2), 4 million tons of brick (Alternative-3), and 3.3 million tons of wood (Alternative-4) would be obtained for reuse after the earthquake. In addition, 10 kg plastics (Alternative-5), 2 kg glass (Alternative-6), five electronic devices (Alternative-7), two items of household goods (Alternative-8) and 3 kg other items (Alternative-9) were considered to be recoverable from debris per building.

Six criteria were identified for the evaluation of the mentioned alternatives. Separability (C-1): materials/raw materials that can be considered to be economically suitable for separation from debris. Transportation (C-2): the distance between the debris zone and the segregation area is important. After the disaster, damage to transportation roads, the shortest/suitable route, etc., situations will likely increase the cost of transportation. These situations should not prevent economic gain from separation. Contribution to the economy (C-3): this criterion expresses the revenue obtained by bringing the materials/raw materials into the economy and reconstruction. Remanufacture/refresh time (C-4): some of the separated materials can be processed for reuse, so this criterion refers to the additional time after the separation of debris. Amount of materials required (C-5): How much material will be needed in the reconstruction process and how much of this need will come from the debris? This criterion corresponds to the amount of materials needed. Costs (C-6): this criterion covers all costs, from the creation of segregation sites to the completion of segregation works and sending them to the place where they are needed.

After these determinations, it was ensured that the decision-makers ranked the criteria in order of importance according to their own thoughts and then the criteria were evaluated (assigning s_j values) by pair-wise comparisons. After this process, all steps of the SWARA method were applied and the weights of each decision-maker's evaluations were determined and are presented in Table 15.3. For the group decision, the final weights were determined by the average of the individual values.

Weights were determined as follows: separability (C1), 0.266; transportation (C2), 0.121; contribution to the economy (C3), 0.284; remanufacture/refresh time (C4), 0.131; amount of materials required (C5), 0.091; costs (C-6), 0.107.

TABLE 15.3
SWARA Method Calculations

	DM-1					DM-2					DM-3			
	Sj	Kj	Vj	Wj		Sj	Kj	Vj	Wj		Sj	Kj	Vj	Wj
C3		1.000	1.000	0.343	C1		1.000	1.000	0.343	C1		1.000	1.000	0.299
C1	0.700	1.700	0.588	0.202	C3	0.350	1.350	0.741	0.254	C3	0.300	1.300	0.769	0.230
C2	0.200	1.200	0.490	0.168	C4	0.450	1.450	0.511	0.175	C2	0.300	1.300	0.592	0.177
C4	0.400	1.400	0.350	0.120	C6	0.500	1.500	0.341	0.117	C4	0.250	1.250	0.473	0.141
C6	0.200	1.200	0.292	0.100	C5	0.700	1.700	0.200	0.069	C6	0.450	1.450	0.326	0.098
C5	0.500	1.500	0.195	0.067	C2	0.650	1.650	0.121	0.042	C5	0.750	1.750	0.187	0.056

	DM-4					DM-5			
	Sj	Kj	Vj	Wj		Sj	Kj	Vj	Wj
C3		1.000	1.000	0.356	C1		1.000	1.000	0.249
C1	0.500	1.500	0.667	0.237	C3	0.050	1.050	0.952	0.237
C2	0.600	1.600	0.417	0.148	C5	0.200	1.200	0.794	0.198
C4	0.200	1.200	0.347	0.124	C6	0.300	1.300	0.611	0.152
C6	0.800	1.800	0.193	0.069	C4	0.600	1.600	0.382	0.095
C5	0.050	1.050	0.184	0.065	C2	0.400	1.400	0.273	0.068

According to the expert opinions, the weights of separability (C1) and contribution to the economy (C3) were similar. The reason for this may be due to the fact that all the experts were civil engineers.

After determining the weights, the data obtained from the Union of Chambers of Turkish Engineers and Architects about alternatives, which is the initial CoCoSo method information/table, are presented in Table 15.4.

After implementing the CoCoSo method, the priority order of alternatives was as presented in Table 15.5.

When the findings are examined, it was determined that concrete, brick, and steel materials have priority and can be separated from debris for use in the reconstruction phase. In other words, it would be appropriate to sort these materials first and it would be a beneficial gain to the economy after a possible Istanbul earthquake.

TABLE 15.4
Information/Initial Table of CoCoSo Method

Criterion	C1(tons)	C2 (unit/m3)	C3 (unit/$)	C4 (unit/ months)	C5 (tons)	C6 (unit/$)
Alternatives	Max	Min	Max	Min	Max	Min
Weight	0.266	0.121	0.284	0.131	0.091	0.107
Concrete	7.10	6.33	2.540	0.00	7.00	3.00
Steel	3.50	5.90	3.600	0.10	12.00	4.00
Brick	4.00	6.00	1.727	0.00	12.00	1.00
Wood	3.30	2.50	1.100	1.00	9.00	0.85
Plastics	10.00	8.90	1.560	2.20	1.00	1.80
Glass	2.00	7.00	1.000	3.00	1.00	5.00
Electronic devices	5.00	4.00	1.77e8	6.00	1.00	3.00
Household goods	2.00	4.00	1.234	6.00	0.90	4.00
Others	3.00	3.50	900	7.00	0.80	1.00

TABLE 15.5
The Priority Order of Alternatives

	k_a	Ranks	k_b	Ranks	k_c	Ranks	k	Final Ranks
Concrete	0.143	1	5.566	1	0.926	1	3.115	1
Steel	0.131	3	4.840	2	0.850	3	2.682	3
Brick	0.122	4	4.653	3	0.791	4	2.668	2
Wood	0.121	7	4.316	5	0.781	7	2.424	4
Plastics	0.118	2	4.145	4	0.766	2	2.411	5
Glass	0.110	5	4.025	7	0.714	5	2.367	6
Electronic devices	0.096	8	3.040	8	0.623	8	1.820	7
Household goods	0.094	6	2.846	6	0.606	6	1.726	8
Others	0.064	9	2.294	9	0.415	9	1.318	9

TABLE 15.6
Sensitivity Analysis

	k	Original Ranks	k	Analysis-1 Ranks	k % 30	k % 40	k % 50	k % 60	Analysis-2 Ranks
Concrete	3.115	1	3.099	1	2.965	3.157	3.255	3.265	1
Steel	2.682	3	2.795	3	2.756	2.862	2.850	2.856	2
Brick	2.668	2	2.851	2	2.738	2.729	2.714	2.715	3
Wood	2.424	4	2.521	5	2.515	2.518	2.522	2.527	5
Plastics	2.411	5	2.523	4	2.412	2.529	2.531	2.532	4
Glass	2.367	6	2.367	6	2.358	2.359	2.360	2.361	6
Electronic devices	1.820	7	1.829	8	2.187	2.190	2.197	2.201	7
Household goods	1.726	8	2.005	7	1.605	1.890	1.918	2.005	8
Others	1.318	9	1.219	9	1.320	1.322	1.325	1.326	9

To increase the reliability and robustness of the study, sensitivity analysis of the change in weight values was done. This was first carried out by evaluating criteria weights as "interval" (± % 10 changes in weight parameters), and the results were very close to the results of the original solution.

Another sensitivity analysis was implemented by clustering the criteria as cost and benefit (changing the criteria weight parameters). Separability (C1), transportation (C2), and costs (C-6) were accepted as "cost"; contribution to the economy (C3), remanufacture/refresh time (C4), and amount of materials required (C5), were accepted as "benefit" criteria. In this analysis, weights of 30%, 40%, 50%, and 60%, respectively were divided equally for the criteria determined in the "cost" cluster; the remaining ratio was equally divided in the "benefit" cluster. The determined results were consistent with the original results. In both sensitivity analyses (changes of weights and changes by clustering), the results changed, but the sorting did not change in the first three alternatives. Overall assessment of the sensitivity analysis and original results are provided in Table 15.6.

15.6 CONCLUSIONS

Disasters end lives, bring economies and education to a halt, and change lifestyles and even living spaces, depending on their size. Although there are activities to be done beforehand in dealing with disasters, it is very important to normalize life afterwards. Many activities take place before disasters. These are primarily technical arrangements to ensure that building/facility structures incur minimal or no damage, if possible, and to train people in how to protect their lives during disasters. After major disasters, life often comes to a standstill. Particularly in the first stage, precautions are taken to save lives and to allow people to temporarily continue their lives. Production may stop at this stage, and aid materials may not be sent for various reasons. However, after the first stage, the disaster site or a new location must be reconstructed. In this study, the debris formed after an earthquake in İstanbul was evaluated by MCDM methods.

The study concluded that the possible debris after an earthquake could be used for reconstruction. In the solution, experts decided that materials such as concrete, brick, and steel could be used after separation/recycling from the debris during the reconstruction phase. According to the data of the "Probable Earthquake Loss Estimates of Istanbul Province Updating Project", there would be 25 million tons of debris after an Mw 7.5 earthquake, and all activities after an earthquake would cost $21.9 million. A saving of approximately $10 million could be achieved by recycling and reusing the debris. To summarize, it is necessary to be prepared for the aftermath of a disaster, as well as to prepare for them occurring in the first place.

In the framework of the criteria weighted with expert opinions, it was found that the recycling of materials obtained from debris will both contribute to the economy and accelerate the post-disaster reconstruction process. As a result, it would be beneficial to add debris/recycling management to post-disaster planning activities. In this study, it was determined that, within the scope of the decision-makers' preferences and available data, concrete, steel, and bricks should be separated and brought into the economy. This study can be extended for future studies, renewed with other methods, or re-evaluated with mathematical models.

REFERENCES

Avdar, R. 2017. 1999–2011 between period arriving economic effects of earthquakes in Turkey. *Econder International Academic Journal* 1:1, 53–63.

Aktaş, Z. 2015. Bina Yıkım Atıklarının Altyapı Projelerinde Değerlendirilmesi, M.S. Thesis, İlbank A.Ş., Turkey.

Basnayake, B.F.A., Chiemchaisri, C., and Mowjood, M.I.M. 2005. Solid wastes arise from the Asian Tsunami disaster and their rehabilitation activities: case study of affected coastal belts in Sri Lanka and Thailand. In: *Tenth International Waste Management and Landfill Symposium*, Sardinia.

Baycan, F. and Petersen, M. 2002. Disaster waste management—C&D waste. In: *Annual Conference of the International Solid Waste Association*, Istanbul, Turkey.

Baycan, F. 2004. Emergency planning for disaster waste: a proposal based on the experience of the Marmara Earthquake in Turkey. In: *International Conference and Student Competition on post-disaster reconstruction* "Planning for reconstruction", Coventry, UK.

Brown, C., Milke, M. and Seville, E. 2011a. Disaster waste management following the 2009 Victorian Bushfires. *Journal of Emergency Management* 26: 2, 17–22.

Brown, C., Milke, M., and Seville, E. 2011b. Disaster waste management: a review article. *Waste Management* 6: 1085–1098.

Brown, C. and Milke, M. 2016. Recycling disaster waste: Feasibility, method and effectiveness. *Resources, Conservation and Recycling* 106: 21–32.

EM-DAT. 2021. Centre for Research on the Epidemiology of Disasters, EM-DAT (021, General Classification, www.emdat.be/classification (accessed December 7, 2021).

Eroğlu, Ö. and Gencer, C. 2021. Classification on SWARA method and an application with SMAA-2. *Journal of Polytechnic* 24: 1707–1718.

EPA. 2019. Planning for natural disaster debris. United States Environmental Protection Service, www.epa.gov/homeland-security-waste/guidance-about-planning-natural-disaster-debris (accessed December 7, 2021).

Fetter, G. and Rakes, T. 2012. Incorporating recycling into post-disaster debris disposal. *Socio-Economic Planning Sciences* 46: 1, 14–22.

Gulledge, J. H. 1995. Calabasa landfill earthquake debris recovery programme. In *Earthquake Waste Symposium* (pp. 12–13).

Güler, Ü. A. 2018. Sürdürülebilir afet yönetiminde atık yönetimi. *Doğal Afetler ve Çevre Dergisi* 4: 2, 236–246.

Howell, J. and Elliott, J. R. 2018. As disaster costs rise, so does inequality. *Socius* 4: 2378023118816795.

İBB and Deprem ve Zemin Şube Müdürlüğü. 2021. İstanbul İli Olası Deprem Kayıp Tahminlerinin Güncellenmesi Projesi. Deprem ve Zemin İnceleme Müdürlüğü. https://depremzemin.ibb.istanbul/calismalarimiz/tamamlanmis-calismalar/istanbul-ili-olasi-deprem-kayip-tahminlerinin-guncellenmesi-projesi/ (accessed December, 16 2021).

Jones, D. 1996. Beirut, capital of Lebanon: waste management after the war. *Warmer Bulletin* 2: 140–161.

Keršuliene, V., Zavadskas, E. K., and Turskis, Z. 2010. Selection of rational dispute resolution method by applying new step-wise weight assessment ratio analysis (SWARA). *Journal of Business Economics and Management* 11: 22, 243–258.

Kobayashi, Y. 1995. Disasters and the problems of wastes—institutions in Japan and issues raised by the Great Hanshin-Awaji earthquake. In *Earthquake Waste Symposium*, 12–13 June, Osaka.

Kuramoto N. 1995. The actual state of damage and measures under taken in Hyogo prefecture. In: *Earthquake Waste Symposium*, 12–13 June, Osaka.

Lorca, Á., Çelik, M., Ergun, Ö., and Keskinocak, P. 2015. A decision-support tool for post-disaster debris operations. *Procedia Engineering* 107: 154–167.

UNDP. 2015. Signature product: guidance note on debris management. https://www.undp.org/publications/signature-product-guidance-note-debris-management (accessed December 17, 2021).

UNDP. 2017. Haiti technical guide for debris management. https://www.undp.org/publications/haiti-technical-guide-debris-management (accessed December 17, 2021).

Yazdani, M., Zarate, P., Zavadskas, E. K., and Turskis, Z. 2019. A Combined Compromise Solution (CoCoSo) method for multi-criteria decision-making problems *Management Decision* 57: 9.

16 A GIS-Based Multi-Criteria Decision Analysis Framework for Evaluation of Emergency Assembly Points

Ahmet Bahadır Şimşek
University of Gümüşhane

CONTENTS

16.1 Introduction 235
16.2 Decision Analysis Framework for Evaluation of EAPs 237
16.2.1 GIS Overview 238
16.2.2 CoCoSo Method 239
16.3 Implementation of the Proposed Decision Analysis Framework 240
16.4 Conclusions 246
References 248

16.1 INTRODUCTION

Urbanization increases disaster risk as it causes the population density to be concentrated in a small area. In parallel with the world urbanization rate, which was 29.6% in 1950, 56.15% in 2020, and expected to be 68.4% in 2050, an increasing number of people are exposed to disaster risk every year (United Nations 2020; World Bank 2020). Urbanization and disaster risk have been the subject of many studies in worldwide, and these studies have emphasized that another side effect of increasing urbanization that disaster operations have become more complex (Klose 2011; Zhang, Ma, and Wang 2008; He et al. 2016). Oden et al. (2012) classified disaster operation challenges into four categories: coordination, communication, information, and logistics. To alleviate these challenges, the approach of gathering in predetermined places is used effectively, especially in disasters that force people to leave their homes. Although gathering places are determined under certain assumptions within the possibilities, for some reasons they may not fulfill their intended functions when needed. To minimize the difficulties and take the necessary precautions, it is important to evaluate the suitability of gathering places.

DOI: 10.1201/9781003212904-16

Disaster management has three concepts based on gathering people at certain places: i) emergency assembly points (EAPs); ii) temporary shelters; and iii) permanent shelters. This chapter focuses on EAPs and compares them as follows. The infrastructure requirement of EAPs is low as they are generally planned to be used for less than a day, while others are constructed from tents or containers for long-term use and have high infrastructure requirements. Aside from the differences in use, a key feature that distinguishes EAPs from others is that *location* and *capacity* decisions cannot be changed right after the event. People are expected to gather in EAPs at the start of an the event, so there is no way to change the locations of the EAPs. The same is true for the capacity decision. Since open areas surrounded by structures such as parks in urban settlements are designated as EAPs, it is not possible to expand these areas if the capacity is insufficient. Given post-disaster timelines, EAPs save time by reducing the requirement for the construction of long-term shelters. In this way, even if a decision has been made about the location and capacity of long-term shelters, it can be changed according needs. The noted differences distinguish EAPs from others, and underline the importance of decisions regarding the determination of suitable EAPs. An additional motivation that requires assessment of EAP compliance is that EAPs that fail to fulfill their expected roles cause people to relocate rather than wait for rescue teams, thus disrupting disaster operations. Knowing the suitability scores of EAPs can play an active role in determining intervention policies.

The reason why not every EAP is suitable can be explained as follows. In a recent study, Savaş et al. (2021) presented a set of criteria to establish standards for ideal EAPs as follows: capacity; position/accessibility; size; connection; function; infrastructure; support units; ownership. For example, an ideal EAP should be larger than 500 m^2, and, at most, 500 m or 15 minutes' walking distance for people. Accordingly, new settlements will have ideal EAPs if open areas are blueprinted by considering relevant criteria in the master plan. However, it can be difficult to find open areas that satisfy the standards in older settlements. For this reason, authorities have to determine the available open areas for EAPs, even if they do not comply with the standards. Although standards exist for ideal EAPs, concerns about whether open areas satisfy standards motivate evaluation of the suitability of an EAP.

Recent studies on the identification and evaluation of EAPs highlight the attraction of the subject. Savaş et al. (2021) proposed a multi-criteria decision analysis formulated with the geographic information system (GIS)-based analytic hierarchy process method to determine the most suitable areas for EAPs. The proposed approach envisaged the evaluation of existing EAPs in terms of location, capacity, resources, victim allocation, and evacuation principles. It also covered the elimination of unsuitable EAPs, and then the identification of new EAPs if the remaining EAPs were not sufficient for the region. Şenik & Uzun (2021) evaluated emergency assembly points in Düzce, Turkey, which has a history of disasters, in terms of size, accessibility, and other site selection criteria. In the study, besides determining whether the EAPs were sufficient or not, EAPs were categorized as very suitable, suitable, or unsuitable. While data production was provided by GIS, the suitability of EAPs was determined by scoring based on criteria. Chrysoulidis (2019) focused on identifying evacuation assembly points for the Latsia region of Cyprus. In the first stage of the study, the Corine land cover, hydrological map, and slope maps of the region were classified as suitable and unsuitable in line with the criteria in the literature, and the suitable areas throughout the region were determined by overlapping the maps via GIS. In the second stage, suitable areas were evaluated in terms of accessibility and road network, and sufficient areas were selected according to the population of the region.

The studies mentioned focused on either identifying new suitable EAPs or identifying unsuitable EAPs for replacement. However, knowledge of the relative suitability score is significant, and not just whether the EAPs are suitable. This information can provide notable benefits to decision-makers when determining response or evacuation policy on many issues such as guiding teams, allocating resources, and specifying evacuation priority. The suitability of EAPs has not previously been investigated in terms of the managerial insight they will provide to decision-makers.

This chapter presents a multi-criteria decision analysis framework based on GIS for the relative evaluation of existing EAPs. It consists of three customizable stages: i) digitizing the urban area and generating data; ii) evaluating the EAPs; and iii) visualizing the results. GIS was used in the first and third stages because the evaluation of EAPs requires the processing of georeferenced data such as size and distance to buildings. The fact that the evaluation criteria contradict each other necessitates the search for a compromise. In the evaluation of EAPs, poor performance in any criterion cannot be tolerated with superior performance in other criteria. For example, an open area with a small size cannot gain a high suitability score, regardless of its performance in other criteria. To reflect this understanding in the evaluation, the combined compromise solution (CoCoSo) method was preferred in the second stage. This chapter also demonstrates the implementation of the proposed decision analysis framework through a case study and the managerial insights it provides to decision-makers.

The major contributions of this chapter are as follows. i) It focuses on the managerial outcomes that suitability scores provide to decision-makers, rather than whether EAPs are suitable. In this regard, it fills a gap in the literature and enables decision-makers to develop detailed policies by increasing their dominance over EAPs. ii) With the flexible structure of the proposed decision analysis framework, it will appeal to all practitioners and researchers interested in different types of disasters anywhere in the world. Its plain and open-to-development structure allows it to be used in different disaster management problems. iii) To our knowledge, it is the first study to integrate the CoCoSo method into GIS. In this respect, it enriches the multi-criteria decision analysis literature by encouraging interdisciplinary studies.

This chapter is organized as follows. Section 16.2 introduces the proposed decision analysis framework. Section 16.3 exhibits a demonstration of the implementation of the proposed decision analysis framework. Conclusions are provided in Section 16.4.

16.2 DECISION ANALYSIS FRAMEWORK FOR EVALUATION OF EAPS

This section outlines the stages of decision analysis. The proposed three-stage decision analysis framework (Figure 16.1) has the flexibility to adapt to a variety of disasters, criteria, and evaluation methods. The stages can be described as follows.

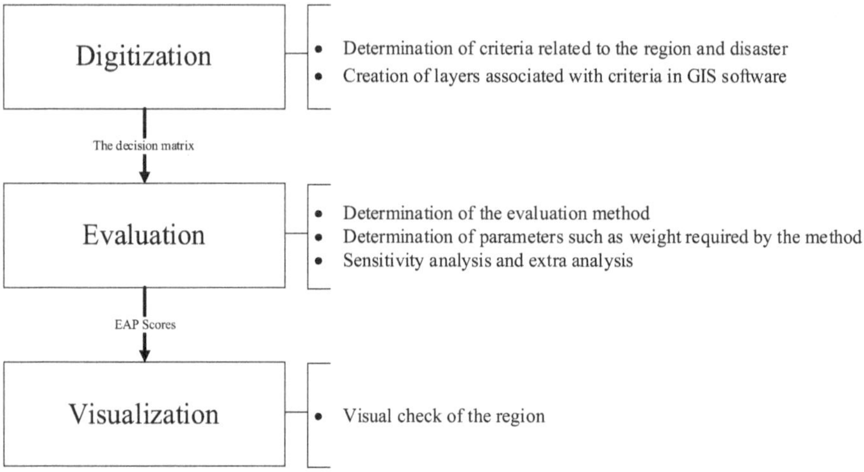

FIGURE 16.1 Decision analysis framework stages for the evaluation of EAPs.

The digitization stage involves digitizing the region to produce the decision matrix. Many features of EAPs, such as size, distance to buildings or facilities, slope, and so on, are georeferenced. With GIS software that allows spatial analysis, geographical, infrastructure, and other relevant factor layers are created that will enable the production of the decision matrix. *The evaluation stage* involves processing the decision matrix. Given the impact of EAP suitability scores on operational policy decisions and the problems that can arise from improper scoring, it is expected that all criteria will consent to the suitability score assigned to an EAP. This drives the need to evaluate the relative suitability of EAPs with a compromise solution. Another issue that requires attention during the evaluation is the difference in opinions among stakeholders. For example, proximity to the main road is more important to relief teams, while the size is more important to the population. Similar differences are reflected in the evaluation with weights. A weight set can only reflect a single stakeholder's opinion. Therefore, stakeholder bias can be avoided by combining scores obtained with weight sets representing different stakeholders. Finally, in addition to the EAP evaluation scores, clustering EAPs and repeating the analysis with a virtual EAP can provide useful hints to decision-makers. *The visualization stage* covers a visual analysis of the location and buffer zones of EAPs. The visualization also facilitates the detection of areas of neglect or overlap throughout the region.

16.2.1 GIS Overview

GIS is a kind of information system that can store, analyze, and report spatially referenced data (Fischer and Nijkamp 1992). It is a tool that provides processed data to decision-makers and is widely used in different fields such as health care (Dudko et al. 2018), urban management (Masser 2001), and waste management (Erfani et al. 2017).

Considering that disasters have intense spatial factors by nature, GIS can be used in critical tasks within the scope of disaster management. GIS provides support to decision-makers at every stage of the disaster cycle such as shelter location evaluations (Sanyal and Lu 2009), risk/hazard assessment (Sari et al. 2021; Wang et al. 2021; Van Westen 2013), risk mapping (Tran et al. 2009), evacuation (Uno and Kashiyama 2008), and resilience (Liu et al. 2019).

16.2.2 CoCoSo Method

The CoCoSo method was introduced by Yazdani et al. (2019). It is based on the combination of three compromise score functions and can thus provide solutions with highly reliable results (Torkayesh et al. 2021). The method is often preferred for problems seeking a compromise solution (Lai et al. 2020; Lahane and Kant 2021; Zolfani et al. 2020). The steps of the method are as follows (Yazdani et al. 2019).

1. The decision matrix and basic notations.
   m alternatives $(A_1, A_2, \ldots, A_m)$, n criteria $(C_1, C_2, \ldots, C_n)$ and $W = \{w_1, \ldots, w_n\} w_j$ is the weight of criterion C_j.

$$x_{ij} = \begin{bmatrix} x_{11} & \cdots & x_{1n} \\ \vdots & \ddots & \vdots \\ x_{m1} & \cdots & x_{mn} \end{bmatrix}; \ i = 1,2,\ldots,m; \ j = 1,2,\ldots,n$$

2. Normalize the decision matrix.

$$r_{ij} = \left\{ \left(\frac{x_{ij} - \min_i x_{ij}}{\max_i x_{ij} - \min_i x_{ij}} \mid j \in I \right) or \left(\frac{\max_i x_{ij} - x_{ij}}{\max_i x_{ij} - \min_i x_{ij}} \mid j \in J \right) \right\}; j = 1,2,\ldots,n$$

   Sets I and J cover the benefit and cost criteria, respectively.

3. Calculate the sum of weighted normalized decision matrix (S_i) and power-weighted normalized decision matrix sequences (P_i) for each alternative.

$$S_i = \sum_{j=1}^{n} (w_j r_{ij}); P_i = \sum_{j=1}^{n} \left(r_{ij}\right)^{w_j}; i = 1,2,\ldots,m$$

4. Calculate the value in three different aggregation strategies (a, b, c) for each alternative.

$$k_{ia} = \frac{P_i + S_i}{\sum_{i=1}^{m} \left(P_i + S_i\right)}$$

$$k_{ib} = \frac{S_i}{\min_i S_i} + \frac{P_i}{\min_i P_i}$$

$$k_{ic} = \frac{\lambda(S_i) + (1-\lambda)(P_i)}{\lambda \max_i S_i + (1-\lambda)\max_i P_i}; \; 0 \le \lambda \le 1$$

a expresses the average of sums of S and P, b states a sum of relative scores of S and P compared to the best, c provides the trade-off between S and P with λ. λ is generally accepted to be 0.5.

5. Determine the evaluation scores for each alternative.

$$k_i = \left(k_{ia} k_{ib} k_{ic}\right)^{\frac{1}{3}} + \frac{1}{3}\left(k_{ia} + k_{ib} + k_{ic}\right)$$

k is sorted in descending order to obtain the order of the alternatives.

16.3 IMPLEMENTATION OF THE PROPOSED DECISION ANALYSIS FRAMEWORK

This section presents a case analysis to show how the proposed decision framework works and what benefits it offers in terms of managerial insight. It is explained step by step so that the evaluation can be easily adapted, considering different disaster types and areas.

The case analysis covers EAPs in the Çankaya region, one of the major districts of Ankara, the capital of Turkey. Çankaya, which has been a district since 1936, has a population of 925,828 with an area of 483 km^2 and a population density of 1,917 per km^2 as of 2020 (Çankaya Municipality 2015; TURKSTAT 2022). Çankaya has a total of 124 neighborhoods mixed with old and new settlements, and has been expanding dramatically every year. Ankara Governorship has determined 60 EAPs in Çankaya (Figure 16.2).

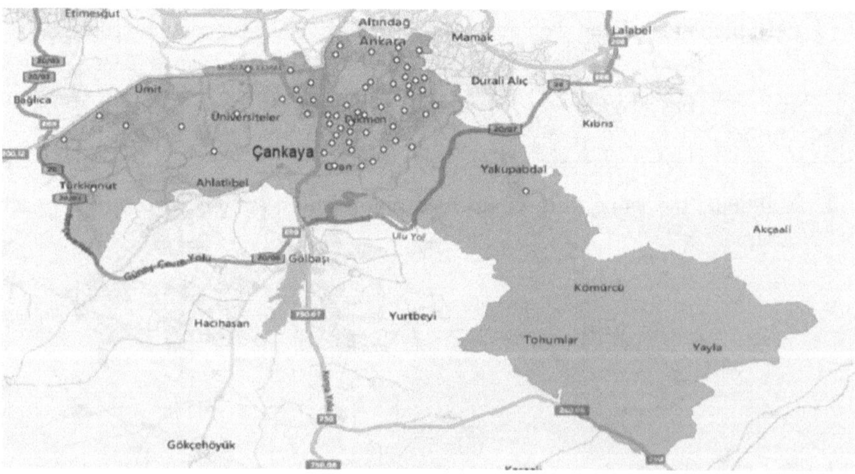

FIGURE 16.2 Çankaya district boundaries and current EAPs.

The suitability of the existing EAPs is doubtful, as it is not known by what criteria the open areas were created. In addition, since there is no alternative to open areas, especially in old settlements, it is not possible to eliminate them and to determine new EAPs. All these factors motivated the investigation of the relative suitability of EAPs in the region to determine effective policies during the preparedness, response, or evacuation phases.

Step 1. Digitization

This step began with determination of the evaluation criteria for EAPs. Considering the conditions of the region and the perspectives of the main stakeholders (the population, relief teams, and disaster management authority), the evaluation criteria were determined as follows. *i) Distance to the nearest building (C1)*: nearby buildings collapse during earthquakes or after aftershocks, posing a risk to EAPs. It is therefore desirable that the nearest building be as far away as possible. The disaster-management authority gives importance to this criterion, as it bears the primary responsibility for determining risky open areas as EAPs. *ii) Size (C2)*: the size of an EAP is directly proportional to the number of people that can gather there. Small EAPs designated for large numbers of people (especially those in old settlements) motivate people to move rather than wait. EAPs are expected to be as large as possible. This criterion is essential for people. *iii) Distance to a multilane road (C3)*: since multilane roads are wider than streets, it should be assumed that relief teams can only move on the multilane road network, considering the possibility that streets will be unusable after an earthquake. Therefore, the closer the EAPs are to a multilane road, the easier it is for the teams to reach them. This criterion is critical for relief teams.

After the criteria were determined, a decision matrix was created that included the performance of the EAPs according to each criterion. For this, EAPs and buildings in the region were created in polygon and roads in polyline format as a separate layer with the help of a base map. Figure 16.3 shows a snapshot of the ArcGIS software. The decision matrix (Table 16.1) was created by performing geometric operations and proximity analyses on the relevant layers.

Step 2. Evaluation

To eliminate stakeholder bias and evaluate the EAP scores objectively, the evaluation was repeated with four different weight sets (Table 16.2). Each of the W1–W3 sets is designated to represent a stakeholder's perspective. In these, one criterion was dominant (50%) and the other two were weighted equally (25%). In W4, each criterion was of equal (33%) importance. The final score was obtained by calculating the geometric mean of the four CoCoSo scores of each EAP (Table 16.3). The variance of scores in each weight set was calculated as an indicator of stakeholder agreement on the final EAP score. Accordingly, a low variance indicates that the stakeholders agreed on the EAP score, and a high variance indicated disagreement. Basic statistics of all calculated (Table 16.4) scores were set as a benchmark. The ranking according to the final suitability scores is shown in Table 16.5.

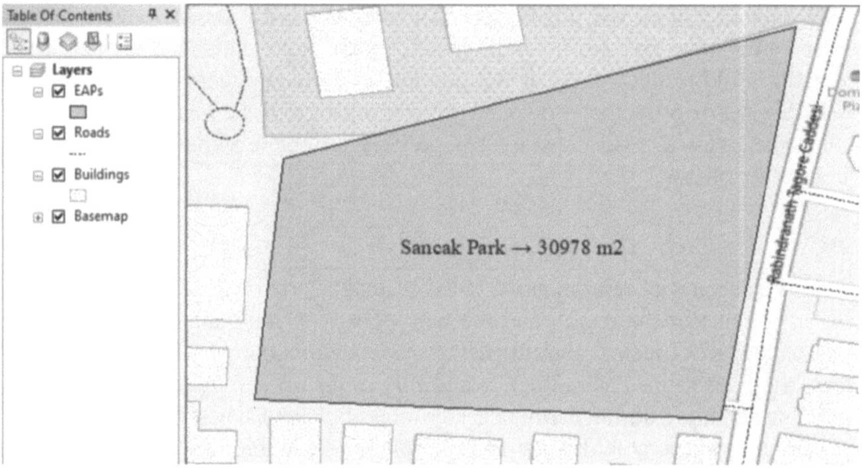

FIGURE 16.3 A snapshot of ArcGIS software.

TABLE 16.1
The Decision Matrix

EAP	C1	C2	C3	EAP	C1	C2	C3	EAP	C1	C2	C3
#1	17.25	18528	12.43	**#21**	646.65	6170	34.19	**#41**	14.40	8819	33.42
#2	13.82	8286	0.00	**#22**	334.02	3278	16.48	**#42**	31.33	16619	54.61
#3	30.70	32497	0.00	**#23**	0.00	3413	116.69	**#43**	54.49	6782	0.00
#4	103.22	5771	14.37	**#24**	59.39	16270	0.00	**#44**	19.74	5546	7.54
#5	10.07	3206	38.44	**#25**	22.62	10639	51.40	**#45**	51.43	4353	115.24
#6	8.00	10116	138.98	**#26**	39.44	4837	0.00	**#46**	65.79	2843	14.99
#7	28.42	26824	0.00	**#27**	12.59	130697	0.00	**#47**	39.74	4688	209.19
#8	2.06	14888	9.80	**#28**	95.40	54011	0.00	**#48**	24.69	6063	89.26
#9	12.50	9747	0.00	**#29**	40.11	29633	0.00	**#49**	57.39	29924	0.00
#10	8.91	197468	0.00	**#30**	127.96	8464	0.00	**#50**	18.70	6019	10.39
#11	0.08	155726	87.37	**#31**	10.20	1141	0.00	**#51**	89.80	9635	27.91
#12	468.66	21464	0.00	**#32**	83.74	5148	0.00	**#52**	1.40	3575	0.00
#13	94.08	10854	99.65	**#33**	49.76	2058	35.89	**#53**	69.58	56859	143.60
#14	17.86	1902	0.00	**#34**	1.75	6590	65.21	**#54**	6.98	12051	0.00
#15	5.19	4333	0.00	**#35**	16.50	12851	432.17	**#55**	8.45	24041	77.06
#16	37.28	5419	0.00	**#36**	55.21	3566	106.60	**#56**	39.27	9844	6.15
#17	36.21	4914	0.00	**#37**	141.90	4288	0.00	**#57**	92.94	25876	103.76
#18	22.08	2577	120.99	**#38**	81.26	19358	0.00	**#58**	178.68	9718	136.54
#19	8.57	5571	0.00	**#39**	18.72	18836	0.00	**#59**	434.05	12255	173.03
#20	146.57	10703	132.19	**#40**	5.34	8629	64.77	**#60**	124.48	5367	599.59

Note: C1: Minimum Distance to Nearest Building (m); C2: size (m^2); C3: minimum distance to nearest multilane road (m).

TABLE 16.2
Weight Sets

Weight Sets	C1	C2	C3
W1	50%	25%	25%
W2	25%	50%	25%
W3	25%	25%	50%
W4	33.3%	33.3%	33.3%

TABLE 16.3
EAP Evaluation Scores by Weight Sets

EAP	W1	W2	W3	W4	Variance	Geo. Mean	EAP	W1	W2	W3	W4	Variance	Geo. Mean
#1	2.2	3.0	4.6	3.1	0.75	3.12	#31	1.8	2.5	4.2	2.6	0.77	2.66
#2	2.1	2.8	4.5	2.9	0.77	2.98	#32	2.4	3.0	4.8	3.2	0.79	3.26
#3	2.5	3.4	5.0	3.4	0.81	3.49	#33	2.1	2.7	4.3	2.8	0.66	2.90
#4	2.5	3.1	4.8	3.2	0.73	3.34	#34	1.9	2.4	4.0	2.5	0.61	2.63
#5	1.9	2.5	4.2	2.6	0.72	2.71	#35	1.3	1.6	1.9	1.5	0.05	1.58
#6	1.8	2.4	3.7	2.4	0.48	2.52	#36	2.0	2.6	4.0	2.7	0.53	2.75
#7	2.4	3.3	4.9	3.3	0.81	3.39	#37	2.6	3.2	5.0	3.4	0.79	3.46
#8	2.1	2.8	4.5	2.9	0.77	2.98	#38	2.6	3.4	5.0	3.4	0.76	3.54
#9	2.1	2.8	4.6	2.9	0.84	3.00	#39	2.3	3.1	4.7	3.1	0.76	3.22
#10	3.4	6.3	6.5	5.0	1.54	5.18	#40	1.9	2.5	4.1	2.6	0.66	2.69
#11	2.9	5.1	5.4	4.1	0.96	4.31	#41	2.1	2.7	4.3	2.8	0.66	2.90
#12	4.1	4.6	6.2	4.7	0.62	4.89	#42	2.2	2.9	4.4	3.0	0.64	3.04
#13	2.4	2.9	4.3	3.0	0.49	3.10	#43	2.3	3.0	4.7	3.1	0.77	3.19
#14	2.0	2.6	4.4	2.8	0.79	2.84	#44	2.1	2.7	4.5	2.9	0.79	2.94
#15	2.0	2.6	4.4	2.7	0.80	2.84	#45	2.0	2.5	3.9	2.6	0.49	2.69
#16	2.2	2.8	4.6	3.0	0.79	3.05	#46	2.2	2.8	4.6	3.0	0.79	3.05
#17	2.2	2.8	4.6	3.0	0.79	3.05	#47	1.8	2.2	3.3	2.3	0.31	2.36
#18	1.8	2.3	3.7	2.4	0.49	2.48	#48	2.0	2.5	4.0	2.6	0.55	2.71
#19	2.0	2.7	4.4	2.8	0.77	2.87	#49	2.6	3.5	5.0	3.5	0.74	3.57
#20	2.5	3.0	4.3	3.1	0.44	3.18	#50	2.1	2.7	4.5	2.9	0.79	2.94
#21	4.5	4.6	6.3	4.9	0.52	5.07	#51	2.5	3.1	4.7	3.2	0.66	3.31
#22	3.3	3.7	5.4	3.9	0.63	4.04	#52	1.9	2.5	4.3	2.7	0.79	2.73
#23	1.7	2.0	3.5	2.2	0.47	2.44	#53	2.6	3.6	4.5	3.4	0.46	3.48
#24	2.5	3.2	4.9	3.3	0.77	3.40	#54	2.1	2.8	4.5	2.9	0.77	2.98
#25	2.1	2.8	4.3	2.8	0.65	2.94	#55	2.1	2.9	4.2	2.9	0.57	2.95
#26	2.2	2.8	4.6	3.0	0.79	3.05	#56	2.3	3.0	4.7	3.1	0.77	3.29
#27	3.0	5.2	5.9	4.3	1.18	5.54	#57	2.5	3.2	4.5	3.2	0.52	3.30
#28	2.9	4.1	5.5	3.9	0.86	4.03	#58	2.6	3.1	4.4	3.2	0.44	3.29
#29	2.5	3.4	5.0	3.4	0.81	2.92	#59	3.6	3.8	4.9	3.9	0.25	4.06
#30	2.7	3.3	5.0	3.4	0.72	2.98	#60	0.9	0.9	1.0	0.9	0.00	0.90

TABLE 16.4
Basic Statistics of EAP Scores

	W1	W2	W3	W4	Variance	Geo. Mean
Average	2.3293	3.0443	4.5128	3.0693	0.6776	3.1424
Max	4.4911	6.3130	6.4612	4.9804	1.5245	5.1253
Min	0.9470	0.9222	0.9552	0.9360	0.0002	0.9400

TABLE 16.5
EAP Ranks

EAP	Rank	EAP	Rank	EAP	Rank	EAP	Rank	EAP	Rank	EAP	Rank
#10	1	#38	11	#32	21	#26	31	#25	41	#5	51
#21	2	#29	12	#58	22	#42	32	#33	42	#40	52
#12	3	#37	13	#39	23	#17	33	#41	43	#31	53
#27	4	#53	14	#20	24	#9	34	#19	44	#34	54
#11	5	#3	15	#43	25	#54	35	#14	45	#6	55
#22	6	#7	16	#56	26	#2	36	#15	46	#18	56
#59	7	#24	17	#1	27	#8	37	#52	47	#47	57
#28	8	#4	18	#13	28	#55	38	#36	48	#23	58
#49	9	#57	19	#46	29	#44	39	#45	49	#35	59
#30	10	#51	20	#16	30	#50	40	#48	50	#60	60

The evaluation results allowed decision-makers to evaluate the relative suitability of EAPs at the macro and micro levels, and customize preparedness, response, or evacuation policies on an EAP basis. The main managerial insights that the evaluation results provided to the decision-makers are as follows.

- The ranking could be an excellent prioritization guide for decision-makers in situations that require the selection of some EAPs for any relief, response, or evacuation action. Prioritization should be read in reverse order, as EAPs with low sustainability scores are given lower ranks. For example, if there is an action to improve the suitability of the EAPs, such as the placement of a disaster and emergency/emergency container, the last five EAPs (#18, #47, #23, #35, #60) can be prioritized for this action.
- EAPs that require special precautions can be determined by analyzing the change in suitability scores according to the rankings of the consecutive EAPs. When EAPs are of similar suitability, the difference in suitability scores across the ranking is expected to be low. A high difference indicates a difference between the respective EAPs that is worth examining. With $x > y$, let Δ_x^y show the difference in suitability scores of the xth and yth EAPs. The average difference between consecutive EAPs in the ranking is 0.07 (calculated from Table 16.3), while Δ_{58}^{57} is 0.10 and Δ_{59}^{58} is 0.67. Δ_{59}^{58} is about seven times higher than Δ_{58}^{57} and about 10 times more than the average difference. There is a notable difference between

the 58th and 59th EAPs. Therefore, specific policies can be set for EAP 35 and EAP 60, which are ranked 59th and 60th, respectively. For example, measures such as preparing a training program and making informative announcements can be taken for those expected to gather in the last two EAPs.

- Keeping in mind that each weight set is dominated by a different criterion, the strengths and weaknesses of EAPs can be determined by examining the variances. For example, EAP 10, which has the highest variance, obtained a high suitability score in the W2 and W3, but a low score in the W1. The weakness of EAP 10 is the risk of collapse of the surrounding buildings.
- Similarly, ranking average suitability scores across weight sets can provide information on the strengths and weaknesses of the overall EAPs. For example, when the average suitability scores of weight sets are ranked, the descending order of the criteria for which the EAPs are strong is as follows: C3, C2, C1. In general, the strength of EAPs in the region is that they are close to the main road, while their weakness is that they are close to buildings. This can also be interpreted as if there is a possibility of retrofitting buildings, retrofitting should be applied first to buildings around EAPs with low scores in W3.

Step 2.1. Extra Analysis

- Clustering
  Putting all EAPs in the same equation can lead to erroneous interpretations. Therefore, EAPs should also be considered in homogeneous clusters so that achievable policies can be determined. Also, with clustering, similar EAPs can be labeled, and preparedness and response policies differentiated on a label-by-tag basis.

  For this purpose, EAPs were clustered using the k-means algorithm. How many clusters the EAPs will be divided into was determined by testing the algorithm with different cluster numbers. As a result of the test, while the number of clusters was 3, the silhouette score[1] value (0.694), which shows the cluster separation performance of the algorithm, reached the highest value. Clustering results are shown in Table 16.6.

TABLE 16.6
EAP Clustering Results

EAP	Cluster	EAP	Cluster	EAP	Cluster	EAP	Cluster	EAP	Cluster	EAP	Cluster
#10	A	#38	B	#32	B	#26	B	#25	B	#5	B
#21	A	#29	B	#58	B	#42	B	#33	B	#40	B
#12	A	#37	B	#39	B	#17	B	#41	B	#31	B
#27	A	#53	B	#20	B	#9	B	#19	B	#34	B
#11	A	#3	B	#43	B	#54	B	#14	B	#6	B
#22	A	#7	B	#56	B	#2	B	#15	B	#18	B
#59	A	#24	B	#1	B	#8	B	#52	B	#47	B
#28	A	#4	B	#13	B	#55	B	#36	B	#23	B
#49	B	#57	B	#46	B	#44	B	#45	B	#35	C
#30	B	#51	B	#16	B	#50	B	#48	B	#60	C

The segregation of EAPs into three clusters indicates that not all should be treated the same. When clusters are analyzed in terms of suitability scores of EAPs, it is seen that the two lowest-scoring EAPs are in cluster C, the eight highest-scoring EAPs are in cluster A, and the others are in cluster B. To balance its disadvantages, cluster C can be given priority in communicating and assisting. B and A follow cluster C. Cluster B is the one with the highest number of suitable EAPs according to standard policy due to its high number and average evaluation scores. Cluster A consists of EAPs with relatively high suitability. EAPs in cluster A can tolerate operational disruptions more than others.

- Virtual EAP

  Virtual EAP allows for the analysis of results by including in the evaluation of an EAP that does not exist, which can be defined with threshold values based on criteria for a specific purpose. For example, if criteria thresholds for an unsuitable EAP can be defined, a virtual EAP with these values will be included in the evaluation and those below the virtual EAP may be interpreted as unsuitable.

  The thresholds of the virtual EAP can be determined optionally, using a literature review, or using the basic statistics of the decision matrix. For example, the criteria-based values of the virtual EAP, which represents the suitability threshold, were determined as C1:30, C2:500, and C3:500 by using the published literature (Şenik and Uzun 2021; Savaş, Cenani, and Çağdaş 2021). With the inclusion of a virtual EAP, a total of 61 EAPs were re-evaluated and only EAP60 was considered below virtual EAP. In this case, it was concluded that EAP60 is unsuitable and the others were above the specified suitability threshold. The virtual EAP approach can be extended to cover specific cases that the decision-makers wish to investigate.

Step 3. Visualization

This stage provided a visual analysis of EAP locations and detected overlaps in coverage and also areas not covered by any EAP. For this purpose, buffer zones formed at certain levels of EAPs were created with GIS software.

Buffer zones with a radius of 500, 1000, and 2000 m for the Çankaya region are shown in Figure 16.4. In the central and northern parts, it was seen that EAPs are close to each other and therefore their coverage areas overlap. There are many alternative EAPs where people can gather after a disaster. While this may sound good, it can cause chaos in terms of coordination. Some EAPs may be in excessive demand, while others may be idle. It is noteworthy that EAPs are less frequent in the western part of the region and there are areas not covered by any EAP. Moreover, the locations of new EAPs can be roughly determined by visual analysis.

16.4 CONCLUSIONS

Statistics show that urbanization is constantly increasing, and scientific studies examining the relationship between urbanization and disaster risk reveal that

FIGURE 16.4 EAP buffer analysis.

urbanization exposes more people to disaster risk. EAPs play a critical role in disaster operations performed in urban areas. For the efficiency of operations, people need to gather in certain areas and wait for professional teams. Among shelters aimed at keeping people together, EAPs differ in location and capacity decisions due to their unique circumstances. Locations or capacities of EAPs cannot be flexible to demand. For this reason, evaluation of the suitability of EAPs is important for the effectiveness of disaster operations and in determining operational policies for pre- and post-disaster.

Studies in the literature have focused on identifying new EAPs, or on determining whether existing EAPs are suitable for replacement. However, even if they are not suitable, especially in old settlements, there is no possibility of changing existing EAPs because there are no alternatives. This reality necessitates evaluation of EAPs. In this chapter, a multi-criteria decision analysis framework based on GIS for the relative evaluation of existing EAPs was presented. This decision analysis framework, which can be adapted to different disaster and geographic conditions, can provide information to decision-makers for the determination of disaster preparedness, response, and evacuation policies.

The proposed decision analysis framework, consisting of digitization, evaluation, and visualization stages, uses GIS software in the first and third stages, and uses the CoCoSo method in the second stage. While GIS is frequently used in studies within the scope of disaster management, the CoCoSo method was used for the first time in this study. The fact that evaluation scores can guide disaster-management policies and the problems that can be caused by incorrect scoring requires precision in the selection of the evaluation method. CoCoSo provides reliable results to compromise solution problems and is therefore preferred in many studies. How the proposed decision analysis framework works was shown with a case study, and the information provided to decision-makers exemplified.

NOTE

1 Silhouette Score is an indicator used to calculate the quality of clustering. It takes a value between –1 and 1. 1: means that the clusters are relatively widely apart and easily identified.

REFERENCES

Çankaya Municipality. 2015. “History of Çankaya.” 2015. https://en.cankaya.bel.tr/Our-Cankaya/Cankaya-History.

Chrysoulidis, Savvas. 2019. “Identifying Evacuation Assembly Points: An Application at Latsia, Nicosia, Cyprus.” *European Journal of Human Security* 3 (3): 48–73. https://doi.org/10.18485/FB_EJHS.2019.3.3.

Dudko, Yevgeni, Dennis E. Robey, Estie Kruger, Marc Tennant, Yevgeni Dudko, Dennis E. Robey, Estie Kruger, and Marc Tennant. 2018. “Selecting a Location for a Primary Healthcare Facility: Combining a Mathematical Approach with a Geographic Information System to Rank Areas of Relative Need.” *Australian Journal of Primary Health* 24 (2): 130–34. https://doi.org/10.1071/PY17093.

Erfani, Seyed Mohammad Hassan, Shahnaz Danesh, Seyed Mohsen Karrabi, and Rouzbeh Shad. 2017. “A Novel Approach to Find and Optimize Bin Locations and Collection Routes Using a Geographic Information System.” *Waste Management and Research* 35 (7): 776–85. https://doi.org/10.1177/0734242X17706753.

Fischer, Manfred M, and Peter Nijkamp. 1992. “Geographic Information Systems and Spatial Analysis.” *The Annals of Regional Science* 26: 3–17.

He, Chunyang, Qingxu Huang, Yinyin Dou, Wei Tu, and Jifu Liu. 2016. “The Population in China’s Earthquake-Prone Areas has Increased by Over 32 Million Along with Rapid Urbanization.” *Environmental Research Letters* 11 (7): 074028. https://doi.org/10.1088/1748-9326/11/7/074028.

Klose, Christian D. 2011. “Evidence for Higher Tropical Storm Risks in Haiti Due to Increasing Population Density in Hazard Prone Urban Areas.” *Environmental Research Letters* 6 (4): 044020. https://doi.org/10.1088/1748-9326/6/4/044020.

Lahane, Swapnil and Ravi Kant. 2021. “A Hybrid Pythagorean Fuzzy AHP – CoCoSo Framework to Rank the Performance Outcomes of Circular Supply Chain Due to Adoption of its Enablers.” *Waste Management* 130 (July): 48–60. https://doi.org/10.1016/J.WASMAN.2021.05.013.

Lai, Han, Huchang Liao, Zhi Wen, Edmundas Kazimieras Zavadskas, and Abdullah Al-Barakati. 2020. “An Improved CoCoSo Method with a Maximum Variance Optimization Model for Cloud Service Provider Selection.” *Engineering Economics* 31 (4): 411–24. https://doi.org/10.5755/J01.EE.31.4.24990.

Liu, Bin, Xudong Chen, Zhongli Zhou, Min Tang, and Shiming Li. 2019. “Research on Disaster Resilience of Earthquake-Stricken Areas in Longmenshan Fault Zone Based on GIS.” *Environmental Hazards* 19 (1): 50–69. https://doi.org/10.1080/17477891.2019.1671785.

Masser, Ian. 2001. “Managing Our Urban Future: The Role of Remote Sensing and Geographic Information Systems.” *Habitat International* 25 (4): 503–12. https://doi.org/10.1016/S0197-3975(01)00021-2.

Oden, Razia V.N., Laura G. Militello, Karol G. Ross, and Christen E. Lopez. 2012. “Four Key Challenges in Disaster Response.” *Proceedings of the Human Factors and Ergonomics Society Annual Meeting* 56 (1): 488–92. https://doi.org/10.1177/1071181312561050.

Sanyal, J. and X. X. Lu. 2009. “Ideal Location for Flood Shelter: A Geographic Information System Approach.” *Journal of Flood Risk Management* 2 (4): 262–71. https://doi.org/10.1111/J.1753-318X.2009.01043.X.

Sari, Indah Cahyaning, Fadly Usman, Keisuke Murakami, and I. Nyoman Suluh Wijaya. 2021. "Geographic Information System and Weighting Technique for Tsunami Risk Assessment in Coastal Villages of Jember Regency, Indonesia." *Disaster Advances*, July, 38–50. https://doi.org/10.25303/148DA3821.

Savaş, Sezer, Şehnaz Cenani, and Gülen Çağdaş. 2021. "Selection of Emergency Assembly Points: A Case Study for the Expected Istanbul Earthquake." In *Contributions to Management Science*, 37–67. Springer, Cham. https://doi.org/10.1007/978-3-030-52406-7_2.

Şenik, Berfin, and Osman Uzun. 2021. "An Assessment on Size and Site Selection of Emergency Assembly Points and Temporary Shelter Areas in Düzce." *Natural Hazards* 105 (2): 1587–602. https://doi.org/10.1007/s11069-020-04367-0.

Torkayesh, Ali Ebadi, Fatih Ecer, Dragan Pamucar, and Çağlar Karamaşa. 2021. "Comparative Assessment of Social Sustainability Performance: Integrated Data-Driven Weighting System and CoCoSo Model." *Sustainable Cities and Society* 71 (August): 102975. https://doi.org/10.1016/J.SCS.2021.102975.

Tran, Phong, Rajib Shaw, Guillaume Chantry, and John Norton. 2009. "GIS and Local Knowledge in Disaster Management: A Case Study of Flood Risk Mapping in Viet Nam." *Disasters* 33 (1): 152–69. https://doi.org/10.1111/J.1467-7717.2008.01067.X.

TURKSTAT. 2022. "Turkish Statistical Institute." 2022. https://www.tuik.gov.tr/.

United Nations. 2020. "World Urbanization Prospects – Population Division – United Nations." *United Nations*. https://population.un.org/wup/DataQuery/.

Uno, Keisuke and Kazuo Kashiyama. 2008. "Development of Simulation System for the Disaster Evacuation Based on Multi-Agent Model Using GIS." *Tsinghua Science and Technology* 13 (SUPPL. 1): 348–53. https://doi.org/10.1016/S1007-0214(08)70173-1.

Van Westen, C. J. 2013. "3.10 Remote Sensing and GIS for Natural Hazards Assessment and Disaster Risk Management." *Treatise on Geomorphology* 3 (January): 259–98. https://doi.org/10.1016/B978-0-12-374739-6.00051-8.

Wang, Si, Lin Mu, Zhenfeng Yao, Jia Gao, Enjin Zhao, and Lizhe Wang. 2021. "Assessing and Zoning of Typhoon Storm Surge Risk with a Geographic Information System (GIS) Technique: A Case Study of the Coastal Area of Huizhou." *Natural Hazards and Earth System Sciences* 21 (1): 439–62. https://doi.org/10.5194/NHESS-21-439-2021.

World Bank. 2020. "Urban Population Statistics – Urban Population (% of Total)." The World Bank. 2020. https://data.worldbank.org/indicator/SP.URB.TOTL.IN.ZS.

Yazdani, Morteza, Pascale Zarate, Edmundas Kazimieras Zavadskas, and Zenonas Turskis. 2019. "A Combined Compromise Solution (CoCoSo) Method for Multi-Criteria Decision-Making Problems." *Management Decision* 57 (9): 2501–19. https://doi.org/10.1108/MD-05-2017-0458.

Zhang, Hao, Wei-chun Ma, and Xiang-rong Wang. 2008. "Rapid Urbanization and Implications for Flood Risk Management in Hinterland of the Pearl River Delta, China: The Foshan Study." *Sensors* 8 (4): 2223–39. https://doi.org/10.3390/S8042223.

Zolfani, Sarfaraz Hashemkhani, Morteza Yazdani, Ali Ebadi Torkayesh, and Arman Derakhti. 2020. "Application of a Gray-Based Decision Support Framework for Location Selection of a Temporary Hospital during COVID-19 Pandemic." *Symmetry* 12 (6): 886. https://doi.org/10.3390/SYM12060886.

17 A Novel Interval Valued Neutrosophic AHP-WASPAS Methodology for Emergency Supply Depot Location Selection Problems

Ertugrul Ayyildiz[1] *and Alev Taskin*[2]

[1]Karadeniz Technical University

[2]Yildiz Technical University

CONTENTS

17.1 Introduction 251
17.2 The Proposed Interval Valued Neutrosophic AHP-WASPAS Methodology 253
17.2.1 Criteria Determination 253
17.2.2 Interval Valued Neutrosophic Sets and Preliminaries 253
17.2.3 IVN-AHP 256
17.2.4 IVN-WASPAS 258
17.3 Application to Istanbul 259
17.3.1 Main Criteria and Sub-Criteria Weights 259
17.3.2 Alternative Comparison by IVN-WASPAS 261
17.4 Conclusions 264
References 264

17.1 INTRODUCTION

Natural, man-made, or technological events that cause physical, social, and economical losses to society, and interrupt or stop the flow of human activities and normal life, are defined as disasters (Buluş Kırıkkaya and Gerdan 2018). Due to its geological features and climate characteristics, Turkey is a country where disasters occur frequently (Kürüm Varolgüneş 2021). Landslides, floods, rockfalls, avalanches, and especially earthquakes are the most common types of natural disaster in Turkey (Baris 2009).

DOI: 10.1201/9781003212904-17

Natural disasters are events that cannot be predicted clearly and cannot be prevented, and therefore being ready for all stages of disasters is crucial in terms of savings lives and properties (Flores and Smith 2013). After a disaster, victims may face significant problems such as not being able to meet their most basic needs. In such cases, it is vital to systematically meet the basic needs of the disaster victims to minimize the loss of life. Disaster management is defined as planning, directing, coordinating, supporting, and effectively implementing the necessary measures and studies before, during, and after a disaster to reduce and prevent damage (Turan 2012). Disaster management consists of three stages: pre-disaster, during the disaster, and post-disaster. Pre-disaster management involves planning, purchasing, transportation, warehouse management, reporting, and human resources development activities (Tanyaş et al. 2013). At this stage, the aim is to reduce the need and cost of work during and after the disaster by reducing material and moral losses.

After major natural disasters such as earthquakes, people first evacuate from risky areas and move to safer assembly areas. People often stay at assembly areas for less than three days until they are settled at safer points such as shelters and tent cities in assembly areas, which are also called preliminary evacuation areas (Yildiz et al. 2020). To deliver emergency aid such as fresh water, food, medicines, and medical supplies to the disaster victims gathered at these assembly areas is an important activity to be done as quickly as possible after the disaster. Effective and efficient distribution of emergency aid to people affected by natural disasters is critical to the emergency response. In this context, the location of emergency supply depots should be determined carefully. Thus, managers should determine the location of the emergency supply depot by considering more than one criterion.

In this chapter, a hierarchical criteria structure has been constructed based on the disaster-management literature and expert views, to handle the problem of emergency supply depot location selection. Different from other humanitarian relief studies, the location selection of an emergency supply depot rather than a distribution center was handled as a multi-criteria decision-making (MCDM) problem. Then, for the first time, weighted aggregated sum product assessment (WASPAS) was extended with analytic hierarchy process (AHP) to develop criteria weight determination under an interval valued neutrosophic (IVN) environment. The criteria weights were determined by IVN-AHP; these weights were then used to evaluate candidate locations by interval valued neutrosophic WASPAS (IVN-WASPAS). A real case study was performed for the pilot region to show the applicability of the proposed methodology.

According to a detailed MCDM literature review based on neutrosophic logic, no study that extended IVN-WASPAS with IVN-AHP was found. Therefore, a novel neutrosophic solution methodology, consisting of IVN-AHP and IVN-WASPAS, is proposed for the first time to solve MCDM problems. In addition, using neutrosophic logic for the first time in determining emergency depot locations is one of the novelties of the study. Therefore, both methodological and practical innovations are proposed in this study.

17.2 THE PROPOSED INTERVAL VALUED NEUTROSOPHIC AHP-WASPAS METHODOLOGY

IVN-WASPAS was integrated with IVN-AHP to evaluate an emergency supply depot location selection problem. First, criteria were determined based on a comprehensive disaster-management literature review and expert opinions. Second, a two-level criteria hierarchy was structured. IVN-AHP was employed to weight main and sub-criteria for each level and alternatives were evaluated by IVN-WASPAS with respect to these criteria weights. Five alternatives on the Anatolian side of Istanbul were evaluated as a real case study. The methodology is explained with details in following sub-sections.

17.2.1 Criteria Determination

In the first stage of the proposed methodology, the main and sub-criteria were defined to determine the best location for an emergency supply depot. A two-level hierarchical structure was constructed considering the criteria and sub-criteria. These criteria were determined by literature review and through consultation with anonymous experts. Published studies on disaster management were comprehensively searched for and analyzed from an MCDM perspective. Web of Science, Google Scholar, and Scopus were searched via keywords such as "disaster management", "MCDM", "emergency supply", "distribution center", "warehouse ", and "location selection". The criteria considered in this study are given in Table 17.1.

17.2.2 Interval Valued Neutrosophic Sets and Preliminaries

Zadeh introduced fuzzy theory into the literature to handle uncertainty (Zadeh 1965). Subsequently, many approaches have been proposed. In traditional fuzzy logic, fuzzy sets are represented by only a truth-membership function. Intuitionistic fuzzy sets include a falsity-membership function and a truth-membership function (Atanassov 1986). Smarandache (1998) developed neutrosophic logic as an advanced version of intuitionistic fuzzy logic to better cope with ambiguity and vagueness by adding a new function called "uncertainty" (Kahraman, Oztaysi, and Cevik Onar 2020). In this sub-section, neutrosophic sets are briefly explained and then interval-valued neutrosophic numbers are explained via definitions.

Definition 1: $\tilde{\ddot{A}}$ is the neutrosophic number that defined in universe E. $\tilde{\ddot{A}}$ is defined by $T_A(x)$, $F_A(x)$, and $I_A(x)$, which represented truth, falsity, and indeterminacy membership functions, respectively (Bolturk and Kahraman 2018b; Biswas, Pramanik, and Giri 2014).

$$\tilde{\ddot{A}} = \{x, (T_A(x), I_A(x), F_A(x)) >: x \in E, (T_A(x), I_A(x), F_A(x) \in]^{-}0, 1[^{+})\} \quad (17.1)$$

TABLE 17.1
Main and Sub-Criteria

Main Criteria	Sub-Criteria	Source
Traffic	Airway	(Yılmaz and Kabak 2020; Roh, Jang, and Han 2013)
	Sea route	(Yılmaz and Kabak 2020; Roh, Jang, and Han 2013)
	Land route	(Yılmaz and Kabak 2020; Roh, Jang, and Han 2013)
	Congestion	(Timperio et al. 2017)
	Railway	Expert view
Coverage	Access to affected zone	(Timperio et al. 2017; Sari et al. 2020; Tian et al. 2013)
	Access to corridor	(Timperio et al. 2017)
	Delivery time	(Sari et al. 2020; Yılmaz and Kabak 2020)
	Access to non-governmental organizations	(Sari et al. 2020; Roh, Jang, and Han 2013)
	Access to suppliers	(Sari et al. 2020; Tian et al. 2013)
Infrastructure	Capacity	(He et al. 2017)
	Water	(Timperio et al. 2017; Yılmaz and Kabak 2020; Tian et al. 2013)
	Electric	(Timperio et al. 2017; Yılmaz and Kabak 2020; Tian et al. 2013)
	Internet	(Timperio et al. 2017; Yılmaz and Kabak 2020; Tian et al. 2013)
	Gas	(Timperio et al. 2017; Yılmaz and Kabak 2020; Tian et al. 2013)
Economic	Investment	(Sari et al. 2020; Yılmaz and Kabak 2020)
	Distribution	(Sari et al. 2020; Roh, Jang, and Han 2013)
	Operating	(Sari et al. 2020; Timperio et al. 2017)
	Storing	(Yılmaz and Kabak 2020; Roh, Jang, and Han 2013)
	Transportation	(Yılmaz and Kabak 2020)
Social	Crime level	(Sari et al. 2020)
	Employee availability	(Sari et al. 2020; Yılmaz and Kabak 2020)
	Security	(Yılmaz and Kabak 2020)
	Political support	(Yılmaz and Kabak 2020; Timperio et al. 2017)
Environment	Rainfall	(Sari et al. 2020)
	Humidity	(Sari et al. 2020)
	Temperature	Expert view
	Disaster-free location	(Yılmaz and Kabak 2020; Roh, Jang, and Han 2013)

$T_A(x)$, $I_A(x)$, and $F_A(x)$ are elements of $]^{-}0,1[^{+}$, and the sum of them can take values between $0^{-} \leq T_A(x)+I_A(x)+, F_A(x) \leq 3^{+}$.

Definition 2: $\tilde{\tilde{N}}$ is the interval valued neutrosophic number defined in universe E. $\tilde{\tilde{N}}$ is defined by $T_N(x)$, $F_N(x)$, and $I_N(x)$, which represent truth membership, falsity membership, and indeterminacy membership functions, respectively (Karaşan, Bolturk, and Kahraman 2020).

$$\ddot{\tilde{N}} = \left\{x, \left[T_N^L(x), T_N^U(x)\right], \left[I_N^L(x), I_N^U(x)\right], \left[F_N^L(x), F_N^U(x)\right] \; x \epsilon E\right\} \tag{17.2}$$

where

$$T_N(x) = \left[T_N^L(x), T_N^U(x)\right] \subseteq [0,1]; \; T_N^L(x) \leq T_N^U(x) \tag{17.3}$$

$$I_N(x) = \left[I_N^L(x), I_N^U(x)\right] \subseteq [0,1]; \; I_N^L(x) \leq I_N^U(x) \tag{17.4}$$

$$F_N(x) = \left[F_N^L(x), F_N^U(x)\right] \subseteq [0,1]; \; F_N^L(x) \leq F_N^U(x) \tag{17.5}$$

Definition 3: $\ddot{\tilde{N}}$ is deneutrosophicated as below (Bolturk and Kahraman 2018a).

$$\mathfrak{D}\left(\ddot{\tilde{N}}\right) = \left(\frac{\left(T_N^L(x) + T_N^U(x)\right)}{2} + \left(I_N^U(x)\right)\left(1 - \frac{\left(I_N^L(x) + I_N^U(x)\right)}{2}\right) - \left(1 - F_N^U(x)\right)\left(\frac{F_N^L(x) + F_N^U(x)}{2}\right)\right) \tag{17.6}$$

Definition 4: The sum and difference of two interval valued neutrosophic numbers $\left(\ddot{\tilde{N}}_1 = \left\langle\left[T_{N_1}^L, T_{N_1}^U\right], \left[I_{N_1}^L, I_{N_1}^U\right], \left[F_{N_1}^L, F_{N_1}^U\right]\right\rangle; \ddot{\tilde{N}}_2 = \left\langle\left[T_{N_2}^L, T_{N_2}^U\right], \left[I_{N_2}^L, I_{N_2}^U\right], \left[F_{N_2}^L, F_{N_2}^U\right]\right\rangle\right)$:

$$\ddot{\tilde{N}}_1 \oplus \ddot{\tilde{N}}_2 = \left\langle\left[T_{N_1}^L + T_{N_2}^L - T_{N_1}^L T_{N_2}^L, T_{N_1}^U + T_{N_2}^U - T_{N_1}^U T_{N_2}^U\right], \left[I_{N_1}^L I_{N_2}^L, I_{N_1}^U I_{N_2}^U\right], \left[F_{N_1}^L F_{N_2}^L, F_{N_1}^U F_{N_2}^U\right]\right\rangle \tag{17.7}$$

$$\ddot{\tilde{N}}_1 \ominus \ddot{\tilde{N}}_2 = \left\langle\left[T_{N_1}^L - F_{N_2}^U, T_{N_1}^U - F_{N_2}^L\right], \left[\max\left(I_{N_1}^L, I_{N_2}^L\right), \max\left(I_{N_1}^U, I_{N_2}^U\right)\right], \left[F_{N_1}^L - T_{N_2}^U, F_{N_1}^U - T_{N_2}^L\right]\right\rangle \tag{17.8}$$

Definition 5: The multiplication of two interval valued neutrosophic numbers $\left(\ddot{\tilde{N}}_1 \text{ and } \ddot{\tilde{N}}_2\right)$

$$\ddot{\tilde{N}}_1 \otimes \ddot{\tilde{N}}_2 = \left\langle\left[T_{N_1}^L T_{N_2}^L, T_{N_1}^U T_{N_2}^U\right], \left[I_{N_1}^L + I_{N_2}^L - I_{N_1}^L T_{N_2}^L, I_{N_1}^U + I_{N_2}^U - I_{N_1}^U I_{N_2}^U\right], \left[F_{N_1}^L + F_{N_2}^L - F_{N_1}^L F_{N_2}^L, F_{N_1}^U + F_{N_2}^U - F_{N_1}^U F_{N_2}^U\right]\right\rangle \tag{17.9}$$

Definition 6: The power and multiplication by positive scalar:

$$\lambda\left(\tilde{\dddot{\mathrm{N}}}_1\right)=\left\langle\left[1-\left(1-\mathrm{T}_{\mathrm{N}_1}^{\mathrm{L}}\right)^{\lambda},1-\left(1-\mathrm{T}_{\mathrm{N}_1}^{\mathrm{U}}\right)^{\lambda}\right],\left[\left(\mathrm{I}_{\mathrm{N}_1}^{\mathrm{L}}\right)^{\lambda},\left(\mathrm{I}_{\mathrm{N}_1}^{\mathrm{U}}\right)^{\lambda}\right],\left[\left(\mathrm{F}_{\mathrm{N}_1}^{\mathrm{L}}\right)^{\lambda},\left(\mathrm{F}_{\mathrm{N}_1}^{\mathrm{U}}\right)^{\lambda}\right]\right\rangle \tag{17.10}$$

$$\left(\tilde{\dddot{\mathrm{N}}}_1\right)^{\lambda}=\left\langle\left[\left(\mathrm{T}_{\mathrm{N}_1}^{\mathrm{L}}\right)^{\lambda},\left(\mathrm{T}_{\mathrm{N}_1}^{\mathrm{U}}\right)^{\lambda}\right],\left[\left(\mathrm{I}_{\mathrm{N}_1}^{\mathrm{L}}\right)^{\lambda},\left(\mathrm{I}_{\mathrm{N}_1}^{\mathrm{U}}\right)^{\lambda}\right],\left[1-\left(1-\mathrm{F}_{\mathrm{N}_1}^{\mathrm{L}}\right)^{\lambda},1-\left(1-\mathrm{F}_{\mathrm{N}_1}^{\mathrm{U}}\right)^{\lambda}\right]\right\rangle \tag{17.11}$$

Definition 7: A score $\mathrm{S}\left(\tilde{\dddot{\mathrm{N}}}_1\right)\in[0,1]$ function (Xu, Wei, and Wei 2017):

$$\mathrm{S}\left(\tilde{\dddot{\mathrm{N}}}_1\right)=\frac{\left(2+\mathrm{T}_{\mathrm{N}_1}^{\mathrm{L}}-\mathrm{I}_{\mathrm{N}_1}^{\mathrm{L}}-\mathrm{F}_{\mathrm{N}_1}^{\mathrm{L}}\right)+\left(2+\mathrm{T}_{\mathrm{N}_1}^{\mathrm{U}}-\mathrm{I}_{\mathrm{N}_1}^{\mathrm{U}}-\mathrm{F}_{\mathrm{N}_1}^{\mathrm{U}}\right)}{6} \tag{17.12}$$

17.2.3 IVN-AHP

Saaty (1988) developed the AHP method based on the study by Myers and Alpert (1977). Saaty systematized the method according to the logic of constructing a decision-making problem in a hierarchical level through pairwise comparisons of components (Thor, Ding, and Kamaruddin 2013). However, AHP sometimes fails to reflect ambiguity in human opinions. To cope with this, AHP was extended with neutrosophic logic and IVN-AHP emerged. Many researchers have used IVN-AHP and neutrosophic AHP in different MCDM problems (Bolturk and Kahraman 2018a), (Radwan, Senousy, and Riad 2016), (Abdel-Basset, Mohamed, and Sangaiah 2018). In this study, the criteria weights for each level affecting the selection of an emergency supply depot location were determined by IVN-AHP. The application steps of the IVN-AHP methodology are given below.

Step 1. The pairwise comparison matrix $(\tilde{\dddot{P}})$ is constructed. To construct the matrix, the linguistic terms given in Table 17.2 were used.

Step 2. Pairwise comparison matrix $(\tilde{\dddot{P}})$ converted interval-valued neutrosophic comparison matrix constructed using Table 17.2.

$$\tilde{\dddot{P}}=\begin{bmatrix}\langle[T_{11}^L,T_{11}^U],[I_{11}^L,I_{11}^U,[F_{11}^L,F_{11}^U]]\rangle & \langle[T_{12}^L,T_{12}^U],[I_{12}^L,I_{12}^U,[F_{12}^L,F_{12}^U]]\rangle & \cdots & \langle[T_{1n}^L,T_{1n}^U],[I_{1n}^L,I_{1n}^U,[F_{1n}^L,F_{1n}^U]]\rangle \\ \langle[T_{21}^L,T_{21}^U],[I_{21}^L,I_{21}^U,[F_{21}^L,F_{21}^U]]\rangle & \ddots & \ddots & \vdots \\ \vdots & \ddots & \ddots & \vdots \\ \langle[T_{n1}^L,T_{n1}^U],[I_{n1}^L,I_{n1}^U,[F_{n1}^L,F_{n1}^U]]\rangle & \cdots & \cdots & \langle[T_{nn}^L,T_{nn}^U],[I_{nn}^L,I_{nn}^U,[F_{nn}^L,F_{nn}^U]]\rangle\end{bmatrix} \tag{17.13}$$

TABLE 17.2
Linguistic Terms

Linguistic Terms		T^L	T^U	I^L	I^U	F^L	F^U
Absolutely more important	AMI	0.9	0.95	0	0.05	0.05	0.15
Strongly more important	SMI	0.8	0.9	0.05	0.1	0.1	0.2
More important	MI	0.7	0.8	0.15	0.25	0.2	0.3
Weakly more important	WMI	0.6	0.7	0.25	0.35	0.3	0.4
Equal importance	EI	0.5	0.5	0.5	0.5	0.5	0.5
Weakly less important	WLI	0.4	0.5	0.55	0.65	0.5	0.6
Less important	LI	0.3	0.4	0.65	0.75	0.6	0.7
Strongly less important	SLI	0.2	0.3	0.75	0.85	0.7	0.8
Absolutely less important	ALI	0.1	0.2	0.9	0.95	0.8	0.9

Step 3. $\tilde{\dddot{P}}$ is deneutrosophicated to check for consistency. The neutrosophic matrix is evaluated as consistent only if the deneutrosophicated matrix is determined to be consistent.

Step 4. The criteria weights are normalized (Vafaei, Ribeiro, and Camarinha-Matos 2016).

$$\tilde{\dddot{N}}_{ij} = \left[\frac{T_{kj}^L}{\sum_{k=1}^n T_{kj}^U}, \frac{T_{kj}^U}{\sum_{k=1}^n T_{kj}^U}\right], \left[\frac{I_{kj}^L}{\sum_{k=1}^n I_{kj}^U}, \frac{I_{kj}^U}{\sum_{k=1}^n I_{kj}^U}\right], \left[\frac{F_{kj}^L}{\sum_{k=1}^n F_{kj}^U}, \frac{F_{kj}^U}{\sum_{k=1}^n F_{kj}^U}\right] \tag{17.14}$$

where n represents the number of criteria.

Step 5. The mean of each row is calculated to determine the neutrosophic weight vector.

$$\tilde{\dddot{W}}_j = \left[\frac{\sum_{k=1}^n \frac{T_{1j}^L}{\sum_{k=1}^n T_{kj}^U}}{n}, \frac{\sum_{k=1}^n \frac{T_{1j}^U}{\sum_{k=1}^n T_{kj}^U}}{n}\right], \left[\frac{\sum_{k=1}^n \frac{I_{1j}^L}{\sum_{k=1}^n I_{kj}^U}}{n}, \frac{\sum_{k=1}^n \frac{I_{1j}^U}{\sum_{k=1}^n I_{kj}^U}}{n}\right], \left[\frac{\sum_{k=1}^n \frac{F_{1j}^L}{\sum_{k=1}^n F_{kj}^U}}{n}, \frac{\sum_{k=1}^n \frac{F_{1j}^U}{\sum_{k=1}^n F_{kj}^U}}{n}\right] \tag{17.15}$$

Step 6. Equation 17.6 is used to determine criteria weights.
Step 7. The weights are normalized to determine final weights $\left(w_j\right)$.

17.2.4 IVN-WASPAS

WASPAS is used to select best among all options (Zavadskas et al. 2012). The method combines two well-known methods, weighted total and weighted multiplication to enable a more consistent decision to be made (Tumsekcali, Ayyildiz, and Taskin 2021). In this study, AHP was used to determine criteria weights to use in WASPAS with interval valued neutrosophic numbers. IVN-WASPAS was performed to better represent ambiguity and hesitancy in information. The IVN-WASPAS methodology steps are given below.

Step 1. The alternative evaluation matrix $\left(\tilde{\dddot{X}}\right)$ is constructed by the linguistic terms shown in Table 17.3. Next, linguistic terms are converted to interval valued neutrosophic numbers.

Let $\tilde{\dddot{x}}_{ij}$ be the value of alternative location i according to criterion j. m represents the number of alternative locations.

$$\tilde{\dddot{X}} = \begin{bmatrix} \tilde{\dddot{x}}_{11} & \tilde{\dddot{x}}_{12} & \dots & \tilde{\dddot{x}}_{1n} \\ \tilde{\dddot{x}}_{21} & \tilde{\dddot{x}}_{22} & \dots & \tilde{\dddot{x}}_{2n} \\ \vdots & \dots & \ddots & \vdots \\ \tilde{\dddot{x}}_{m1} & \tilde{\dddot{x}}_{m2} & \dots & \tilde{\dddot{x}}_{mn} \end{bmatrix} \tag{17.16}$$

Step 2. The weighted sum model $(\tilde{Q}^1)$ is calculated for each alternative. Equation 17.7 and Equation 17.10 are used to calculate $\tilde{Q}_i^1$.

TABLE 17.3
Scale for the IVN-WASPAS Evaluations

Linguistic Terms	T^L	T^U	I^L	I^U	F^L	F^U
Certainly low (CL)	0.05	0.2	0.4	0.6	0.85	1
Very low (VL)	0.15	0.3	0.3	0.5	0.75	0.9
Low (L)	0.25	0.4	0.2	0.4	0.65	0.8
Below average (BA)	0.35	0.5	0.1	0.3	0.55	0.7
Average (A)	0.45	0.6	0	0.2	0.45	0.6
Above average (AA)	0.55	0.7	0.1	0.3	0.35	0.5
High (H)	0.65	0.8	0.2	0.4	0.25	0.4
Very high (VH)	0.75	0.9	0.3	0.5	0.15	0.3
Certainly high (CH)	0.85	1	0.4	0.6	0.05	0.2

$$\tilde{Q}_i^1 = \sum_{j=1}^{n} \ddot{\tilde{x}}_{ij} w_j \tag{17.17}$$

Step 3. The weighted sum model ($\tilde{Q}^2$) is calculated for each alternative. Equation 17.9 and Equation 17.11 are used to calculate $\tilde{Q}_i^2$.

$$\tilde{Q}_i^2 = \prod_{j=1}^{n} \left(\ddot{\tilde{x}}_{ij}\right)^{w_j} \tag{17.18}$$

Step 4. Threshold value (λ) is specified to aggregate $\tilde{Q}^1$ and $\tilde{Q}^2$.

$$\tilde{Q}_i = \lambda \tilde{Q}_i^1 + (1-\lambda)\tilde{Q}_i^2 \tag{17.19}$$

where λ takes value in the range of $|0,1|$ and usually it is taken as 0.5 (Zavadskas et al. 2012).

Step 5. Interval valued neutrosophic numbers are deneutrosophicated to determine the final scores.
Step 6. The alternatives are ranked with decreasing scores.

17.3 APPLICATION TO ISTANBUL

We evaluate an emergency supply depot location selection problem. First, the criteria were weighted in a hierarchical structure; second, alternative locations in the Anatolian side of Istanbul were ranked with respect to these weights. Istanbul is located close to the one of the fastest moving and most active faults in the world, namely the North Anatolian Fault. The city is considered to be a second-degree earthquake zone (Hamiyet 1996). In considering major events, starting with the Erzincan earthquake in 1939 and continuing until 1967, it was seen that the earthquakes on the North Anatolian Fault systematically migrated from east to west along the fault. The eastern side of the North Anatolian Fault, which is one of the most active faults in Turkey, was destroyed by severe earthquakes and discharged its energy. However, on the west side, energy has accumulated to produce large earthquakes. The Marmara Region, especially Istanbul, is one of the riskiest areas in terms of earthquakes occurring (Taymaz, Eyidoğan, and Jackson 1991). In this context, it is essential to prepare for natural disasters such as earthquakes in the region. Therefore, Istanbul was chosen as the application area for the proposed methodology.

17.3.1 Main Criteria and Sub-Criteria Weights

First, predetermined criteria were weighted by IVN-AHP according to expert opinions. Three experts were asked for their opinions on criteria for an emergency

TABLE 17.4
Pairwise Comparison Matrices

Expert 1

	Traffic	Coverage	Infrastructure	Economic	Social	Environment
Traffic	EI	SLI	WLI	WMI	SMI	MI
Coverage	SMI	EI	SMI	AMI	AMI	AMI
Infrastructure	WMI	SLI	EI	MI	SMI	MI
Economic	WLI	ALI	LI	EI	MI	EI
Social	SLI	ALI	SLI	LI	EI	WLI
Environment	LI	ALI	LI	EI	WMI	EI

Expert 2

	Traffic	Coverage	Infrastructure	Economic	Social	Environment
Traffic	EI	SLI	WMI	MI	MI	SMI
Coverage	SMI	EI	AMI	AMI	AMI	AMI
Infrastructure	WLI	ALI	EI	WMI	WMI	MI
Economic	LI	ALI	WLI	EI	WLI	WMI
Social	LI	ALI	WLI	WMI	EI	WMI
Environment	SLI	ALI	LI	WLI	WLI	EI

Expert 3

	Traffic	Coverage	Infrastructure	Economic	Social	Environment
Traffic	EI	MI	SMI	WMI	LI	AMI
Coverage	LI	EI	WMI	LI	ALI	MI
Infrastructure	SLI	WLI	EI	LI	ALI	WMI
Economic	WLI	MI	MI	EI	LI	SMI
Social	MI	AMI	AMI	MI	EI	AMI
Environment	ALI	LI	WLI	SLI	ALI	EI

supply depot location selection problem. Two experts worked on humanitarian relief supply chain operations; the other was an academic studying humanitarian relief. Comparison matrices were constructed via interviews with experts to evaluate the criteria. Experts used the linguistic terms presented in Table 17.2. Table 17.4 presents the pairwise comparison matrices of six main criteria for each expert.

Pairwise comparison matrices were checked for consistency (for details, see Gulum, Ayyildiz, and Taskin Gumus 2021) and determined to be consistent for each expert. After pairwise comparison matrices were determined as consistent, the IVN-AHP methodology was applied, and the main criteria weights were determined as given in Table 17.5.

The criteria weights for three experts were aggregated via the geometric mean and normalized, and thus the final weights were calculated. The main criteria were ordered as "Coverage" > "Traffic" > "Social" > "Infrastructure" > "Economic" > "Environment" according to their importance, from highest to the lowest. "Coverage" was determined as the most significant criterion with the 0.206 importance weight.

TABLE 17.5
The Weights for Main Criteria

Main Criteria	Expert -1	Expert -2	Expert -3	Final Weight	Rank
Traffic	0.174	0.182	0.193	0.185	2
Coverage	0.238	0.240	0.146	0.206	1
Infrastructure	0.184	0.160	0.134	0.160	4
Economic	0.145	0.143	0.184	0.158	5
Social	0.122	0.147	0.229	0.162	3
Environment	0.137	0.129	0.115	0.128	6

The size of the service area that the warehouse can serve is important. It is necessary to deliver emergency supplies to people living in the coverage area. Therefore, factors such as population density in the coverage area and infrastructure facilities should be considered. "Traffic" is another important main criterion, because the traffic density that may occur after a disaster may make it difficult to deliver the emergency supplies to the assembly areas. "Environment" is the least important among the six main criteria. After determining the main criteria weights, pairwise comparison matrices were constructed for each main criterion to determine the weights of its sub-criteria. All matrices were determined to be consistent. Then, the weights of the sub-criteria were calculated by IVN-AHP considering each expert. The sub-criteria weights were multiplied by corresponding main criteria weights, and the sub-criteria weights were aggregated to determine the final weights. The final weights of the sub-criteria are presented in Table 17.6.

"Political support" was determined to be the most significant sub-criterion. Political support is essential in constructing an emergency supply depot, because the construction of these warehouses is costly and this cost should be covered by politicians; the support of local government is also essential for building a warehouse in the most suitable place. Ensuring the security of warehouses is also the responsibility of politicians. "Delivery time" was also an important sub-criterion, with an importance rate 5.3%. Delivery time represents the time from when emergency supplies leave the warehouse to the time they reach the victims. A short delivery time is important for the effective execution of post-earthquake activities. As the delivery time increases, it can become more difficult to meet the immediate needs of victims, which can sometimes lead to chaos. "Access to affected zone" is the distance between depot and disaster points. A short distance is important in terms of fast delivery of needs. In addition, due to the deterioration of infrastructures after a disaster, transportation to the region may be difficult. Therefore, easy access to these areas is important in terms of carrying out post-disaster activities.

17.3.2 Alternative Comparison by IVN-WASPAS

The alternatives were evaluated with respect to the predetermined criteria. Different districts in the Anatolian side of Istanbul were evaluated as alternative locations to

TABLE 17.6
The Importance Weights for Sub-Criteria

Sub-Criteria	E -1	E-2	E-3	Weight	Sub-Criteria	E -1	E-2	E-3	Weight
Airway	0.036	0.035	0.041	0.038	Investment	0.025	0.022	0.048	0.030
Sea route	0.030	0.030	0.044	0.035	Distribution	0.036	0.040	0.035	0.038
Land route	0.044	0.045	0.052	0.048	Operating	0.025	0.033	0.044	0.034
Congestion	0.040	0.049	0.036	0.042	Storing	0.029	0.023	0.029	0.028
Railway	0.025	0.024	0.020	0.023	Transportation	0.030	0.023	0.028	0.028
Access to affected zone	0.061	0.063	0.034	0.052	Crime level	0.023	0.037	0.053	0.036
Access to corridor	0.044	0.035	0.022	0.033	Employee availability	0.031	0.020	0.042	0.030
Delivery time	0.065	0.056	0.038	0.053	Security	0.026	0.042	0.057	0.041
Access to non-governmental organizations	0.033	0.037	0.020	0.029	Political support	0.043	0.047	0.077	0.055
Access to suppliers	0.036	0.049	0.032	0.039	Rainfall	0.040	0.037	0.033	0.037
Capacity	0.045	0.041	0.037	0.042	Humidity	0.024	0.033	0.028	0.029
Water	0.032	0.033	0.027	0.031	Temperature	0.024	0.017	0.015	0.019
Electric	0.045	0.037	0.030	0.038	Disaster-free location	0.049	0.041	0.039	0.044
Internet	0.038	0.029	0.020	0.028					
Gas	0.024	0.019	0.020	0.021					

show the applicability of the proposed methodology. Five alternative locations were determined by interviewing the same experts. Then, the proposed neutrosophic decision-making methodology was utilized to determine the best location for an emergency supply depot. In the application discussed here, the candidate locations are shown in Figure 17.1.

First, the same experts evaluated alternative locations using the linguistic terms given in Table 17.3 via a questionnaire. Table 17.7 shows the evaluation values of the alternative locations.

The weighted sum and product model were calculated according to criteria-alternative evaluations, with respect to criteria weights obtained by IVN-AHP. Then, the threshold value (λ) was fixed to 0.5 and the weighted sum and product model combined. Last, interval valued neutrosophic numbers were deneutrosophicated to determine the final score of each alternative.

Then, the alternatives were sorted in descending order. Table 17.8 shows the final scores of the alternatives. Kartal was determined to be the best location for an emergency supply depot according to the results of the IVN-AHP integrated IVN-WASPAS methodology. It stands out with its coverage and traffic-related factors. Kartal also has better infrastructure to help during a disaster. Tuzla is another good option. Çekmeköy was the worst location in which to construct an emergency supply depot, particularly because of its coverage.

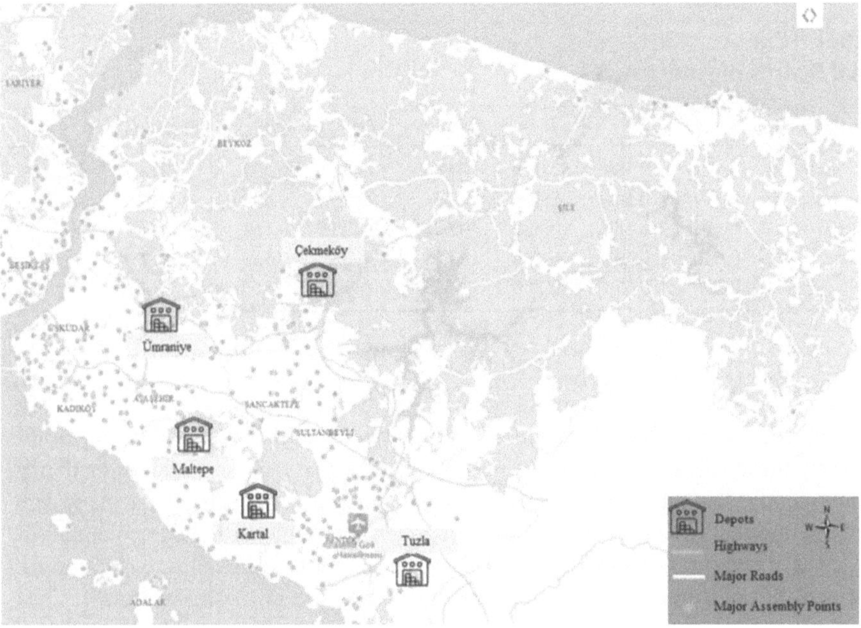

FIGURE 17.1 Alternative locations.

TABLE 17.7
Alternative Evaluation of Experts

Alternative	C11	C12	C13	C14*	C15	C21	C22	C23	C24	C25	C31	C32	C33	C34
Çekmeköy	L	VL	BA	L	BA	L	BA	L	AA	AA	VH	AA	BA	BA
Ümraniye	VL	BA	AA	AA	BA	BA	BA	BA	VH	H	H	AA	BA	BA
Maltepe	L	AA	H	VH	H	H	AA	H	H	VH	AA	AA	AA	AA
Kartal	H	H	H	H	H	VH	H	H	H	H	BA	AA	AA	AA
Tuzla	VH	VH	BA	BA	H	AA	AA	AA	L	BA	H	BA	AA	BA

Alternative	C35	C41*	C42*	C43*	C44*	C45*	C51*	C52	C53	C54	C61*	C62	C63	C64
Çekmeköy	BA	VL	H	BA	H	VH	VH	BA	BA	H	AA	AA	H	VH
Ümraniye	BA	BA	AA	AA	H	H	H	AA	H	H	AA	AA	H	H
Maltepe	AA	VH	BA	H	AA	BA	BA	VH	H	L	BA	BA	BA	BA
Kartal	AA	H	L	H	BA	L	L	H	H	L	BA	BA	BA	BA
Tuzla	AA	L	BA	BA	AA	BA	VL	H	AA	AA	BA	BA	BA	L

* Cost criterion.

TABLE 17.8
Final Scores of Alternatives

Alternative	T^L	T^U	I^L	I^U	F^L	F^U	Final Score	Rank
Çekmeköy	0.421	0.589	0.398	0.606	0.468	0.635	0.605	5
Ümraniye	0.449	0.609	0.368	0.584	0.445	0.605	0.628	4
Maltepe	0.493	0.656	0.389	0.599	0.399	0.562	0.667	3
Kartal	0.530	0.691	0.405	0.608	0.363	0.523	0.700	1
Tuzla	0.518	0.678	0.360	0.579	0.376	0.534	0.693	2

17.4 CONCLUSIONS

This study focused on the problem of an emergency supply depot location. In this context, the criteria that are effective in decision-making were determined first. Subsequently, Istanbul was selected as the application area due to its many strategic features such as its proximity to fault zones with high earthquake potential, the presence of important industrial establishments in and near its borders, and the fact that one-fifth of Turkey's population live in the city. The IVN-AHP integrated IVN-WASPAS methodology was then structured; the problem was solved by this methodology for Istanbul and the best alternative was determined to be "Kartal".

The contributions of this study to the disaster-management literature are as follows. (1) The criteria that play a key role in selecting the best location for an emergency supply depot were determined and grouped. (2) The most important criteria were determined with neutrosophic logic for the first time in the literature. (3) A real case application for Istanbul was performed to show the reliability and applicability of the proposed hybrid MCDM methodology. (4) Five different locations were evaluated according to the criteria determined, and the best one selected. (5) This study also presents the first integration of IVN-WASPAS with IVN-AHP. (6) The aim was that the proposed IVN-AHP integrated IVN-WASPAS can be used as a guide by governmental organizations and non-governmental organizations to improve their disaster and post-disaster strategies.

Despite the aforementioned contributions, this study had some limitations. The first limitation was the evaluation of a limited number of alternative districts within the scope of the study and the difficulties experienced in collecting data to evaluate these districts. Second, each expert, whose opinions were consulted during the implementation stages of IVN-AHP and IVN-WASPAS, had equal importance. Experts can be evaluated according to their experience and coefficients can be assigned to experts. In this context, the site-selection model proposed in this study could be enriched with these extensions for future studies.

REFERENCES

Abdel-Basset, Mohamed, Mai Mohamed, and Arun Kumar Sangaiah. 2018. "Neutrosophic AHP-Delphi Group Decision Making Model Based on Trapezoidal Neutrosophic Numbers." *Journal of Ambient Intelligence and Humanized Computing*. doi:10.1007/s12652-017-0548-7.

Atanassov, Krassimir T. 1986. "Intuitionistic Fuzzy Sets." *Fuzzy Sets and Systems*. doi:10.1016/S0165-0114(86)80034-3.

Baris, M. E. 2009. "Effectiveness of Turkish Disaster Management System and Recommendations." *Biotechnology & Biotechnological Equipment* 23 (3): 1391–98. doi:10.1080/13102818.2009.10817677.

Biswas, Pranab, Surapati Pramanik, and Bibhas C Giri. 2014. "A New Methodology for Neutrosophic Multi-Attribute Decision Making with Unknown Weight Information." *Neutrosophic Sets and Systems* 3: 42–52.

Bolturk, Eda and Cengiz Kahraman. 2018a. "Interval-Valued Neutrosophic AHP with Possibility Degree Method." *International Journal of the Analytic Hierarchy Process*. doi:10.13033/ijahp.v10i3.545.

Bolturk, Eda and Cengiz Kahraman. 2018b. "A Novel Interval-Valued Neutrosophic AHP with Cosine Similarity Measure." *Soft Computing* 22 (15): 4941–58. doi:10.1007/s00500-018-3140-y.

Buluş Kırıkkaya, Esma and Serpil Gerdan. 2018. "Engelli ve Engelli Adayı Bireylerin Bir Afet Anında Nasıl Davranacaklarına İlişkin Görüşleri." *Resilience* 2 (2): 123–29. doi:10.32569/RESILIENCE.477814.

Flores, Alejandro Quiroz and Alastair Smith. 2013. "Leader Survival and Natural Disasters." *British Journal of Political Science* 43 (4): 821–43. doi:10.1017/S0007123412000609.

Gulum, Pelin, Ertugrul Ayyildiz, and Alev Taskin Gumus. 2021. "A Two Level Interval Valued Neutrosophic AHP Integrated TOPSIS Methodology for Post-Earthquake Fire Risk Assessment: An Application for Istanbul." *International Journal of Disaster Risk Reduction* 61 (July): 102330. doi:10.1016/j.ijdrr.2021.102330.

Hamiyet, Sezer. 1996. "1894 İstanbul Depremi Hakkında Bir Rapor Üzerine İnceleme." *Ankara Üniversitesi Dil ve Tarih-Coğrafya Fakültesi Tarih Bölümü Tarih Araştırmaları Dergisi*. doi:10.1501/tarar_0000000097.

He, Junkang, Chenpeng Feng, Dan Hu, and Liang Liang. 2017. "A Decision Model for Emergency Warehouse Location Based on a Novel Stochastic MCDA Method: Evidence from China." *Mathematical Problems in Engineering*. doi:10.1155/2017/7804781.

Kahraman, Cengiz, Başar Oztaysi, and Sezi Cevik Onar. 2020. "Single Interval-Valued Neutrosophic AHP Methods: Performance Analysis of Outsourcing Law Firms." *Journal of Intelligent and Fuzzy Systems* 38 (1): 749–59. doi:10.3233/JIFS-179446.

Karaşan, Ali, Eda Bolturk, and Cengiz Kahraman. 2020. "An Integrated Interval-Valued Neutrosophic AHP and TOPSIS Methodology for Sustainable Cities' Challenges." *Advances in Intelligent Systems and Computing* 1029: 653–61. doi:10.1007/978-3-030-23756-1_79.

Kürüm Varolgüneş, Fatma. 2021. "Post-Disaster Permanent Housing: The Case of the 2003 Bingöl Earthquake in Turkey." *Disaster Prevention and Management: An International Journal* 30 (2): 77–93. doi:10.1108/DPM-09-2019-0307/FULL/PDF.

Myers, James H. and Mark I. Alpert. 1977. "Semantic Confusion in Attitude Research: Salience vs. Importance vs. Determinance." *ACR North American Advances* 4 (1): 106–110.

Radwan, Nouran M., M. Badr Senousy, and Alaa El Din M. Riad. 2016. "Neutrosophic AHP Multi Criteria Decision Making Method Applied on the Selection of Learning Management System." *International Journal of Advancements in Computing Technology* 8 (5): 95–105.

Roh, Sae Yeon, Hyun Mi Jang, and Chul Hwan Han. 2013. "Warehouse Location Decision Factors in Humanitarian Relief Logistics." *Asian Journal of Shipping and Logistics* 29 (1): 103–20. doi:10.1016/j.ajsl.2013.05.006.

Saaty, Thomas. 1988. "Axiomatic foundation of the analytic hierarchy process." *Management Science* 32 (7): 841–55. doi:10.1287/mnsc.32.7.841.

Sari, Diana Puspita, Purnawan Adi Wicaksono, Naniek Utami Handayani, and Yessica Efirani Siahaan. 2020. "Integrating AHP, Cluster Analysis, and Fuzzy TOPSIS to Evaluate Emergency Warehouse Locations of Mount Merapi Eruption Victims." *International Journal on Advanced Science, Engineering and Information Technology* 10 (5): 1839–45. doi:10.18517/ijaseit.10.5.6923.

Smarandache, F. 1998. "Neutrosophy: Neutrosophic Probability, Set, and Logic: Analytic Synthesis & Synthetic Analysis." Rehoboth, NM: American Research Press.

Tanyaş, Mehmet, Yavuz Günalay, Levent Aksoy, and Burak Küçük. 2013. "Afet Lojistik Yönetiminde Rize İline Yönelik Yeni Model Önerisi." In *II. Rize Kalkınma Sempozyumu*. Rize, Turkey.

Taymaz, Tuncay, Haluk Eyidoğan, and James Jackson. 1991. "Source Parameters of Large Earthquakes in the East Anatolian Fault Zone (Turkey)." *Geophysical Journal International*. doi:10.1111/j.1365-246X.1991.tb06328.x.

Thor, Jureen, Siew-hong Ding, and Shahrul Kamaruddin. 2013. "Comparison of Multi Criteria Decision Making Methods From The Maintenance Alternative Selection Perspective." *International Journal Of Engineering And Science (IJES)* 2 (6): 27–34.

Tian, Weidong, Haiqiu Jiang, Hongjuan Zhou, Dan Guo, Wenbo Li, and Yang Lu. 2013. "A Decision-Making Model for Distribution Center Location during Earthquake Responses." *Advanced Materials Research* 629: 808–14. doi:10.4028/www.scientific.net/AMR.629.808.

Timperio, Giuseppe, Gajanan Bhanudas Panchal, Avinash Samvedi, Mark Goh, and Robert De Souza. 2017. "Decision Support Framework for Location Selection and Disaster Relief Network Design." *Journal of Humanitarian Logistics and Supply Chain Management* 7 (3): 222–45. doi:10.1108/JHLSCM-11-2016-0040

Tumsekcali, Ecem, Ertugrul Ayyildiz, and Alev Taskin. 2021. "Interval Valued Intuitionistic Fuzzy AHP-WASPAS Based Public Transportation Service Quality Evaluation by a New Extension of SERVQUAL Model: P-SERVQUAL 4.0." *Expert Systems with Applications* 186: 115757. doi:10.1016/J.ESWA.2021.115757

Turan, Menaf. 2012. "Lessons Learnt from Van and Erciş Earthquakes 2011, Turkey: An Evaluation of Disaster Management." *International Journal of Business and Social Science* 3 (22): 42–52.

Vafaei, Nazanin, Rita A. Ribeiro, and Luis M. Camarinha-Matos. 2016. "Normalization Techniques for Multi-Criteria Decision Making: Analytical Hierarchy Process Case Study." *IFIP Advances in Information and Communication Technology* 470: 261–68. doi:10.1007/978-3-319-31165-4_26

Xu, Dong-Sheng, Cun Wei, and Gui-Wu Wei. 2017. "TODIM Method for Single-Valued Neutrosophic Multiple Attribute Decision Making." *Information* 8 (4): 125. doi:10.3390/info8040125

Yildiz, Aslihan, Ertuğrul Ayyildiz, Coşkun Özkan, and Alev Taşkin-Gümüs. 2020. "Drinking Water Distribution Problem for Assembly Areas after Earthquake: A Case Study in Sakarya, Turkey." *Turk Hijyen ve Deneysel Biyoloji Dergisi* 77 (Supplement 4): 21–34. doi:10.5505/TurkHijyen.2020.28000

Yılmaz, Hafize and Özgür Kabak. 2020. "Prioritizing Distribution Centers in Humanitarian Logistics Using Type-2 Fuzzy MCDM Approach." *Journal of Enterprise Information Management* 33 (5): 1199–1232. doi:10.1108/JEIM-09-2019-0310

Zadeh, L. A. 1965. "Fuzzy Sets." *Information and Control* 8 (3): 338–53. doi:10.1016/S0019-9958(65)90241-X

Zavadskas, E. K., Z. Turskis, J. Antucheviciene, and A. Zakarevicius. 2012. "Optimization of Weighted Aggregated Sum Product Assessment." *Elektronika Ir Elektrotechnika* 122 (6): 3–6. doi:10.5755/j01.eee.122.6.1810

18 UAV Selection for Post-Disaster Medical Supply Distribution Using a Two-Level Interval-Valued Pythagorean Fuzzy AHP Integrated Interval-Valued Pythagorean Fuzzy VIKOR Methodology, and an Application

Pelin Gulum Tas,[1] Bahar Yalcin Kavus,[2] Ertugrul Ayyildiz,[3] and Alev Taskin[4]

[1]Delft University of Technology, Faculty of Technology, Policy and Management

[2]Nisantasi University, bahar.yalcin@nisantasi.edu.tr

[3]Karadeniz Technical University

[4]Yildiz Technical University

CONTENTS

18.1 Introduction 268
18.2 Evaluation Criteria 269
18.3 The Proposed Methodology 270
  18.3.1 Interval Valued Pythagorean Fuzzy Sets 271
  18.3.2 Interval-Valued Pythagorean Fuzzy Analytic Hierarchy Process (IVPF-AHP) 272

DOI: 10.1201/9781003212904-18

18.3.3 Interval-Valued Pythagorean Fuzzy Vikor (IVPF-VIKOR)..........273
18.4 Case Study 275
18.5 Conclusion 276
References 278

18.1 INTRODUCTION

In parallel with technological developments, post-disaster operations are transforming and improving. Traditional methods give way to new, faster, and more effective methods that offer better (without a) performance. Unmanned aerial vehicles (UAVs) have started to attract attention in carrying out different operations in disaster-affected areas (Chowdhury et al. 2021). Compared to helicopters or airplanes, they are more compact, easy to control, and do not require a special area for take-off and landing activities. These features make UAVs an increasingly popular vehicles (Nur et al. 2020). They are becoming more frequently preferred, especially in areas where transportation is almost impossible after a disaster, for purposes such as damage detection (Ezequiel et al. 2014), image collection (Mishra et al. 2020), and signal reinforcement by acting as a base station (Erdelj et al. 2017). In addition to these activities, especially for developing countries, the task of using UAVs for transportation in post-disaster response operations is promising (Iqbal et al. 2015). Although UAV delivery for commercial purposes and some medical supplies has been discussed, there are still limited studies in post-disaster transportation operations (Panda et al. 2019). Since post-disaster regions may have many features hindering the task of transporting medical supplies with UAVs, the preferred one should be chosen meticulously. Although there are studies in the literature of UAV selection problems for classical transportation or disaster-response missions, no research has been found on UAVs that will carry medical supplements specifically. For instance, Nur et al. (2020) investigated which UAV configuration would be most appropriate in UAV delivery missions using fuzzy TOPSIS, one of the multi-criteria decision-making (MCDM) methods. In their study, they defined a wide hierarchy of criteria and performed applications for different scenarios after expert evaluations. Similarly, Iqbal et al. (2015), addressed the problem of UAV selection for post-disaster recovery and compared five models according to different specifications. They decided that the rotary wing system is the most appropriate type of UAV for given specific purposes.

This study deals with the selection of a UAV model that will specifically perform the task of transporting medical supplies in disaster areas. First, a wide hierarchy of criteria affecting the performance of a UAV in disaster areas is presented. Second, the importance weights are revealed with the help of expert evaluations using the interval-valued Pythagorean fuzzy analytic hierarchy process (IVPF-AHP) and interval-valued Pythagorean fuzzy VIKOR (IVPF-VIKOR) methods. Then, four popular UAV models in the market, most often preferred in transport missions, were evaluated with the proposed methodology instead of method. The most suitable UAV model for transportation between medical equipment warehouses and areas of need post-disaster is defined.

18.2 EVALUATION CRITERIA

In this study, the main and sub-criteria were determined via a comprehensive literature review and then formed by listening to expert opinions. In order to represent the problem, a two-level criteria hierarchy was preferred, and a total of 16 sub-criteria related to three main criteria were included. The criterion hierarchy is provided in Table 18.1.

First, the criteria are classified into three groups: *technical specifications*, *performance*, and *reliability and security*. *Technical specifications*, which has six sub-criteria, represents the physical and technical characteristics of the UAV. *Speed limit*

TABLE 18.1
Main and Sub-Criteria

Main Criteria	Sub-Criterion	References
1. Technical Specifications	1.1 Speed limit	(Nur et al. 2020; Yaacoub et al. 2020; Iqbal et al. 2015; Braun et al. 2019; Scott and Scott 2017; Fu et al. 2021)
	1.2 Weight	(Nur et al. 2020; Yaacoub et al. 2020; Iqbal et al. 2015; Fu et al. 2021; EUCHI 2021)
	1.3 Range	(Nur et al. 2020; Iqbal et al. 2015; Fu et al. 2021; EUCHI 2021)
	1.4 Total recharge/refuel time	(Nur et al. 2020; Fu et al. 2021; Hong, Kuby, and Murray 2017)
	1.5 Cargo capacity	(Nur et al. 2020; Yaacoub et al. 2020; Iqbal et al. 2015; Braun et al. 2019; Scott and Scott 2017; Fu et al. 2021)
	1.6 Propeller types (fixed, rotary…)	(Nur et al. 2020; Yaacoub et al. 2020; Iqbal et al. 2015; Braun et al. 2019; Fu et al. 2021; EUCHI 2021)
2. Performance	2.1 Charge endurance	(Nur et al. 2020; EUCHI 2021)
	2.2 Location accuracy	(Nur et al. 2020; Yaacoub et al. 2020; Scott and Scott 2017)
	2.3 Traceability and remote-control convenience	(Nur et al. 2020; Yaacoub et al. 2020; Iqbal et al. 2015; Braun et al. 2019; Fu et al. 2021)
	2.4 Maneuverability and flexibility	(Nur et al. 2020; Yaacoub et al. 2020; Braun et al. 2019; EUCHI 2021; Xiao et al. 2016)
	2.5 Loading–unloading practicality	(Nur et al. 2020; Iqbal et al. 2015)
	2.6 Maintenance and operating costs	(Nur et al. 2020; Yaacoub et al. 2020; Iqbal et al. 2015; Fu et al. 2021)
3. Reliability and Security	3.1 Handling and delivery security	(Nur et al. 2020; Fu et al. 2021; EUCHI 2021)
	3.2 Communication and cyber security	(Nur et al. 2020; Yaacoub et al. 2020; Braun et al. 2019; Shakeri et al. 2019)
	3.3 Emergency landing and collision safety system	(Nur et al. 2020; Yaacoub et al. 2020; Fu et al. 2021)
	3.4 Disturbance effect	(Iqbal et al. 2015)

indicates the maximum speed limit of the UAV in transport missions (Nur et al. 2020; Braun et al. 2019; Scott and Scott 2017). The *weight* of the UAV is critical as it is directly related to factors such as agility, power consumption, and speed (Nur et al. 2020; EUCHI 2021). UAVs have different weights depending on the material they are made of and their structure, but UAVs used for transportation are usually weigh between 5 and 8.6 kg (Nur et al. 2020). The *range* represents the maximum range of the UAV (Iqbal et al. 2015). The *total recharge/refuel time* indicates the time to recharge or refuel the UAV and must be performed as quickly as possible to avoid disruption to transport operations (Nur et al. 2020). *Cargo capacity* represents the maximum payload that can be carried by the UAV and varies depending on many factors such as model, loading mechanism, power, etc. (Iqbal et al. 2015; Braun et al. 2019). While the *propeller types* criterion, which represents the type of UAVs, affects different variables such as speed, fuel consumption, and maneuverability, it has been determined that some models are more suitable for transport tasks than others (Nur et al. 2020; Iqbal et al. 2015).

The *performance* criterion represents all the capabilities and features of the UAVs in fulfilling the transport task and includes six sub-criteria, detailed below. The *charge endurance* indicates the carrying performance and endurance of UAVs over a single charge period (Nur et al. 2020; Fu et al. 2021). The *location accuracy* refers to the ability of the UAV to accurately detect the location of the delivery point (Nur et al. 2020; Fu et al. 2021). *Traceability and remote-control convenience* is important so that the UAV performing the transport task can be coordinated and continuously monitored from a remote location such as the disaster coordination center (Yaacoub et al. 2020; Scott and Scott 2017; EUCHI 2021). *Maneuverability and flexibility* refers to the maneuverability of the UAV in motion and its ability to adapt to rapid changes in route or location (Nur et al. 2020; Yaacoub et al. 2020). *Loading–unloading practicality* refers to the speed of loading and unloading of the cargo. *Maintenance and operating costs* indicates the costs incurred for the periodic maintenance of the UAV and keeping it up and running (Yaacoub et al. 2020). *Reliability and security*, a two-dimensional main criterion, represents both the reliability of the UAV while performing the task and the risk of causing danger in the region while doing so. The *Handling and delivery security* sub-criterion refers to the ability of a UAV to pick up loads and deliver them undamaged and safely (Nur et al. 2020; Scott and Scott 2017; Hong, Kuby, and Murray 2017). *Communication and cyber security* refers to the safe and correct communication of the UAV with the central point, and its security against information loss due to cyber-attacks (Yaacoub et al. 2020). *Emergency landing and collision safety system* refers to the UAV's security system in landing safely in the event of a collision or other emergency (Yaacoub et al. 2020; Fu et al. 2021). Finally, *disturbance effect* refers to factors such as noise and wind, and that the UAV does not disturb other people and processes in the disaster area during transportation.

18.3 THE PROPOSED METHODOLOGY

In this chapter, IVPF-AHP and IVPF-VIKOR methods are proposed to select a suitable UAV to serve not go between a medical supply warehouse and disaster areas after an earthquake. This section defines the proposed methodology.

18.3.1 Interval Valued Pythagorean Fuzzy Sets

Pythagorean fuzzy sets (PFS) have often been applied by many researchers to various areas since they were introduced by Yager (2014). Similar to intuitionistic fuzzy sets, PFS include a membership, non-membership, and hesitancy degree. However, the degrees of membership and non-membership can exceed 1 in PFS (Gul et al. 2021).

Definition 1: Assume X is a fixed set and PFS is denoted by P (Zhang and Xu 2014)

$$P=\left\{x,\mu_{\tilde{P}}(x),\ v_{\tilde{P}}(x);x\in X\right\} \tag{18.1}$$

where the degree of membership of the element $x\in X$ to P is defined as $\mu_P(x): X\mapsto[0,1]$. However, the degree of non-membership of the element $x\in X$ to P is defined as $v_P(x): X\mapsto[0,1]$ and for every $x\in X$, it takes:

$$0\le\mu_P(x)^2+v_P(x)^2\le1 \tag{18.2}$$

For any set P and $x\in X$, the hesitancy degree is given by Eq. (18.3):

$$\pi_P^2(x)=1-(\mu_P(x))^2-(v_P(x))^2 \tag{18.3}$$

Definition 2: Let X be a non-empty finite set. The interval-valued PFS (IVPFS) is defined as (Garg 2016):

$$P=\left\{x,\left[\mu^L(x),\mu^U(x)\right],\left[v^L(x),v^U(x)\right],x\in X\right\} \tag{18.4}$$

where $\mu(x)$ and $v(x)$ are the degree of membership and non-membership of the $x\in X$, respectively, where $(\mu^U(x))^2+(v^U(x))^2\le1$. The hesitancy degree is equal to

$$\pi(x)=\left[\pi^L(x),\pi^U(x)\right]=\left[\sqrt{1-(\mu^L(x))^2-(v^L(x))^2},\ \sqrt{1-(\mu^U(x))^2-(v^U(x))^2}\right] \tag{18.5}$$

An interval-valued Pythagorean fuzzy number (IVPFN) is defined as $\left[\mu_A^L(x),\mu_A^U(x)\right],\left[v_A^L(x),v_A^U(x)\right]$. For simplicity, an IVPFN can be denoted by $\alpha=[a,b],[c,d]$. The score function of an IVPFN is defined as:

$$S(\alpha)=\frac{a^2+b^2-c^2-d^2}{2} \tag{18.6}$$

Definition 3: Let $\alpha=[a,b],[c,c]$, $\alpha_1=\left[a_1,b_1\right],\left[c_1,d_1\right]$, and $\alpha_2=\left[a_2,b_2\right],\left[c_2,d_2\right]$ be three IVPFNs, and $\lambda>0$. Then, the operations of them are defined as (Peng and Yang 2016):

$$\alpha_1 \oplus \alpha_2 = \left\langle \left[\sqrt{a_1^2 + a_2^2 - a_1^2 a_2^2}, \sqrt{b_1^2 + b_2^2 - b_1^2 b_2^2} \right], \left[c_1 c_2, d_1 d_2 \right] \right\rangle \tag{18.7}$$

$$\alpha_1 \otimes \alpha_2 = \left\langle \left[a_1 a_2, b_1 b_2 \right], \left[\sqrt{c_1^2 + c_2^2 - c_1^2 c_2^2}, \sqrt{d_1^2 + d_2^2 - d_1^2 d_2^2} \right] \right\rangle \tag{18.8}$$

$$\alpha^\lambda = \left\langle \left[a^\lambda, b^\lambda \right], \left[\sqrt{1-\left(1-c^2\right)^\lambda}, \sqrt{1-\left(1-d^2\right)^\lambda} \right] \right\rangle \tag{18.9}$$

$$\lambda\alpha = \left\langle \left[\sqrt{1-\left(1-a^2\right)^\lambda}, \sqrt{1-\left(1-b^2\right)^\lambda} \right], \left[c^\lambda, d^\lambda \right] \right\rangle \tag{18.10}$$

Definition 4: The distance between two IVPFNs $\alpha_1 = \left[a_1, b_1\right], \left[c_1, d_1\right]$ and $\alpha_2 = \left[a_2, b_2\right], \left[c_2, d_2\right]$ can be defined as follows (Wu, Huang, and Geng 2021):

$$d\left(\alpha_1, \alpha_2\right) = \frac{1}{4}\Big(\left|a_1^2 - a_2^2\right| + \left|b_1^2 - b_2^2\right| + \left|c_1^2 - c_2^2\right| + \left|d_1^2 - d_2^2\right| + \left|\left(\pi_{\alpha_1}^L\right)^2 - \left(\pi_{\alpha_2}^L\right)^2\right| + \left|\left(\pi_{\alpha_1}^U\right)^2 - \left(\pi_{\alpha_2}^U\right)^2\right|\Big) \tag{18.11}$$

18.3.2 Interval-Valued Pythagorean Fuzzy Analytic Hierarchy Process (IVPF-AHP)

In this section, the IVPF-AHP method is presented step by step (Rosanisah, Mohd, and Abdullah 2017).

Step 1. First, in order to achieve linguistic pairwise comparisons, expert opinions were sought.

Then, a linguistic pairwise comparison matrix $R = (rjt)_{mxn}$ was formed according to the decision model and expert opinions. Table 18.2 demonstrates the linguistic pairwise comparison scale preferred in this study.

Step 2. The differences matrix, defined as $D = (d_{ij})_{mxm}$, was calculated using Eqs. (18.12) and (18.13). These equations present the difference between lower and upper points of the membership function and non-membership functions, respectively.

$$d_{ij}^L = \mu_{ij}^{L2} - v_{ij}^{L2} \tag{18.12}$$

$$d_{ij}^U = \mu_{ij}^{U2} - v_{ij}^{U2} \tag{18.13}$$

Step 3. The interval multiplicative matrix was defined as $S = (s_{ij})_{mxm}$. This matrix was calculated via Eqs. (18.14) and (18.15):

TABLE 18.2
Linguistics Pairwise Comparison Scale for the IVPF-AHP Evaluations

	IVPF Number			
Linguistic Terms	μ_L	μ_U	v_L	v_u
AHI—Absolutely High Importance	0.90	1.00	0.00	0.00
VHI—Very High Importance	0.80	0.90	0.10	0.20
HI—High Importance	0.65	0.80	0.20	0.35
AAI—Above Average Importance	0.55	0.65	0.35	0.45
EI—Equal Importance	0.45	0.55	0.45	0.55
BAI—Below Average Importance	0.35	0.45	0.55	0.65
LI—Low Importance	0.20	0.35	0.65	0.8
VLI—Very Low Importance	0.10	0.2	0.8	0.9
ALI—Absolutely Low Importance	0.00	0.00	0.90	1.00

$$s_{ij}^{L} = \sqrt{1000^{d_{ij}^{L}}} \tag{18.14}$$

$$s_{ij}^{U} = \sqrt{1000^{d_{ij}^{U}}} \tag{18.15}$$

Step 4. The indeterminacy value was defined as $H = (h_{ij})_{mxm}$ of the r_{jt} and was calculated by using Eq. (18.16):

$$h_{ij} = 1 - \left(\mu_{ij}^{U2} - \mu_{ij}^{L2}\right) - \left(v_{ij}^{U2} - v_{ij}^{L2}\right) \tag{18.16}$$

Step 5. To unnormalized weights $T = (\tau_{ij})_{mxm}$ via Eq. (18.17), the indeterminacy values were multiplied with the $S = (s_{ij})_{mxm}$ matrix to calculate the matrix of:

$$t_{ij} = \left(\frac{s_{ij}^{L} + s_{ij}^{U}}{2}\right) h_{ij} \tag{18.17}$$

Step 6. The fuzzy weights for criteria of dimensions w_i were found via Eq. (18.18).

$$w_i = \frac{\sum_{i=1}^{m} t_{ij}}{\sum_{i=1}^{m}\sum_{j=1}^{m} t_{ij}} \tag{18.18}$$

18.3.3 Interval-Valued Pythagorean Fuzzy Vikor (IVPF-VIKOR)

VIKOR, presented by Opricovic in (2004), is a convenient method used to solve complex MCDM problems. Under the defined criteria the method allows determination

TABLE 18.3
Linguistic Terms for IVPF-VIKOR Evaluations

	IVPF Number			
Linguistic Terms	μ_L	μ_U	v_L	v_u
VB—Very Bad	0.00	0.10	0.85	0.95
B—Bad	0.10	0.25	0.70	0.85
MB—Moderately Bad	0.25	0.40	0.55	0.70
M—Moderate	0.40	0.55	0.40	0.55
MG—Moderately Good	0.55	0.70	0.25	0.40
G—Good	0.70	0.85	0.10	0.25
VG—Very Good	0.85	0.95	0.00	0.10

of consensus-based ranking and for a compromise solution to be reached. With the VIKOR method, alternatives are evaluated as stated by the pre-determined criteria and then, by determining the positive and negative ideal solutions, the closeness values of the alternatives to the ideal solutions are calculated, and thus the most appropriate ranking of the alternatives is obtained. However, sometimes crisp sets fail to reflect the ambiguity of human thoughts. In order to avoid this uncertainty, in this study IVPF-VIKOR was used to better represent the fuzziness in human thoughts. The IVPF-VIKOR method steps are given below (Wu, Huang, and Geng 2021).

Step 1. Based on the linguistic terms given in Table 18.3, the alternative evaluation decision matrix was constructed. $\tilde{R}_{ij}$ shows the Pythagorean fuzzy evaluation value of alternative i according to criterion j.

Step 2. The positive $\left(P_j^+\right)$ and negative ideal solutions $\left(P_j^-\right)$ were determined based on the score values for each criterion. The score value of $\tilde{X}_{ij}$ was calculated by Equations (18.19) and (18.20).

$$P_j^+ = \max_i\left\{\tilde{\mathrm{R}}_{\mathrm{ij}}\right\} \tag{18.19}$$

$$P_j^+ = \min_i\left\{\tilde{\mathrm{R}}_{\mathrm{ij}}\right\} \tag{18.20}$$

Step 3. The group utility (S_i) and individual regret (R_i) values were determined. In this step criteria weights calculated by IVPF-AHP were used.

$$S_i = \sum_{j=1}^{n} \frac{w_j d\left(R_{ij}, P_j^+\right)}{d\left(P^+, P_j^-\right)} \tag{18.21}$$

$$R_i = \max \frac{w_j d\left(R_{ij}, P_j^+\right)}{d\left(P^+, P_j^-\right)} \tag{18.22}$$

Step 4. The compromise (Q_i) value was calculated for each alternative:

$$Q_i = v\frac{(S_i - S^+)}{(S^- - S^+)} + (1-v)\frac{(R_i - R^+)}{(R^- - R^+)} \tag{18.23}$$

where $S^+ = \min_i S_i; S^- = \max_i S_i$ and $R^+ = \min_i R_i; R^- = \max_i R_i$. v is the weight of strategy of maximum.

Step 5. Alternatives were ordered by the values of S_i, R_i, and Q_i.

Step 6. A compromise solution for the given set of criteria, the alternative (a') which is ranked as the first by the minimum Q_i was proposed if the following two conditions were satisfied (Tzeng, Lin, and Opricovic 2005):

C1: "Acceptable advantage": $Q(a'') - Q(a') \geq D(Q)$; (a'') is the second alternative in the ranking by Q_i and $D(Q) = 1/(m-1)$;

C2: "Acceptable stability in decision-making": alternative (a') must be best ranked by S_i and/or R_i.

If one condition is not satisfied, then a set of compromise solutions was proposed (San Cristóbal 2011):

- If C2 was not satisfied, Alternative (a') and Alternative (a'');
- If C1 was not satisfied, Alternative (a'), Alternative (a'') ... Alternative (Z) was determined by the relation $Q(Z) - Q(a') < D(Q)$ for maximum Z.

18.4 CASE STUDY

First, all the criteria in Table 18.1 were evaluated by experts and a linguistic evaluation matrix was obtained. After that, the main and sub-criteria weights were calculated by applying the IVPF-AHP method to the matrix obtained. Finally, four different UAV alternatives were evaluated according to the given criterion, and one was selected as the most suitable model for the medical supply-carrying task. The weights for three main and 16 sub-criteria were calculated as in Table 18.4 by applying the steps of the proposed methodology.

While the main criterion with the highest importance was *technical specifications* with a weight of 0.6549, the most important among the sub-criteria were *charging endurance* and *emergency landing*. The criteria weights obtained were used in the IVPF-VIKOR method in the next step to evaluate four different types of UAVs: UAV-1, UAV-2, UAV-3, and UAV-4. The abbreviated UAV names are used for privacy reasons, but Table 18.5 presents the general attributes of each, in order to give a rough idea of their main features. Not all features and criteria are included, so that the privacy of the brands and models is not violated. The intention is to give a general impression of the models and their capabilities instead of all criteria values.

TABLE 18.4
Calculated Criteria Weights

Criterion	Weight	Criterion	Weight	Criterion	Weight
C1	0.6549	C2	0.2785	C3	0.0668
C11	0.1931	C21	0.5763	C31	0.4496
C12	0.1344	C22	0.1572	C32	0.0529
C13	0.0557	C23	0.6033	C33	0.4729
C14	0.1024	C24	0.1155	C34	0.0248
C15	0.3341	C25	0.0604		
C16	0.1806	C26	0.0277		

TABLE 18.5
UAV Models and Some of Their Main Features

UAV	Type	Takeoff System	Delivery Method	Altitude	Cargo Capacity	Speed	Range
UAV-1	Fixed wing	VTOL	Automated skyport	7.5 miles	2.56 kg	140 km/h	65 km
UAV-2	Quad copter	VTOL	Automated ground station	6.25 miles	2.2 kg	40 km/h	20 km
UAV-3	Hexacopter	VTOL	Dropped by rope	20 miles	2 kg	90 km/h	32 km
UAV-4	Fixed wing	Runway	Paper parachute	45 miles	1.75 kg	128 km/h	160 km

After the same expert group was consulted for their opinions, an alternative evaluation with respect to each criterion using linguistic terms was done and the results are presented in Table 18.6.

After obtaining criteria-alternative evaluations from the experts, the S_i, R_i, and Q_i values were calculated. In this step, the criteria weights obtained by IVPF-AHP were used and the λ was assumed to be 0.5. The S_i, R_i, and Q_i values for each alternative are presented in Table 18.7.

UAV-1 was awarded first place in each ranking. Two conditions were satisfied for this alternative. Therefore, "UAV-1" was determined to be the most convenient UAV to be used in post-disaster medical supply distribution tasks.

18.5 CONCLUSION

In this chapter, the problem of selecting a UAV for post-disaster medical supply transportation operations is discussed. An IVPF-AHP and IVPF-VIKOR two-level method was proposed to determine the most appropriate criteria hierarchy and to define the most convenient model for transportation services among four different UAVs. The results of the evaluation process, which was created as a result of a wide hierarchy of criteria and expert opinions, were as follows. (1) The most important criterion when choosing a UAV for a medical supply handling task is the

TABLE 18.6
UAVs Evaluation Matrix

UAV	C11	C12	C13	C14	C15	C16
UAV-1	VG	VG	G	VG	VG	G
UAV-2	MB	MG	MB	M	MG	MG
VAV-3	M	B	M	M	M	MG
UAV-4	G	MB	VG	G	MB	MG

UAV	C21	C22	C23	C24	C25	C26
UAV-1	G	VG	VG	M	VG	M
UAV-2	B	G	G	MG	VG	M
UAV-3	MG	G	G	M	G	M
UAV-4	VG	VG	VG	M	G	M

UAV	C31	C32	C33	C34
UAV-1	VG	VG	G	G
UAV-2	MG	VG	VG	MG
UAV-3	G	G	MG	MG
UAV-4	MG	G	MG	G

TABLE 18.7
S_i, R_i, and Q_i Values and Alternative Rankings

UAV	S_i	R_i	Q_i	Rank (S_i)	Rank (R_i)	Rank (Q_i)
UAV-1	0.108	0.048	0.000	1	1	1
UAV-2	0.785	0.160	0.791	3	2	2
UAV-3	0.841	0.181	0.889	4	3	4
UAV-4	0.599	0.219	0.834	2	4	3

technical specifications of the UAV, followed by performance parameters. (2) In terms of sub-criteria, the sub-criterion of highest importance is *charge endurance*. The UAV model chosen, which can last for a number of hours without losing performance, will be more suitable for disaster areas where a time limit is critical. When an emergency landing is required, the second most important criterion is the ability of the UAV to respond to it, while safety is also important during the taking and delivery of the load (third most important criterion). Obviously, it is important that the transportation must be fast and effective, but it is important not to endanger the people around the site. Considering all the criteria, UAV-1 was found to be the most suitable on the market for the transportation mission. The fact that it is currently being used in transporting medical supplies to certain parts of Africa also supports the results (Triche, Greve, and Dubin 2020).

This research can be expanded by increasing the number of criteria and alternatives or applying different fuzzy sets to handle uncertainty. In future studies, parameters such as climatic conditions and disaster type could be included in the selection problem.

REFERENCES

Braun, Jonathan, S. David Gertz, Ariel Furer, Tarif Bader, Hagay Frenkel, Jacob Chen, Elon Glassberg, and DeanNachman. 2019. "The Promising Future of Drones in Prehospital Medical Care and Its Application to Battlefield Medicine." *The Journal of Trauma and Acute Care Surgery* 87 (1S Suppl 1): S28–34. doi:10.1097/TA.0000000000002221

Chowdhury, Sudipta, Omid Shahvari, Mohammad Marufuzzaman, Xiaopeng Li, and Linkan Bian. 2021. "Drone Routing and Optimization for Post-Disaster Inspection." *Computers & Industrial Engineering* 159: 107495. doi:10.1016/J.CIE.2021.107495

Erdelj, Milan, Enrico Natalizio, Kaushik R. Chowdhury, and Ian F. Akyildiz. 2017. "Help from the Sky: Leveraging UAVs for Disaster Management." *IEEE Pervasive Computing*. doi:10.1109/MPRV.2017.11

Euchi, Jalel. 2021. "Do Drones Have a Realistic Place in a Pandemic Fight for Delivering Medical Supplies in Healthcare Systems Problems?" *Chinese Journal of Aeronautics* 34 (2): 182–90. doi:10.1016/j.cja.2020.06.006

Ezequiel, Carlos Alphonso F., Matthew Cua, Nathaniel C. Libatique, Gregory L. Tangonan, Raphael Alampay, Rollyn T. Labuguen, Chrisandro M. Favila, et al. 2014. "UAV Aerial Imaging Applications for Post-Disaster Assessment, Environmental Management and Infrastructure Development." In *2014 International Conference on Unmanned Aircraft Systems, ICUAS 2014 – Conference Proceedings*. doi:10.1109/ICUAS.2014.6842266

Fu, Chen-Hua, Ming-Wen Tsao, Li-Pin Chi, and Zheng-Yun Zhuang. 2021. "On the Dominant Factors of Civilian-Use Drones: A Thorough Study and Analysis of Cross-Group Opinions Using a Triple Helix Model (THM) with the Analytic." *Drones*. doi:10.3390/drones5020046

Garg, Harish. 2016. "A Novel Accuracy Function under Interval-Valued Pythagorean Fuzzy Environment for Solving Multicriteria Decision Making Problem." *Journal of Intelligent and Fuzzy Systems* 31 (1): 529–40. doi:10.3233/IFS-162165

Gul, Muhammet, Suleyman Mete, Faruk Serin, and Erkan Celik. 2021. "Fine–Kinney-Based Occupational Risk Assessment Using Interval-Valued Pythagorean Fuzzy Vikor." *Studies in Fuzziness and Soft Computing* 398: 45–68. doi:10.1007/978-3-030-52148-6_4

Hong, Insu, Michael Kuby, and Alan Murray. 2017. "A Deviation Flow Refueling Location Model for Continuous Space: A Commercial Drone Delivery System for Urban Areas." *Advances in Geographic Information Science*, 125–32. doi:10.1007/978-3-319-22786-3_12

Iqbal, Umair, Syed Irtiza, Ali Shah, Mohsin Jamil, Syed Omer Gilani, Syed Omer Gillani, and Yasar Ayaz. 2015. "Selection of Unmanned Aerial System (UAS) for Disaster Relief Operations: A Comparison Deep Learning for Real-Time Myoelectric Control View Project Flow Separation Detection on a 3D Printed NACA 0012 Airfoil Using SOI CMOS MEMS Sensor and a Data Acquisition System View Project." *Science International (Lahore)* 27 (4): 3199–3203.

Mishra, Balmukund, Deepak Garg, Pratik Narang, and Vipul Mishra. 2020. "Drone-Surveillance for Search and Rescue in Natural Disaster." *Computer Communications* 156: 1–10. doi:10.1016/J.COMCOM.2020.03.012

Nur, Farjana, Ayat Alrahahleh, Reuben Burch, Kari Babski-Reeves, and Mohammad Marufuzzaman. 2020. "Last Mile Delivery Drone Selection and Evaluation Using the Interval-Valued Inferential Fuzzy TOPSIS." *Journal of Computational Design and Engineering* 7 (4): 397–411. doi:10.1093/jcde/qwaa033.

Panda, Kirtan Gopal, Shrayan Das, Debarati Sen, and Wasim Arif. 2019. Design and Deployment of UAV-Aided Post-Disaster Emergency Network." *IEEE Access* 7: 102985–99. doi:10.1109/ACCESS.2019.2931539

Peng, Xindong and Yong Yang. 2016. "Fundamental Properties of Interval-Valued Pythagorean Fuzzy Aggregation Operators." *International Journal of Intelligent Systems* 31 (5): 444–87. doi:10.1002/int.21790

Rosanisah, Wan, Wan Mohd, and Lazim Abdullah. 2017. "Pythagorean Fuzzy Analytic Hierarchy Process to Multi-Criteria Decision Making." *AIP Conference Proceedings*. doi:10.1063/1.5012208

San Cristóbal, J. R. 2011. "Multi-Criteria Decision-Making in the Selection of a Renewable Energy Project in Spain: The Vikor Method." *Renewable Energy* 36 (2): 498–502. doi:10.1016/j.renene.2010.07.031

Scott, Judy E. and Carlton H. Scott. 2017. "Drone Delivery Models for Healthcare." *Hawaii International Conference on System Sciences (HICSS)*.

Shakeri, Reza, Mohammed Ali Al-Garadi, Ahmed Badawy, Amr Mohamed, Tamer Khattab, Abdulla Khalid Al-Ali, Khaled A. Harras, and Mohsen Guizani. 2019. "Design Challenges of Multi-UAV Systems in Cyber-Physical Applications: A Comprehensive Survey and Future Directions." *IEEE Communications Surveys and Tutorials* 21 (4): 3340–85. doi:10.1109/COMST.2019.2924143

Triche, Ryan M., Ashley E. Greve, and Scott J. Dubin. 2020. "UAVs and their Role in the Health Supply Chain: A Case Study from Malawi." *2020 International Conference on Unmanned Aircraft Systems, ICUAS 2020*, September. Institute of Electrical and Electronics Engineers, 1241–48. doi:10.1109/ICUAS48674.2020.9214064

Tzeng, Gwo Hshiung, Cheng Wei Lin, and Serafim Opricovic. 2005. "Multi-Criteria Analysis of Alternative-Fuel Buses for Public Transportation." *Energy Policy* 33 (11): 1373–83. doi:10.1016/j.enpol.2003.12.014

Wu, Weidong, Peng Huang, and Shuai Geng. 2021. "Application of Interval-Valued Pythagorean Fuzzy VIKOR Approach for Petroleum Sludge Treatment Technology Evaluation and Selection." *Environmental Science and Pollution Research* 28 (36): 50890–907. doi:10.1007/S11356-021-14225-6

Xiao, Xiongwu, Bingxuan Guo, Deren Li, Linhui Li, Nan Yang, Jianchen Liu, Peng Zhang, and Zhe Peng. 2016. "Multi-View Stereo Matching Based on Self-Adaptive Patch and Image Grouping for Multiple Unmanned Aerial Vehicle Imagery." *Remote Sensing* 8 (2). doi:10.3390/rs8020089

Yaacoub, Jean Paul, Hassan Noura, Ola Salman, and Ali Chehab. 2020. "Security Analysis of Drones Systems: Attacks, Limitations, and Recommendations." *Internet of Things* 11: 100218. doi:10.1016/J.IOT.2020.100218

Yager, Ronald R. 2014. "Pythagorean Membership Grades in Multicriteria Decision Making." *IEEE Transactions on Fuzzy Systems* 22 (4): 958–65. doi:10.1109/TFUZZ.2013.2278989

Zhang, Xiaolu and Zeshui Xu. 2014. "Extension of TOPSIS to Multiple Criteria Decision Making with Pythagorean Fuzzy Sets." *International Journal of Intelligent Systems* 29 (12): 1061–78. doi:10.1002/INT.21676

19 A Simulation-Based Approach for Decision-Making in Earthquake Crisis Management

Hassan Ahmadi Choukolaei,[1]
Mustafa Jahangoshai Rezaee,[2]
Samuel Yousefi,[3] *and Morteza Saberi*[4]

[1]Faculty of Industrial Engineering, Urmia University of Technology, Urmia, Iran

[2]Faculty of Industrial Engineering, Urmia University of Technology, Urmia, Iran

[3]School of Engineering, University of British Columbia, Okanagan Campus, Kelowna, Bc V1v 1v7, Canada

[4]School of Information, Systems and Modelling, University of Technology Sydney, Australia

CONTENTS

19.1 Introduction 281
19.2 The Role of Simulation in Preparing for Disasters and Natural Disasters 282
19.3 Investigating the Performance of Simulation Structure in the Problem of Crisis Management in a Tehran Earthquake 284
19.4 Simulation Model for Evaluating Relief Goods 285
19.5 Evaluating Relief Centers in Terms of Housing Priority and Relief Priority for the Allocation of Relief Goods 290
19.6 Conclusion 300
References 300

19.1 INTRODUCTION

Simulation is used in many fields, such as technology simulation for performance tuning or optimization, safety engineering, testing, training, and video games. Computer software or hardware is often used to study simulated models. Today,

DOI: 10.1201/9781003212904-19

simulations have been used by scientific modeling of natural or human systems to gain insight into their performance. Simulation can be used to show the real effects of a phenomenon on the target subject, under controlled and regulated conditions. Simulations are also used in cases where the actual system is not usable, as such a system may not be available or its use may be dangerous or unacceptable. The main issues in a simulation are access to reliable sources of information about the proper selection of characteristics and features of key behaviors of the phenomenon under study. Approximately, simplifying and minimizing the assumptions of the simulation model so that it does not distort the real principles of the phenomenon and does not invalidate it are other aspects of simulation. Therefore, simulation is a scientific activity that requires procedures and protocols related to model validation, refinement, research, and development in technology, especially in a computer simulation. As today's business world becomes more competitive and business processes become more complex, the application of simulation has become more widespread. In a study between 1974 and 1988, Lane, Mansour, and Harpell (1993) found that simulation rankings were one of the three most important operational research tools. The two methods of mathematical and statistical programming are in the next steps. After reviewing 1294 articles from various journals from 1970 to 1992, Gupta and Ferguson (1997) found that simulation ranks second after programming. In recent years, a lot of work has been done in the field of upgrading simulation software, so that today, using simulation software and components that are present by default, complex systems can be simulated. And examined. Attempts have also been made to include reporting tools so that users, after building the model, can receive a variety of reports from each member of the system. With the help of simulation, without incurring a lot of time, money, and possibly life, we can perform trial and error by defining different scenarios and generalizing the results to the real world. With the accurate and appropriate design of the simulation structure, the received output information can be used in better crisis management to increase the speed and performance of support teams. In this chapter, we will refer to applications of simulation in crisis management and get acquainted with some of its basic concepts. We will also address the role of simulation in crisis management and natural disasters such as earthquakes by providing criteria for evaluating goods and relief centers, designing the structure of earthquake effects on urban infrastructure, and using geographic information systems (GIS).

19.2 THE ROLE OF SIMULATION IN PREPARING FOR DISASTERS AND NATURAL DISASTERS

Simulation-based education can be a way to prepare people for natural disasters. Simulators can create and repeat emergencies several times and help people to learn how to respond to hazards. Such an experience may never happen, but preparedness for such catastrophes reduces irreparable damage and destruction. Crisis preparedness simulations may include training in how to deal with terrorist attacks, natural disasters, the spread of an epidemic such as COVID-19, or other life-threatening emergencies. In today's world, preparedness and security in the face of adversity are essential. Due to the increase in population in cities and in urban construction, the structure of cities is more endangered than anything else, and having an open urban space becomes more important (Rapoport 2016). This situation can make cities more dangerous in the event of natural disasters. This issue is so important that developed countries devote an important part of their comprehensive and national planning to it. In recent years, many studies have been conducted in the field of natural disasters (Goodarzian et al. 2022; Ahmadi Choukolaei et al. 2021; Ghasemi et al. 2020). Many researchers have focused specifically on planning and policymaking. Marcelin et al. (2016) used GIS and a p-median-based modeling framework to locate crisis relief facilities based on the age of the victims. The results showed that emergency planners should consider the elderly population and their disaster-related needs in future planning to ensure a fairer distribution system. Chen et al. (2017) developed a systematic planning framework using the opinion of experts and GIS, and based on system theory, to create a temporary settlement as a temporary relief center in Guangzhou. Dabbagh and Chouklaei (2020) evaluated the relief centers considered by the Urmia City Crisis Management. They used the analytic hierarchy process method to weigh the criteria and an evaluation method to prioritize the options. Safety criteria, time of use, and coverage level were the most important research criteria. Borhani et al. (2021) identified multi-purpose shelters and places in critical situations using GIS and multi-criteria decision-making (MCDM) methods. They used the criteria of experts to determine the weight of the criteria, and used simple additive weighting model GIS to integrate the information layers. Ghasemi et al. (2020) used the simulation system to calculate the number of relief goods in the event of an earthquake in different scenarios as random input requests for their mathematical planning model. A Tehran earthquake was studied as a case study. The results of their research showed that the model used can be effective. Lou et al. (2020) evaluated a 19-story hospital at the time of an earthquake using a rescue scenario simulation framework based on the building information model and virtual reality (VR). The results show that the created VR scenario was highly realistic and the effect of smoke on rescue during a fire is more important than fallen debris. Nguyen and Kim (2020) considered parameters such as topography (slope, curvature), soil depth, rainfall, peak ground acceleration, and physical properties of soil using a quasi-static model and a NEWMARK displacement model based on possible landslide risk data. The earth was simulated and evaluated at the time of the earthquake. The simulation results showed certain levels of confidence in the probability of landslides occurring annually. Ghasemi and Khalili-Damghani (2021), using a robust simulation–optimization approach and simulation structure design, evaluated the interactive relationships of a city's critical infrastructure at the time of an earthquake. The simulation system itself calculated relief items, including drinking water, food, blood, medicine, tents, and non-drinking water, in times of crisis. Researchers and crisis-management practitioners are willing to act in decisions that can improve system performance as much as possible. Therefore, the use of simulation can be effective for the validation of models, refinement, research, and development in technology, especially in the field of crisis management (Shirazi, Kia, & Ghasemi 2021). In this research, using information extracted from GIS and a simulation system, in addition to prioritizing relief centers in the study area, the volume of relief goods required during the earthquake was evaluated. Using the opinion of experts and previous studies, and considering the new criteria, the relief centers in the region were also evaluated in terms of allocation priority. The design of a new simulation structure to assess the effects of earthquakes on urban infrastructure and to consider new criteria such as regional risks and type of place to accommodate people are other innovations of this research.

19.3 INVESTIGATING THE PERFORMANCE OF SIMULATION STRUCTURE IN THE PROBLEM OF CRISIS MANAGEMENT IN A TEHRAN EARTHQUAKE

The city of Tehran is located in the foothills of the Alborz Mountains and has many active faults. Area 1 is about 60 km^2. As structures were constructed in Tehran before 1390, engineering criteria and standards were not considered with regard to earthquakes. These old standards make the area vulnerable to earthquakes (Banerjee et al. 2020). The Tehran Crisis Management Organization has invited JICA (Japan International Cooperation Agency) to cooperate in the implementation of preventive measures, preparedness, and response to various crises, especially natural disasters such as earthquakes, to provide operational and tactical conditions to minimize possible earthquake losses and damages. According to the results of JICA research, the amount damage that would be caused by an earthquake in the fault zone in the north of Tehran is estimated to be significant. However, the other three faults, namely the Ray fault, the Mosha fault model, and the floating model, are also of great importance. According to estimates, more than 60,000 buildings in this area would be damaged based on four-fault models (Ray fault model, Mosha fault model, North Tehran fault model, and floating model) (Jica, 2000). The Tehran Crisis Management Organization has identified suitable locations in all 22 districts of Tehran with basic facilities to meet the needs of asylum seekers for 72 hours during a crisis. Figure 19.1 shows the distribution of buildings in Tehran and the study area in terms of the amount of damage incurred in times of crisis. As can be seen in the figure, buildings are divided into five groups, scattered throughout the area: very dangerous, dangerous, neutral (moderate), safe, and very safe. This means that one area is more likely to be severely damaged and other areas less damaged. Crisis-management officials try to minimize the amount of damage by being aware of the amount of damage and number of casualties there will be during a crisis. The main purpose of this study was to present the decision tree of earthquake effects on urban infrastructure using simulation science and MCDM techniques to prioritize relief centers in the region. Estimating the number of relief goods and prioritizing relief centers to accommodate earthquake victims was another goal of this study.

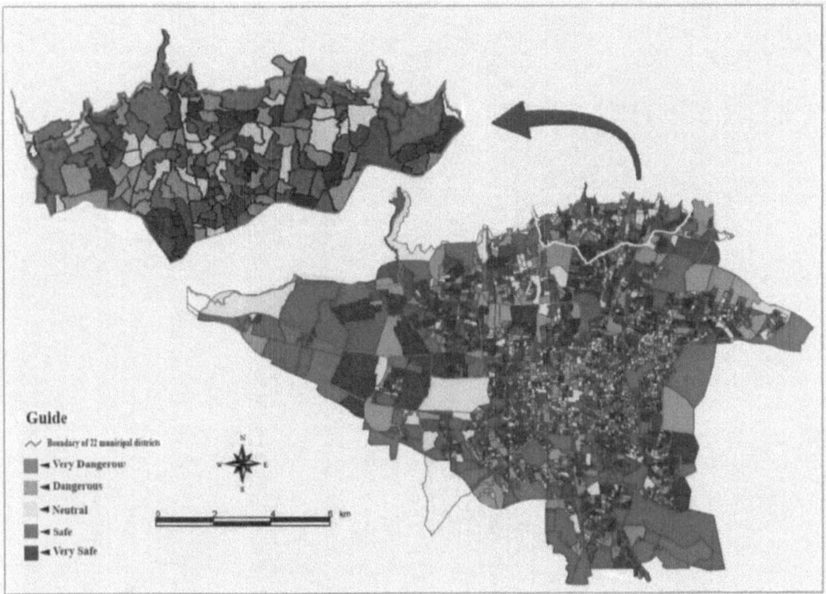

FIGURE 19.1 Dispersion of buildings in tehran in terms of amount of damage.

19.4 SIMULATION MODEL FOR EVALUATING RELIEF GOODS

According to statistics, 78% of infrastructure damage is related to buildings, hospitals, roads, and railways, so this study examined the effects of earthquakes on the destruction of this infrastructure (Ibrion et al. 2015). Figure 19.2 shows the decision tree structure of the impact of an earthquake on infrastructure.

As shown in Figure 19.2, earthquakes affect the infrastructure of buildings, courtyard arteries, hospitals, roads, and railways. The destruction of any of these infrastructures will have reciprocal effects. A building collapse can lead to a "gas station fire", "evacuation", "fire", or "one person killed or injured". However, earthquakes can cause damage to the rail network and roads, and consequently disrupt the transportation network and communication routes, which disrupts relief operations and equipment transfers, ultimately increasing the number of people killed or injured, and the shortage of relief goods increases. The demolition of buildings kills or injures people, and injured people need to be evacuated to hospitals and medical centers. Homeless people must also be transferred to relief centers designated by crisis management after evacuation, in which case they will need relief items that include food packs, bio packs, sanitary water, and drug packs. The simulation output calculates the system demand. The system demand includes relief goods, including food packs, bio packs, sanitary packs of water, and drugs. JICA data and statistics are used for simulation input. For example, Figure 19.3 shows the effect of earthquake damage ratios on different structures due to the Mercalli intensity. This figure is estimated using the ratio of damage incurred during historical earthquakes in Tehran. According to JICA estimates, the most probable magnitude of a Tehran earthquake

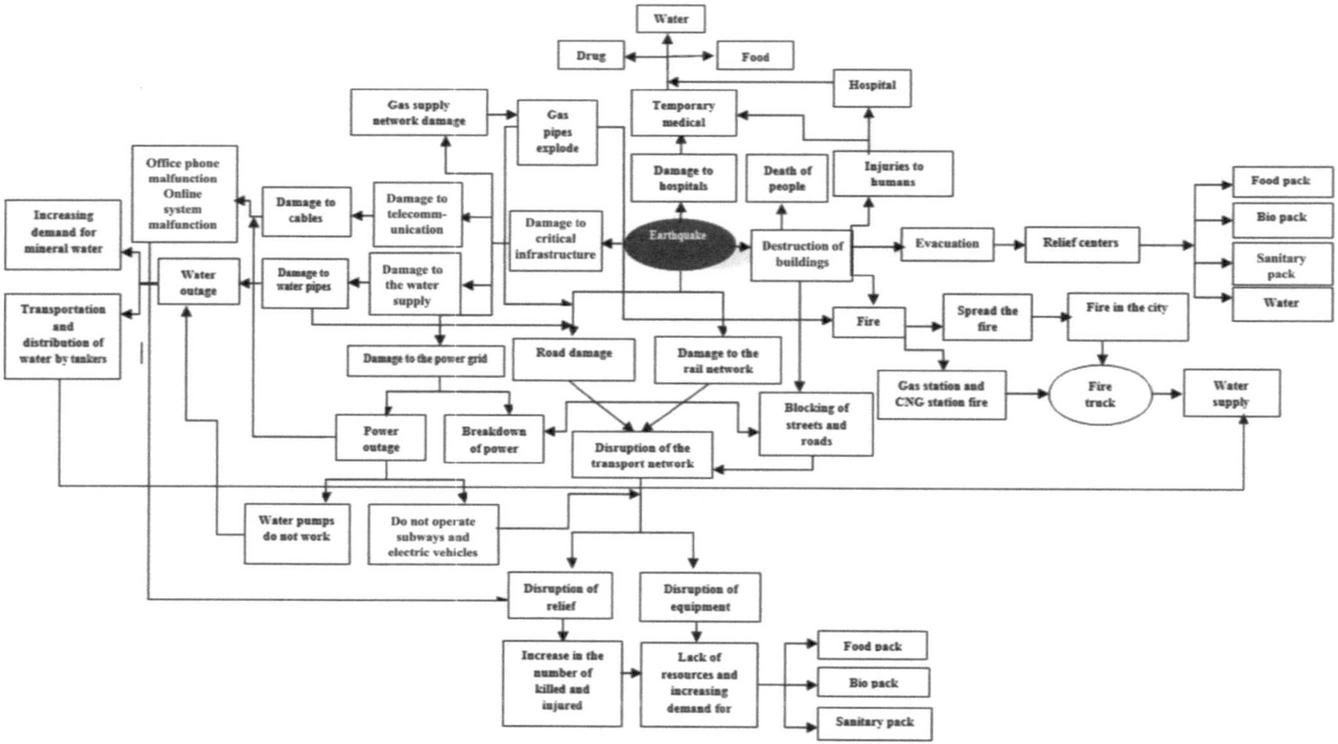

FIGURE 19.2 Network structure of the impact of an earthquake on urban infrastructure.

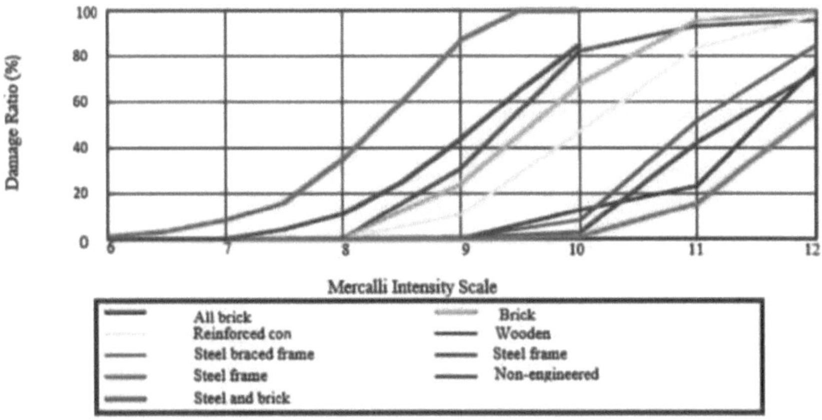

FIGURE 19.3 Comparison of the damage ratio of residential buildings by Mercalli.

is 7, which is the target value of this study used to assess damage and relief goods (Jica, 2000). Figure 19.3 shows the damage function for different types of structures, based on the damage function related to brick and steel structures and by transferring the "damage ratio" curves of that type of structure to the axis of the earthquake intensity scale.

1) Brick and steel structures and stone and steel (the most common type in the city, with thin metal bases). The ratio of this type of structure is considered as the "damage ratio" of the base.
2) Type 1 steel structures (better quality of steel than type 2): the ratio for this type of building is 0.5 units to the right of the "damage ratio" based on the axis of the earthquake intensity scale.
3) Type 2 steel structures (lower quality of steel compared to type 1): the ratio for this type of building is 0.4 units to the right of the "damage ratio" based on the axis of the earthquake magnitude scale.
4) Reinforced concrete structures type zero (the best quality of reinforced concrete structures): the ratio for this type of building is 0.5 units to the right of the "damage ratio" based on the axis of the earthquake intensity scale. In general, most reinforced concrete structures have less earthquake resistance than steel structures.
5) Type 1 reinforced concrete structures (better quality of concrete structures than type 2): the ratio for this type of building is 0.5 units to the right of the "damage ratio" based on the axis of the earthquake intensity scale.
6) Type 2 concrete structures (lower quality of concrete structure than type 1): the ratio for this type of building is 0.7 units to the left of the "damage ratio" based on the axis of the earthquake magnitude scale.
7) Wooden structures. The ratio for this type of building is one unit to the left of the "damage ratio" based on the axis of the earthquake magnitude scale.

8) Cement block with any type of roof, brick and wood or stone and wood, all brick or stone and brick. The ratio for this type of building is 1.2 units to the left of the "damage ratio" based on the axis of the earthquake magnitude scale. For example, the "damage ratio" for the intensity of 7.5 on the Modified Mercalli Intensity scale is assumed to be 0.5.
9) Clay, wood, clay, and mud structures: the ratio for this type of building is two units to the left of the "damage ratio" based on the axis of the earthquake magnitude scale.

Figure 19.4 shows the simulation of the proposed model with Enterprise Dynamic software. The calculated performance measurement values are "AvgContent", as well as the separate performance simulation method; the value of the warm-up period is 100,000 hours and the observation period is 1,000,000 hours. As shown in Figure 19.4, 76 atoms for urban infrastructure design and evaluation of their effects on each other during an earthquake were used. There was one "source" atom, one "sink" atom, and 74 "server" atoms. Five servers called food packs, bio packs, water, and drug health packs were also used to count the estimated demand. The following equation was used to estimate the number of killed and injured people in proportion to the demand for relief goods (Coburn & Spence 2003 and Coburn et al. 1992).

$$* \mathrm{F} = N_1.N_2.N_3.N_4(R + (1 - R).C) \quad (19.1)$$

F = human casualties in residential buildings after an earthquake.
N_1 = number of residents locked in collapsed buildings.
N_2 = percentage of building residents during an earthquake.
N_3 = number of collapsed residential buildings.
N_4 = number of residents in each building based on statistical data.
R = proportion of casualties after the collapse of buildings.
C = casualties after the collapse.

By applying the relation of each of the relief goods in the form of 4Dscript code to the designated servers, their value can be evaluated. For example, seven liters of water per injured person and 2.5 kg per day per injured person is estimated in the atoms of "water" and "Drugs":

Server drug: *Trigger on entry*= 2.5*(AvgContact (AtomByName ([Human Injured], Model)))

Server Water: *Trigger on entry*= 6*(AvgContact (AtomByName ([Human flnjured], Model)))

To validate the simulation model, this model was compared with the actual historical results of similar earthquakes in Iran and the results forecasted by JICA. Figure 19.5 shows the similarity of the simulation results and the real system with a 93% confidence interval. As can be seen, the amount of water, drugs, food packs, bio packs, and health packs are estimated to be 95%, 96%, 90%, 94%, and 90% similar to the real system, respectively. Therefore, according to the simulation results, the results are realistic.

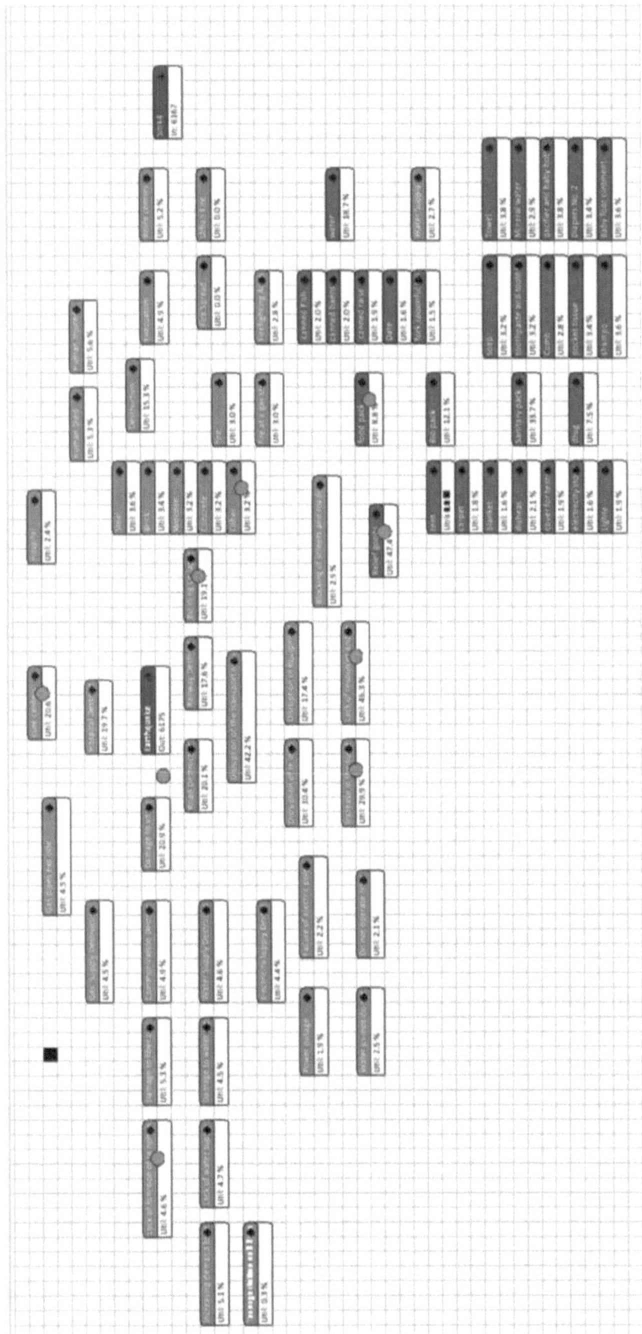

FIGURE 19.4 Structure of a simulation model to evaluate the number of relief goods at the time of an earthquake.

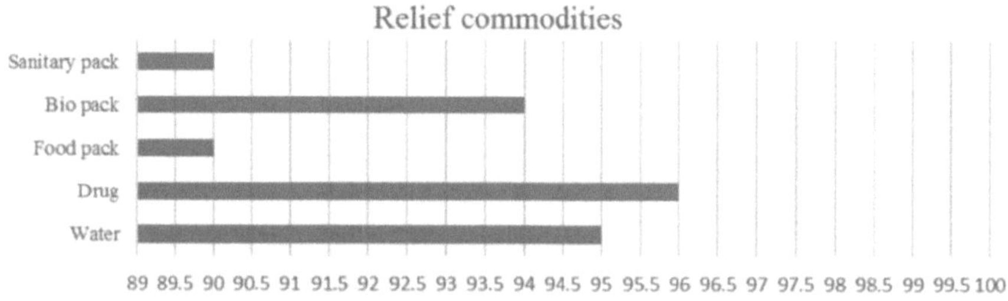

FIGURE 19.5 Ninety-three per cent confidence interval for the results obtained from the simulation structure and the real system for evaluating relief goods.

19.5 EVALUATING RELIEF CENTERS IN TERMS OF HOUSING PRIORITY AND RELIEF PRIORITY FOR THE ALLOCATION OF RELIEF GOODS

After the occurrence of crises and natural disasters such as floods and earthquakes, to create safe conditions for residents and citizens and take them out of dangerous conditions and avoid the dangers of the crisis and its secondary dangers, safe evacuation operations must be carried out. Safe evacuation can be done spontaneously or at the request of relevant authorities. Safe evacuation centers or relief centers are places and spaces where asylum seekers are accommodated in times of crisis and have basic facilities to meet their needs for 72 hours. For this reason, the Tehran Crisis Management Organization identified stadiums, universities, parks, barren lands, etc., in all 22 districts of Tehran among the suitable places for relief centers in crises . Here, relief centers designated by crisis management for housing during an earthquake crisis were evaluated in terms of housing priority and relief priority. First, the necessary criteria to evaluate the centers in terms of accommodation and priority of relief based on previous studies and other additional criteria arrived at in cooperation with experts, are shown in Table 19.1. These criteria were collected by crisis-management experts using the fuzzy Delphi method. The condition for selecting experts is that they had to have more than 10 years of experience and a master's degree. The fuzzy Delphi method was proposed in 1993 by Ishikawa et al. (1993). The advantages of the fuzzy Delphi method over the Delphi method are reversible fuzzy lighting in interviews with experts to obtain logical and appropriate answers. In this method, the calculation process is simple, dealing with multi-level multi-feature decision-making problems and multi-solutions. First, the evaluation criteria were classified by experts and prepared as information layers in ArcGIS. Then the information layers of the analysis and relief centers were evaluated. Table 19.1 shows the number of classes and the value of each class for each of the criteria. The research options included 30 locations designated by Tehran Crisis Management Officers, which included stadiums, sports complexes, universities, barren lands, and parks. Research criteria also include surface area (m^2), distance from Worn texture (m), distance from main roads (m), distance from police (m), distance from the gas station (m), distance from the gas station (m), slope (%), distance from an aqueduct (m), distance from the hospital (m), distance from the fire station (m), distance from the substation (m), population

TABLE 19.1
Classification and Evaluation of Criteria to Evaluate the Prioritization of Centers for Housing

C \ A	Very Good	Good	Average	Bad	Very Bad
Area	+3000	2000–3000	2000	1000–2000	0–1000
Distance from worn texture	0–100	100–200	200–300	300–400	+400
Distance from main ways	0–100	100–200	200–300	300–400	+400
Distance from police stations	0–200	200–400	400–600	600–800	+800
Distance from gas station	+400	200–400	100–200	50–100	0–50
Distance from compressed natural gas station	+400	200–400	100–200	50–100	0–50
Distance from percent slope	1–4	4–6	6–10	10–12	+12
Distance from wells and aqueducts	+300	200–300	100–200	50–100	0–50
Distance from hospital	0–500	500–1000	1000–1500	1500–2000	+2000
Distance from the fire station	0–500	500–1000	1000–1500	1500–2000	+2000
Distance from electricity post	+100	80–100	60–80	30–60	0–30
Population	+120	90–120	60–90	30–60	0–30
Distance from subway	+300	200–300	100–200	50–100	0–50
Distance from fault	+400	200–400	200	100–200	0–100
Distance from dangerous centers	0–200	200–500	500–1000	1000–1500	+1500
Distance from rivers	+700	500–700	200–500	100–200	0–100
Distance from educational centers	0–150	150–300	300–500	500–700	700
Distance from parks and gardens	0–200	200–400	400–600	600–1000	+1000

density (per hectare), distance from metro lines (m), fault lines (m), distance from dangerous centers (m), distance from the river (m), distance from educational centers (mm), and distance from green space (m).

After creating the information layers and integrating them with GIS (geographic information system), the designated relief centers in the region were evaluated in terms of distance and quantity of the ratios of the indicators evaluated in Table 19.1 (position of each optimally extracted area concerning the indicators). Table 19.2 shows the results of this evaluation, where VG, G, A, B, and VB indicate very good, good, average, bad, and very bad, respectively, in accordance with the classification and evaluation criteria in Table 19.1. For example, according to the GIS assessment, the Laleh Park option is in the range of 200–400 in terms of distance from the nearest police station, and according to the information classified by experts (Table 19.1) in the "Good" (G) range.

After evaluating the 30 relief sites in the area and forming a pairwise comparison matrix, the criteria were weighted and prioritized. There are different methods to estimate the weights of criteria. Here the Shannon entropy method was used to determine the weight of the criteria. The entropy method is one of the MCDM methods used to calculate the weight of criteria. In this method, a criterion-option matrix is required. This method was proposed in 1974 by Shannon and Weaver (Gray, 2011). According to the entropy method, the higher the scatter in the values of a criterion, the more important that criterion is. The TOPSIS (Technique for Order of Preference by

TABLE 19.2
Results of the Evaluation of Relief Centers Intended for Accommodation in the Region by GIS According to the Criteria

Criteria / Options	Area	Worn Texture	Main Ways	Police Stations	City Gas Station	Compressed Natural Gas Station	Slope Percent	Wells and Aqueducts	Hospital	Fire Station	Electricity Post	Population	Subway	Fault	Dangerous Centers	Rivers	Educational Centers	Parks and Gardens
Laleh Park	VG	VB	VG	G	VG	VG	A	G	VB	VB	VG	VB	VG	VG	VB	VG	VB	VG
Aseman Park	VG	VG	VG	VB	VG	VG	VB	VG	VB	VB	VG	VB	VG	VG	VB	VG	VB	VG
Morvarid Park	VG	VB	VG	A	VG	VG	VB	VG	VB	VB	VG	VB	VG	VG	VB	VG	B	VG
Aboozar Park	VG	VB	VG	VG	VG	VG	B	VG	VB	A	VG	VB	VG	VG	VB	VG	VG	VG
Fadak Park	VG	VB	G	A	VG	VG	VB	VG	VB	A	VG	VB	VG	VG	VB	VG	VG	VG
Orchid Garden	B	VB	VG	VB	VG	VG	VB	VG	VB	VB	VG	VG	VG	VG	VB	VG	VG	VG
Kowsar Park	VG	VB	VG	G	VG	VG	VB	VG	B	A	VG	VB	VG	VG	VB	VG	VG	VG
Beheshti University	VG	VB	VG	VG	VG	VG	VG	VG	A	VB	VG	VG	G	VG	B	VG	VG	VG
Industry Sports Complex	VG	VG	VG	VG	VG	VG	A	VG	A	G	VG	A	VG	VG	A	G	B	VG
Mehregan Park	VG	VB	VG	VG	VG	VG	VB	VG	VG	G	VG	A	VG	VG	VB	A	VG	VG
Aftab Park	VG	VB	VG	VG	VG	VG	VB	VG	A	G	VG	VB	VG	VG	VB	A	G	VG
Amin Park	VG	VB	A	B	VG	VG	VB	VG	VG	A	VG	VG	VG	VG	VB	A	A	VG
Yas Sport Complex	VG	VB	G	G	VG	VG	VG	VG	G	G	B	VB	A	VG	B	B	G	VG
Darabad Coastal Park	VG	VG	G	G	VG	VG	VB	VG	VG	VB	VG	B	VG	VG	VB	VB	VB	VG
Samen Park	VG	VB	G	G	VG	VG	A	VG	A	VG	VG	VG	VG	VG	VB	VG	VB	VG
Ozgol Park	VG	B	VG	G	VG	VG	VB	VG	VG	A	VG	VB	VG	VG	VB	G	VB	VG
Niavaran Park	VG	VB	VG	VG	VG	VG	VB	VG	G	G	VG	B	VG	VG	VB	VG	B	VG
Baghestan park	VB	VB	G	G	VG	VG	VB	A	VG	VB	VG	VB	VG	VG	VB	B	VG	VG

Gol Mohammadi Park	VG	VB	VG	G	VG	VG	VB	VG	G	A	VG	G	VG	VG	VB	VG	VB	VG
Mehr Park	VG	VB	VG	A	VG	VG	VB	VG	G	A	VG	VB	VG	VG	VB	VG	G	VG
Sadr Park	VG	VB	VG	VB	VG	VG	VB	VG	B	VB	VG	A	VG	VG	B	VG	B	VG
Qeytarieh Park	VG	B	VG	VB	VG	VG	VB	VG	A	A	VG	B	VG	VG	B	VG	A	VG
Shadi Park	B	VB	G	VB	VG	VG	VB	VG	G	VB	VG	VG	VG	VG	VB	G	G	VG
Arezoo Park	VG	G	G	A	VG	VG	A	VG	VG	G	VG	A	VG	VG	G	VB	VG	VG
Zaferaniyeh Park	VG	VB	VG	B	VG	VG	VB	VG	G	VB	VG	A	VG	VB	VB	G	B	VG
Kaj Park	G	VB	VG	A	VG	VG	VB	VG	G	VB	VG	VG	VG	VG	VB	VB	VB	VG
Golzar Park	VG	VB	VG	A	VG	VG	VB	VG	VG	G	VG	B	VG	VB	A	VG	A	VG
Golrizan Park	VG	VB	VG	B	VG	VG	VB	VG	A	G	VG	VG	VG	B	B	VG	G	VG
Negin Park	VG	VB	VG	VB	VG	VG	VB	VG	G	VG	VG	VG	VG	B	B	VG	VG	VG
Wasteland	VG	VB	G	VG	VG	VG	VB	VG	B	A	VG	VB	VG	VG	B	VB	VB	VG

TABLE 19.3
Weight of Calculated Research Criteria by SHANNON ENTROPY Method

Criteria	w_j
Area	0.0559
Distance from worn texture	0.047
Distance from main ways	0.0576
Distance from police stations	0.0556
Distance from gas station	0.0575
Distance from compressed natural gas station	0.0575
Distance from percent slope	0.0514
Distance from wells and aqueducts	0.0573
Distance from hospital	0.0561
Distance from the fire station	0.0542
Distance from electricity post	0.0573
Population	0.0556
Distance from subway	0.0573
Distance from fault	0.0562
Distance from dangerous centers	0.0515
Distance from rivers	0.0589
Distance from educational centers	0.0556
Distance from parks and gardens	0.0575

Similarity to Ideal Solution) technique has also been used to prioritize relief centers. The TOPSIS technique is a MCDM method, first introduced by Wong and Hwang in 1981 (Lai et al. 1994). The TOPSIS technique can be used to rank and determine the distances between options with positive and negative criteria. Comparing criteria with different units of measurement is one of the advantages of this method. Tables 19.3 and 19.4 show the weights of the criteria calculated by the Shannon entropy method and prioritization of research options by the TOPSIS method, respectively.

As can be seen in the weights table (Table 19.3), the criteria of proximity to the river and proximity to main roads have the highest weight values (0.0589 and 0.0576, respectively), and the criteria of worn texture and land slope have the lowest weight values (0.047 and 0.0514, respectively). Relief centers were set up during the crisis. The ranking results also show that based on the criterion of closeness to the positive ideal (CLi*), the criterion of closeness to the ideal of Arezoo Park, Industry Sports Complex and Shahid Beheshti University is higher than other options (0.6655, 0.6499, and 0.564648, respectively). These options were ranked first to third, and the value closest to the positive ideal of Zaferaniyeh Park (0.2782) was less than the other options. Therefore, this place was ranked last as a potential relief center for housing victims in times of crisis. In the next step, the relief centers were examined in terms of the priority of relief work. Criteria for evaluating relief sites in terms of relief priority were divided into three groups: type of place to accommodate people; area risk; and compatible access and proximity. The type of place to accommodate people criterion included the type of relief places that exist in the area and considered by

TABLE 19.4
Prioritization of Crisis Relief Centers in Terms of Accommodation with the TOPSIS Method

Options Issue	CL_i^*	RANK
Laleh Park	0.3471	19
Aseman Park	0.3792	12
Morvarid Park	0.3952	11
Aboozar Park	0.3769	13
Fadak Park	0.3257	25
Orchid Garden	0.3409	22
Kowsar Park	0.3429	21
Shahid Beheshti University	0.5648	3
Industry Sports Complex	0.6499	2
Mehregan Park	0.3979	10
Aftab Park	0.3478	18
Amin Park	0.3662	14
Yas Sport Complex	0.5228	4
Darabad Coastal Park	0.4115	8
Samen Park	0.4580	5
Ozgol Park	0.3433	20
Niavaran Park	0.3613	15
Baghestan park	0.3125	29
Gol Mohammadi Park	0.3572	17
Mehr Park	0.3366	24
Sadr Park	0.3128	28
Qeytarieh Park	0.3592	16
Shadi Park	0.3381	23
Arezoo Park	0.6655	1
Zaferaniyeh Park	0.2782	30
Kaj Park	0.3198	26
Golzar Park	0.4149	7
Golrizan Park	0.4033	9
Negin Park	0.4298	6
Wasteland	0.3139	27

crisis-management officials as relief centers in crises. Experts evaluated these centers according to the level of facilities, and the level of attention and need for relief goods. The evaluation criteria of the relief centers were evaluated in terms of prioritization of relief by crisis-management experts using the best-worst method (BWM) method and the results of this evaluation are presented in Table 19.5. BWM is one of the new MCDM techniques. In this method, the weight of the criteria is obtained by solving a linear optimization model. In BWM, the indicators are determined by the decision-maker and then a minimax problem is formulated and solved to determine the weight of the indicators (Rezaei 2015). Area risk criterion includes vulnerable areas in the region, shown in Figure 19.1, and the location of each relief center (research options)

TABLE 19.5
Evaluation Criteria of Relief Centers in Terms of Priority of Relief in Times of Crisis

Criteria		Weight
Type of place to accommodate people	Universities	0.026
	Wasteland	0.072
	Stadiums	0.016
	Parks and gardens	0.039
Area risk	Very dangerous	0.243
	Dangerous	0.132
	Neutral	0.088
	Safe	0.044
	Very safe	0.032
Compatible access and proximity	Proximity to medical centers	0.096
	Security	0.173
	At-risk population density	0.038

and level of need for relief goods were evaluated by experts. Finally, the criteria of compatible access and proximity, which included proximity to medical centers, security, and population density at risk, were evaluated. Although crisis management strives to be fair in providing relief so that all affected people and affected centers have timely and equal relief, relief centers located near medical centers are the last priority because, in times of crisis, priority is given to centers that are far from medical centers. Also, centers that have higher security, such as stadiums and universities, have priority in providing relief. Centers that are at higher risk in terms of population density have a higher priority than other areas.

Then, using extracted GIS information and according to the opinion of crisis-management experts, the research options were evaluated according to the criteria for determining the priority of relief centers in times of crisis, shown in Table 19.6. Each of the options was evaluated in terms of type of location, and their settlement nature was identified and confirmed. The location of each option was then measured according to the information in Figure 19.1 and one of the defined intervals (very safe, secure, moderate, dangerous, and very dangerous). The three criteria of proximity to medical centers, security, and density of the at-risk population were evaluated qualitatively according to GIS distance information and expert opinions.

After weighing the criteria, according to the pairwise comparison matrix and the information obtained from the evaluation of options (Table 19.6), the research options (quantitative and qualitative) were prioritized using the PROMETHEE (preference ranking organization method for enrichment evaluation) method. The PROMETHEE method performs the evaluation process as a partial or complete ranking on a limited set of options. In this method, each decision criterion is examined based on a separate function and without relation to other criteria (Brans et al 1986). In the PROMETHEE

TABLE 19.6
Couple Comparison Matrix to Determine the Priority of Relief in Crises

Actions	Universities	Wasteland	Stadiums	Parks	Very dangerous	Dangerous	Neutral	Safe	Very safe	Proximity to medical centers	Security	At-risk population density
Weight	0.026	0.072	0.016	0.039	0.234	0.132	0.088	0.044	0.032	0.096	0.0176	0.038
Laleh Park	NO	NO	NO	YES	NO	NO	NO	NO	YES	Very bad	Bad	Very bad
Aseman Park	NO	NO	NO	YES	NO	NO	NO	NO	YES	Very bad	Bad	Very bad
Morvarid Park	NO	NO	NO	YES	NO	NO	NO	NO	YES	Very bad	Bad	Very bad
Aboozar Park	NO	NO	NO	YES	NO	YES	NO	NO	NO	Very bad	Bad	Very bad
Fadak Park	NO	NO	NO	YES	NO	YES	NO	NO	NO	Very bad	Bad	Very bad
Orchid Garden	NO	NO	NO	YES	NO	YES	NO	NO	NO	Very bad	Bad	Very Good
Kowsar Park	NO	NO	NO	YES	NO	YES	NO	NO	NO	Very bad	Bad	Very bad
Golzar Park	NO	NO	NO	YES	NO	NO	YES	NO	NO	Bad	Bad	Very Good
Ozgol Park	NO	NO	NO	YES	NO	NO	YES	NO	NO	Average	Bad	Average
Mehregan Park	NO	NO	NO	YES	NO	NO	YES	NO	NO	Very bad	Average	Average
Aftab Park	NO	NO	NO	YES	YES	NO	NO	NO	NO	Very bad	Bad	Very bad
Amin Park	NO	NO	NO	YES	NO	NO	YES	NO	NO	Very bad	Bad	Very Good
Baghestan park	NO	NO	NO	YES	NO	NO	YES	NO	NO	Bad	Bad	Very bad
Darabad Coastal Park	NO	NO	NO	YES	NO	NO	NO	YES	NO	Very bad	Bad	Bad
Samen Park	NO	NO	NO	YES	NO	NO	YES	NO	NO	Very bad	Bad	Very Good
Industry Sports Complex	NO	NO	YES	NO	NO	NO	NO	YES	NO	Very bad	Good	Very bad
Niavaran Park	NO	NO	NO	YES	NO	NO	NO	YES	NO	Very bad	Bad	Bad
Yas Sport Complex	NO	NO	YES	NO	NO	YES	NO	NO	NO	Very bad	Average	Very bad
Gol Mohammadi Park	NO	NO	NO	YES	YES	NO	NO	NO	NO	Very bad	Very Good	Good
Mehr Park	NO	NO	NO	YES	YES	NO	NO	NO	NO	Very bad	Bad	Very bad
Sadr Park	NO	NO	NO	YES	NO	NO	NO	NO	YES	Bad	Average	Average
Qeytarieh Park	NO	NO	NO	YES	NO	NO	NO	YES	NO	Bad	Bad	Bad

(Continued)

TABLE 19.6 (Continued)
Couple Comparison Matrix to Determine the Priority of Relief in Crises

Actions	Universities	Wasteland	Stadiums	Parks	Very dangerous	Dangerous	Neutral	Safe	Very safe	Proximity to medical centers	Security	At-risk population density
Shadi Park	NO	NO	NO	YES	NO	NO	NO	YES	YES	Very bad	Bad	Very Good
Arezoo Park	NO	NO	NO	YES	NO	NO	YES	NO	NO	Good	Bad	Average
Zaferaniyeh Park	NO	NO	NO	YES	NO	NO	YES	NO	NO	Very bad	Average	Average
Kaj Park	NO	NO	NO	YES	NO	NO	NO	YES	NO	Very bad	Average	Very Good
Shahid Beheshti University	YES	NO	NO	NO	NO	NO	NO	YES	NO	Average	Very Good	Bad
Golrizan Park	NO	NO	NO	YES	NO	YES	NO	NO	NO	Bad	Bad	Very Good
Negin Park	NO	NO	NO	YES	NO	YES	NO	NO	NO	Bad	Bad	Very Good
Wasteland	NO	YES	NO	NO	NO	YES	NO	NO	NO	Bad	Bad	Very bad

TABLE 19.7
Results of Relief Prioritization of Relief Centers in Times of Crisis by PROMETHEE Method

Actions	Phi	Phi+	Phi–	Rank
Wasteland	0.1860	0.3197	0.1336	1
Aftab Park	0.1645	0.3016	0.1371	2
Mehr Park	0.1645	0.3016	0.1371	2
Arezoo Park	0.0989	0.2214	0.1225	4
Baghestan park	0.0906	0.1998	0.1092	5
Gol Mohammadi Park	0.0896	0.2800	0.1904	6
Ozgol Park	0.0870	0.2135	0.1264	7
Golrizan Park	0.0823	0.2116	0.1292	8
Negin Park	0.0823	0.2116	0.1292	8
Aboozar Park	0.0380	0.1605	0.1225	10
Fadak Park	0.0380	0.1605	0.1225	10
Kowsar Park	0.0380	0.1605	0.1225	10
Golzar Park	0.0278	0.1716	0.1438	13
Qeytarieh Park	0.0109	0.1554	0.1444	14
Shahid Beheshti University	0.0043	0.2020	0.1977	15
Yas Sport Complex	–0.0101	0.1689	0.1790	16
Orchid Garden	–0.0248	0.1322	0.1570	17
Sadr Park	–0.0377	0.1351	0.1728	18
Mehregan Park	–0.0754	0.1027	0.1781	19
Zaferaniyeh Park	–0.0754	0.1027	0.1781	19
Amin Park	–0.0793	0.0923	0.1715	21
Samen Park	–0.0793	0.0923	0.1715	21
Laleh Park	–0.0859	0.0736	0.1595	23
Aseman park	–0.0859	0.0736	0.1595	23
Morvarid park	–0.0859	0.0736	0.1595	23
Shadi Park	–0.0942	0.0871	0.1813	26
Darabad Coastal Park	–0.0962	0.0761	0.1722	27
Niavaran Park	–0.0962	0.0761	0.1722	27
Industry Sports Complex	–0.1235	0.0900	0.2136	29
Kaj Park	–0.1535	0.0504	0.2039	30

method, the higher net flow represents the superior option. Table 19.7 shows the results of prioritizing the options. The net flow (Phi) of the options wasteland, Aftab Park, Mehr Park, and Arezoo Park were equal to 0.1860, 0.1645, 0.1645, and 0.0989, respectively (ranks 1–4).

Kaj Park, industry sports complex, and Niavaran Park were ranked last, with a net flow of 0.1535, –0.1235, and –0.0962, respectively. The reason for being in this rank can be considered as non-compliance with some important research criteria. Evaluation results show that parks and Wasteland have a higher priority than sports complexes, universities, and stadiums in terms of receiving relief goods.

19.6 CONCLUSION

One of the issues always considered by responsible organizations in crisis management is planning to prepare and prevent risk reduction. In the twentieth century, urban planning operations in Tehran accelerated without a proper crisis-management system for severe earthquakes, and such an earthquake is predicted to cause an unprecedented catastrophe, so having a comprehensive plan and accurate information to reduce damage is essential. Financial aspects and human life are some of the requirements of Tehran crisis management. In this study, according to the study conducted by JICA and statistical data from the study area, by designing a network structure (decision tree) and investigating the effects of the earthquake on infrastructure, the number of relief goods required through simulation with a 93% confidence interval was calculated. Thirty relief bases in the region, considered by crisis management as relief centers for accommodation during crises, were evaluated in terms of housing priority and priority of sending relief goods. First, the evaluation criteria of relief centers were classified and evaluated through fuzzy Delphi and the opinion of crisis-management experts. Then, using the information extracted by GIS and the entropy method, the weights of the criteria were calculated and the options prioritized using TOPSIS. According to experts and a literature review, criteria for determining the priority of relief centers were considered and weighted by the BWM method, and, finally, research options were prioritized by the PROMETHEE method. The results of the evaluation of relief centers showed that stadiums, universities, and sports centers are suitable places to live in times of an earthquake crisis (Table 19.4). Also, wasteland and parks are of higher priority than sports complexes, universities, and stadiums, and are eligible to receive relief goods (Table 19.7). Also, according to the reliability of the simulation structure (93%), this structure can be used in other areas.

It is suggested that the amount of goods should be evaluated according to different scenarios and different sizes of crisis. It is also suggested that in the event of an epidemic, relief goods and places considered by crisis management should be evaluated in different scenarios. It is suggested that, considering the supply chain, suitable models for the transfer of relief goods to relief centers should be presented in different scenarios. Considering the network structure of earthquake effects, assessing risks in crises and estimating their impact on equipment logistics in crises is also one of the cases suggested. One of the limitations of the research was the lack of access to some layers of information such as sensitive centers and oil and gas pipelines. Also, due to COVID-19 conditions at the time, experts participated in the evaluation of criteria and options remotely and online.

REFERENCES

Ahmadi Choukolaei, Hassan, Mustafa Jahangoshai Rezaee, Peiman Ghasemi, and Morteza Saberi. "Efficient crisis management by selection and analysis of relief centers in disaster integrating GIS and multicriteria decision methods: a case study of Tehran." Mathematical Problems in Engineering 2021 (2021). doi: 10.1155/2021/5944828.

Banerjee, Snehanshu, Md Muhib Kabir, Nashid K. Khadem, and Celeste Chavis. "Optimal locations for bikeshare stations: A new GIS based spatial approach." Transportation Research Interdisciplinary Perspectives 4 (2020): 100101.

Borhani, Kazem, Ashraf Azimzadeh Irany, and Amirhosein Elhami. "Spatial analysis of urban multi-functional land uses integrating MCDM and GIS methods Case study: emergency shelter in Saravan City." Scientific-Research Quarterly of Geographical Data (SEPEHR) 29, no. 116 (2021): 103–118.

Brans, Jean-Pierre, Ph Vincke, and Bertrand Mareschal. "How to select and how to rank projects: The PROMETHEE method." European Journal of Operational Research 24, no. 2 (1986): 228–238.

Chen, Wei, Guofang Zhai, Chenjing Fan, Wenbo Jin, and Ying Xie. "A planning framework based on system theory and GIS for urban emergency shelter system: A case of Guangzhou, China." Human and Ecological Risk Assessment 23, no. 3 (2017): 441–456.

Coburn, Andrew W., Robin JS Spence, and Antonios Pomonis. "Factors determining human casualty levels in earthquakes: mortality prediction in building collapse." In Proceedings of the tenth world conference on earthquake engineering, vol. 10, pp. 5989–5994. Balkema Rotterdam, 1992.

Coburn, Andrew, and Robin Spence. Earthquake protection. John Wiley & Sons (2003).

Dabbagh, Rahim and Hassan Ahmadi Chokalaei. "Optimal location of rescue centers using GIS and multi-criteria decision making methods (case study: Urmia City)." Jorar 12 (2020): 12.

Ghasemi, Peiman and Kaveh Khalili-Damghani. "A robust simulation-optimization approach for pre-disaster multi-period location–allocation–inventory planning." Mathematics and Computers in Simulation 179 (2021): 69–95.

Ghasemi, Peiman, Kaveh Khalili-Damghani, Ashkan Hafezalkotob, and Sadigh Raissi. "Stochastic optimization model for distribution and evacuation planning (a case study of Tehran earthquake)." Socio-Economic Planning Sciences 71 (2020): 100745.

Goodarzian, Fariba, Farzad Bahrami, and Davood Shishebori. "A new location-allocation-problem for mobile telecommunication rigs model under crises and natural disasters: a real case study." Journal of Ambient Intelligence and Humanized Computing 71 (2022): 2565–2583.

Gray, Robert M. Entropy and information theory. Springer Science & Business Media (2011).

Gupta, Akhil and James Ferguson, eds. Culture, power, place: explorations in critical anthropology. Duke University Press (1997).

Ibrion, Michaela, Mohammad Mokhtari, and Farrokh Nadim. "Earthquake disaster risk reduction in Iran: lessons and "lessons learned" from three large earthquake disasters—Tabas 1978, Rudbar 1990, and Bam 2003." International Journal of Disaster Risk Science 6, no. 4 (2015): 415–427.

Ishikawa, Akira, Michio Amagasa, Tetsuo Shiga, Giichi Tomizawa, Rumi Tatsuta, and Hiroshi Mieno. "The max-min Delphi method and fuzzy Delphi method via fuzzy integration." Fuzzy Sets and Systems 55, no. 3 (1993): 241–253.

Lai, Young-Jou, Ting-Yun Liu, and Ching-Lai Hwang. "Topsis for MODM." European Journal of Operational Research 76, no. 3 (1994): 486–500.

Lane, Michael S., Ali H. Mansour, and John L. Harpell. "Operations research techniques: a longitudinal update 1973–1988." Interfaces 23, no. 2 (1993): 63–68.

Lu, Xinzheng, Zhebiao Yang, Zhen Xu, and Chen Xiong. "Scenario simulation of indoor post-earthquake fire rescue based on building information model and virtual reality." Advances in Engineering Software 143 (2020): 102792.

Marcelin, Jean Michael, Mark W. Horner, Eren Erman Ozguven, and Ayberk Kocatepe. "How does accessibility to post-disaster relief compare between the aging and the general population? A spatial network optimization analysis of hurricane relief facility locations." International Journal of Disaster Risk Reduction 15 (2016): 61–72.

Nguyen, Vinh Ba-Quang, and Yun-Tae Kim. "Rainfall-earthquake-induced landslide hazard prediction by Monte Carlo simulation: a case study of MT. Umyeon in Korea." KSCE Journal of Civil Engineering 24, no. 1 (2020): 73–86.

Rapoport, Amos. Human aspects of urban form: towards a man—environment approach to urban form and design. Elsevier (2016).

Rezaei, Jafar. "Best-worst multi-criteria decision-making method." Omega 53 (2015): 49–57.

Shirazi, Hossein, Reza Kia, and Peiman Ghasemi. "A stochastic bi-objective simulation–optimization model for plasma supply chain in case of COVID-19 outbreak." Applied Soft Computing 112 (2021): 107725.

20 Routing of Vehicles to Disaster Relief Tents

Engin Pekel
Department Of Industrial Engineering, Faculty Of Engineering, Hitit University, Çorum, Turkey,

CONTENTS

20.1 Introduction 303
20.2 Mathematical Model 305
20.3 Method 307
20.4 Numerical Result 307
20.5 Conclusion 309
References 310

20.1 INTRODUCTION

In today's world, disasters are an undeniable fact of life. A catastrophe is a devastating circumstance that enormously disrupts the lives of residents in a society. Many studies have entered the literature as a result of disasters being a part of our lives, and the associated painful experiences.

Ahmadi et al. (2015) suggested a model that deals with location and the routing problems considering network troubles, multiple utilizations of tools, and common recovery periods for a multi-depot. The proposed method resolves the location of regional warehouses and the route for the final mile allocation after an earthquake. The method was expanded to a stochastic model performed by random travel time to locate distribution centers. Sharif and Salari (2015) designed a greedy randomized adaptive search procedure (GRASP) to resolve the transportation problems in catastrophe relief circumstances. The problem is to meet the demand of a particular set of customers by constructing available routes employing a delimited number of vehicles in a central warehouse. The authors proposed two mathematical models using some valid inequalities. Bozorgi-Amiri and Khorsi (2016) explored humanitarian assistance logistics where findings are matched pre- and post-disaster by considering a stochastically multi-objective model. Their proposed model has three purposes: Maximum famine duration between affected areas; total visiting time; and total costs before and after disaster. Weimin (2016) dealt with the optimal path choice of public emergencies as an entry point of research to build a post-disaster ad hoc net through wireless sensor technology by using an ant colony algorithm that absorbs the behavior of ants. Uslu et al. (2017) provided a multi-depot vehicle routing

DOI: 10.1201/9781003212904-20

problem with stochastic demand by performing the chance constraint approach. In addition, they performed a case study in Ankara, the capital city of Turkey. Kokaji and Kainuma (2018) proposed a logistics model to provide decent distribution for many victims affected by asynchrony. In their assumption, many routes were inaccessible, and the number of available vehicles was limited because of a shortage of fuel. Their proposed model minimized the time taken for victims to receive relief goods from vehicles. Shi et al. (2018) dealt with the problem of planning relief supply vehicles to maximize their overall weight by proposing a new method called a substitution equilibrium point. They also considered substitution between tools. Vahdani et al. (2018) proposed a new non-linear multi-purpose, multi-period, multi-commodity mathematical integer model to identify distribution centers for timely vital aid delivery to damaged areas, vehicle routing, and emergency road repair operations. Their model minimized the travel time and overall cost, and increased the reliability of routes to solve the designated problem by using two metaheuristic algorithms: non-dominant sequencing genetic algorithm II and multi-objective particle swarm optimization. Gholami-Zanjani et al. (2019) dealt with multiple vehicles allowed to carry products when the vehicles met at demand hubs to better meet late demands in disaster relief. They performed a non-linear mathematical model considering goods transfer between vehicles in dynamic multi-vehicle situations. In addition, a genetic algorithm was developed and performed to achieve the solutions for the problem. Sabouhi et al. (2019) proposed a combined evacuation and allocation logistics design to receive concurrent routing and scheduling of relief vehicles to move people from damaged regions to sanctuaries, and supply them with essential relief supplies. Their proposed method was a linear model to minimize the aggregate arrival times of relief vehicles at damaged regions and hub centers. Shin et al. (2019) proposed a mathematical model, including a mixed integer and metaheuristic algorithm, offering optimal scheduling considering both reconstruction and delivery. They performed the model to schedule delivery of goods, thus reducing bilateral damage following a huge catastrophe. Yi et al. (2019) performed a two-staged solution model to deal with emergency natural disaster relief: the emergency vehicle routing problem by considering relief materials in sudden disasters. The first stage provided the boundary conditions. The second stage constructed an optimization model for relief materials in sudden disasters. Adarang et al. (2020) investigated a non-deterministic location-routing problem for supplying emergency treatments during catastrophes by performing a robust optimization method. Their goals included minimizing rescue time, and the location and routing costs generated by the vehicles. They performed an algorithm, including the ε-constraint and the NSGA-II algorithm, to unravel the problem. Beiki et al. (2020) designed an incorporated relief chain to optimize disaster management's preparedness and reaction phases simultaneously. Judgments to enhance the supply chain comprised uncovering allocation supplies; the quantity of supplies kept in facilities after disaster, locating transient hub points and routing of vehicles used to allocate care, and how to distribute relief services to impacted areas. Sanbouhi et al. (2020) performed two-staged stochastic programming to minimize the expected arrival time of relief vehicles in affected areas, considering the destruction of potential routes and disruptions due to disasters, how the demand of each affected area could be met by multiple vehicles and distribution centers, and vehicles with limited capacity. Akbari et al. (2021) proposed a set of collaborative game theory methods to distribute fairly the cost savings of cooperation to minimize the unmet demands of aid vehicles and transportation costs in affected areas by considering the inherent uncertainties of disasters. Vieira et al. (2021) proposed a two-stage framework to deal with the problem of large-scale water distribution to aridness-affected regions. In the first stage, they used the transportation model to allocate demand hubs to water sources. In the second stage, they modeled the problem as capacitated vehicle routing problems using a metaheuristic hybridized with random variable neighborhood heuristic method considering the total distance and number of vehicles used for large-scale emergency water trucking. Zhan et al. (2021) used a novel concept that replaced conventional disaster relief logistics actions with synchronicity, including demand determination and assignment. In addition, the authors created a dynamic optimization model using the particle swarm optimization algorithm and presented the solution for the case.

The motivation of this study is to propose a mathematical model that enables the delivery of aid to people living in disaster-affected areas in a shorter time after an earthquake, which is considered to be a major disaster. The application of the proposed model plays an active role in routing and scheduling for disasters.

The rest of the chapter is organized as follows. Section 20.2 presents the mathematical model. Section 20.3 presents the methodology that consists of branch and bound algorithms. Section 20.4 presents the numerical results of the applied algorithm for the case study. Section 20.5 summarizes the findings.

20.2 MATHEMATICAL MODEL

This section provides the mathematical model for sending aid vehicles to disaster areas. Let V be a vehicle set, N be all nodes (disaster regions and depot), and N_a is disaster region. The aid vehicles ($k \in V$) have a specific loading capacity, indicated by Q. These vehicles depart from central depots, indicated by d, and try to meet the demands (D) of disaster regions. Let U be the amount of transported aid; U is a continuous variable. Table 20.1 shows the parameters and variables used in the model.

TABLE 20.1
The Descriptions of Parameters and Variables

Parameters	Description
$R_{i,j}$	Transportation cost
$A_{i,j}$	Allocation cost for a relief team
D_i	Demand that is required in region i
Q	Vehicle capacity
Variables	
$X_{i,j,k}$	1 if vehicle k is assigned from region i to region j, otherwise 0
$W_{i,j}$	1 if a team is assigned, otherwise 0
$U_{i,j,k}$	The amount of aid transported from the region i to the region j by vehicle k

$$Min\sum_{i\in N}\sum_{j\in N}\sum_{k\in V}X_{i,j,k}*R_{i,j}+\sum_{i\in N_a}\sum_{j\in N_a}W_{i,j}*A_{i,j} \quad (20.1)$$

$$s.t.\ \sum_{j\in N}X_{i,j,k}=\sum_{j\in N}X_{j,i,k} \qquad \forall i\in N,\forall k\in V \quad (20.2)$$

$$\sum_{j\in N}\sum_{k\in V}X_{i,j,k}=1 \qquad \forall i\in N_a \quad (20.3)$$

$$\sum_{j\in N}X_{d,j,k}=1 \qquad \forall k\in V \quad (20.4)$$

$$\sum_{i\in N}X_{i,d,k}=1 \qquad \forall k\in V \quad (20.5)$$

$$\sum_{k\in V}X_{i,j,k}=W_{i,j} \qquad \forall i\in N_a,\forall j\in N_a \quad (20.6)$$

$$\sum_{i\in N_a}\sum_{j\in N_a}X_{i,j,k}*D_i\le Q \qquad \forall k\in V \quad (20.7)$$

$$\sum_{j\in N}U_{j,i,k}-\sum_{j\in N}U_{i,j,k}=\sum_{j\in N_a}X_{i,j,k}*D_j \qquad \forall i\in N_a,\forall k\in V \quad (20.8)$$

$$\sum_{j\in N_a}U_{d,j,k}=\sum_{i\in N_a}\sum_{j\in N_a}X_{i,j,k}*D_j \qquad \forall i\in N_a,\forall k\in V \quad (20.9)$$

$$U_{i,j,k}\le X_{i,j,k}*M \qquad \forall i\in N,\forall j\in N_a,\forall k\in V \quad (20.10)$$

$$\sum_{i\in N_a}U_{i,d,k}=0 \qquad \forall k\in V \quad (20.11)$$

$$X_{i,j,k}\in\{0,1\} \qquad \forall i\in N,\forall j\in N,\forall k\in V \quad (20.12)$$

$$W_{i,j}\in\{0,1\} \qquad \forall i\in N_a,\forall j\in N \quad (20.13)$$

Equation 20.1 is the objective function that aims to minimize the sum of the allocated distance cost between disaster areas and the cost of the relief team. Equation 20.2 allows a node to have any input and output connections. Any such node cannot remain unconnected. Equation 20.3 allows any node to connect to only one different node, considering all vehicles. Equation 20.4 states that any vehicle starts at the central depot once. Equation 20.5 allows any vehicle to return to the central depot. Equation 20.6 ensures that the assigned nodes are determined by considering all vehicles. Equation

20.7 ensures that the vehicle capacity cannot exceed the specified value Q. Equation 20.8 meets the aid demand for the connected node and provides the aid demand for the next node if a connection is established between any two nodes. Equation 20.9 ensures that a vehicle leaving the central depot meets the aid demand from the disaster zones assigned to it. Let M be a large number, and Equation 20.10 avoids subtours in the assignment process. Equation 20.11 ensures there is no demand for a vehicle to return to the central depot. Equations 20.12 and 20.13 ensure that $X_{i,j,k}$ and $W_{i,j}$ are binary variables.

20.3 METHOD

The branch-and-bound algorithm divides large-sized problems into many subsets and provides a smaller-sized structure, that is, a sub-problem of a tree structure called a branch-and-bound tree. The division mechanism of the algorithm is that S is decomposed into S1 and S2, thus defining two sub-problems. Here, the new sub-problems obtained by decomposition will be decomposed into sub-problems within the following process. To prevent investigation of the whole tree, the algorithm evaluates the node before creating a replacement node within the tree by comparing the worth of the simplest possible solution, which may well be found within the corresponding sub-tree, with the most effective current solution. This process is termed branching. Sub-problems S1 and S2 are called branches created at node S. Subsequently, each sub-problem obtained in solving the matter represents a node within the tree. Branching is not necessarily bidirectional, and multi-directional branching is possible (Kianfar 2020).

An upper bound is updated with the branch-and-bound method and utilized to prune the hubs. Nodes are pruned due to infeasibility, deficiently bound, and the ideal result comes about. If a node is not pruned, it is discovered that the desired circumstances do not occur. In this case, the decomposition process continues into modest sub-problems.

20.4 NUMERICAL RESULT

This section presents the results of different scenarios. Different scenarios have been produced for the problem under consideration. The scenarios produced consider vehicle capacity, the allowed number of vehicles, and the number of vehicles used. Vehicle capacity is 1500, 1000, 750, and 500, five vehicles were allowed, and vehicle numbers are given as 1, 2, 3, 4, and 5. In addition, the amount of demand to be met was 2250 units.

Infeasible results were obtained from the parametric values selected in the first, sixth, seventh, eleventh, twelfth, sixteenth, seventeenth, eighteenth, and nineteenth scenarios. The main reason for obtaining infeasible results was that the total vehicle capacities were insufficient to meet demand. All scenarios are shown in Table 20.2.

Figure 20.1 shows all combinations of average and best total cost change by considering the different vehicle capacities. The point where the average total and best cost is the lowest seems to be the scenario where the vehicle capacity was 1500

TABLE 20.2
The Average and the Best Cost Change with Regard to Vehicle Capacity and Allowed Vehicle Number

Scenario	Vehicle Capacity	Allowed Vehicle Number	Used Vehicle Number	Solution Time (min)	Average Total Cost	Best Total Cost Found
1	1500	1	1	–	Inf	Inf
2	1500	2	2	0.22	562.60	562.60
3	1500	3	3	1.36	560.00	560.00
4	1500	4	4	0.68	557.60	557.60
5	1500	5	5	1.22	563.50	563.50
6	1000	1	1	–	Inf	Inf
7	1000	2	2	–	Inf	Inf
8	1000	3	3	2.78	566.40	566.40
9	1000	4	4	3.10	564.20	564.20
10	1000	5	5	4.09	570.10	570.10
11	750	1	1	–	Inf	Inf
12	750	2	2	–	Inf	Inf
13	750	3	3	7.60	591.50	591.50
14	750	4	4	8.70	585.10	585.10
15	750	5	5	8.47	582.50	582.50
16	500	1	1	–	Inf	Inf
17	500	2	2	–	Inf	Inf
18	500	3	3	–	Inf	Inf
19	500	4	4	–	Inf	Inf
20	500	5	5	15.25	620.00	619.40

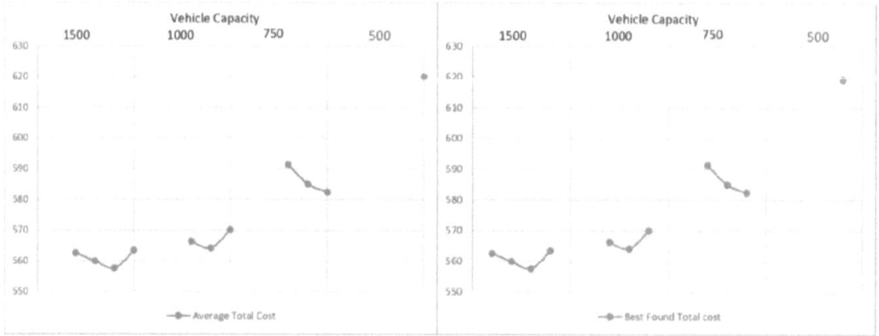

FIGURE 20.1 The change of average and the best total cost concerning vehicle capacity.

units and the number of vehicles used was five units. A detailed examination of Figure 20.1 shows that reducing vehicle capacity increases the average total and best cost. The highest price occurs when the vehicle capacity is reduced to 500 units.

In addition, each vehicle capacity has a different number of vehicles used. Using other vehicle numbers within the same vehicle capacity also causes other costs. The number of vehicles used seems to impact the evaluation of costs significantly.

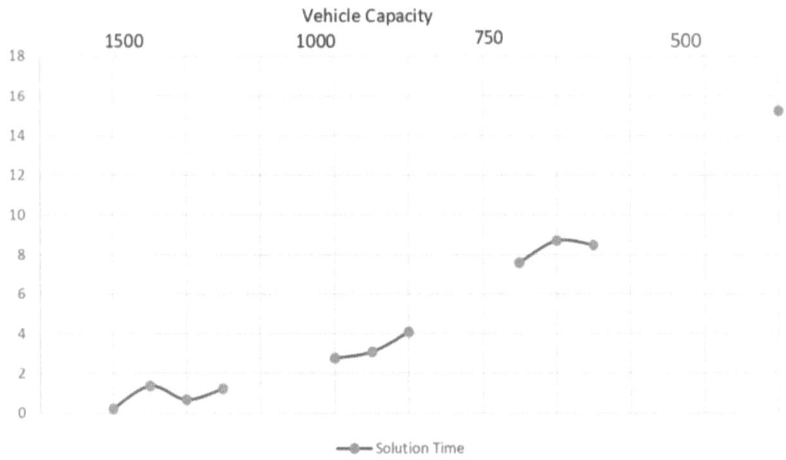

FIGURE 20.2 The change of solution time concerning vehicle capacity.

Figure 20.2 shows all combinations of solution times by considering the different vehicle capacities. The use of varying vehicle capacities seems to have a significant effect on the cost and time taken to obtain the optimal solution values. As the preferred vehicle capacity decreases, the time required to get the best solution to the problem increases significantly. In addition, the use of different numbers of vehicles within the same vehicle capacity also affects the solution process. However, it is not possible to declare that this effect is a positive or negative interaction.

20.5 CONCLUSION

This chapter focused on meeting the material needs of disaster relief tents set up in the case of a possible earthquake by determining relief vehicles and routing them.

In the problem's branch-and-bound solution, one of the algorithms that gives the exact resolution is used, and results obtained according to different parametric values. The results obtained were presented in detail in Section 20.4. When the effects were examined, some conclusions were reached:

- The point where the average total and best cost is the lowest was the scenario where the vehicle capacity was 1500 units and the number of vehicles used was five units.
- Reducing vehicle capacity leads to an increase in the average total and best costs. The highest cost occurred when the vehicle capacity was reduced to 500 units.
- The number of vehicles used seemed to impact the evaluation of costs significantly.
- The use of different vehicle capacities seemed to have a significant effect on the cost and time taken to obtain the optimal solution values.
- The use of different vehicles within the same vehicle capacity also affected the solution process.

The application of larger-scale problems for future studies will play an active role in this proposed method, but the increase in problem size will make it difficult to find solutions with algorithms that give exact solutions. For this reason, approximate solutions can be obtained with strong heuristics or deep learning methods.

REFERENCES

Adarang, H., Bozorgi-Amiri, A., Khalili-Damghani, K., and Tavakkoli-Moghaddam, R. 2020. A robust bi-objective location-routing model for providing emergency medical services. Journal of Humanitarian Logistics and Supply Chain Management, 10: 285–319.

Ahmadi, M., Seifi, A., and Tootooni, B. 2015. A humanitarian logistics model for disaster relief operation considering network failure and standard relief time: A case study on San Francisco district. Transportation Research Part E: Logistics and Transportation Review, 75: 145–163.

Akbari, F., Valizadeh, J., and Hafezalkotob, A. 2021. Robust cooperative planning of relief logistics operations under demand uncertainty: a case study on a possible earthquake in Tehran. International Journal of Systems Science: Operations & Logistics, 1–24. Epub ahead of print.

Beiki, H., Seyedhosseini, S. M., Ghezavati, V. R., and Seyedaliakbar, S. M. 2020. A location-routing model for assessment of the injured people and relief distribution under uncertainty. International Journal of Engineering, 33(7): 1274–1284.

Bozorgi-Amiri, A. and Khorsi, M. 2016. A dynamic multi-objective location–routing model for relief logistic planning under uncertainty on demand, travel time, and cost parameters. The International Journal of Advanced Manufacturing Technology, 85(5): 1633–1648.

Gholami-Zanjani, S. M., Jafari-Marandi, R., Pishvaee, M. S., and Klibi, W. 2019. Dynamic vehicle routing problem with cooperative strategy in disaster relief. International Journal of Shipping and Transport Logistics, 11(6): 455–475.

Kianfar, K. 2010. Branch-and-bound algorithms. Wiley Encyclopedia of Operations Research and Management Science. Hoboken, NJ: Wiley.

Kokaji, K. and Kainuma, Y. 2018. Development of a disaster relief logistics model minimizing the range of delivery time. Operations and Supply Chain Management, 11(2): 66–72.

Sabouhi, F., Bozorgi-Amiri, A., and Vaez, P. 2020. Stochastic optimization for transportation planning in disaster relief under disruption and uncertainty. Kybernetes, 50(2): 2632–2650.

Sabouhi, F., Bozorgi-Amiri, A., Moshref-Javadi, M., and Heydari, M. 2019. An integrated routing and scheduling model for evacuation and commodity distribution in large-scale disaster relief operations: A case study. Annals of Operations Research, 283(1): 643–677.

Sharif, M. T. and Salari, M. 2015. A GRASP algorithm for a humanitarian relief transportation problem. Engineering Applications of Artificial Intelligence, 41: 259–269.

Shi, Q., Yang, W., Yang, Z. M., and Zhao, Q. C. 2018. Relief materials vehicles planning in natural disasters. IEEE/CAA Journal of Automatica Sinica, 5(2): 595–601.

Shin, Y., Kim, S., and Moon, I. 2019. Integrated optimal scheduling of repair crew and relief vehicle after disaster. Computers & Operations Research, 105: 237–247.

Uslu, A., Cetinkaya, C., and İşleyen, S. K. 2017. Vehicle routing problem IN POST-disaster humanitarian relief logistics: A case study in ANKARA. Sigma: Journal of Engineering & Natural Sciences/Mühendislik ve Fen Bilimleri Dergisi, 35(3): 481–499.

Vahdani, B., Veysmoradi, D., Shekari, N., and Mousavi, S. M. 2018. Multi-objective, multi-period location-routing model to distribute relief after earthquake by considering emergency roadway repair. Neural Computing and Applications, 30(3): 835–854.

Vieira, Y. E. M., de Mello Bandeira, R. A., and da Silva Júnior, O. S. 2021. Multi-depot vehicle routing problem for large scale disaster relief in drought scenarios: The case of the Brazilian northeast region. International Journal of Disaster Risk Reduction, 58: 102193.

Weimin, G. 2016. Research the optimal path of postdisaster relief vehicles based on ant colony algorithm and WSN. International Journal of Future Generation Communication and Networking, 9(7): 145–152.

Yi, J. H., Wang, J., and Wang, G. G. 2019. Using Monarch butterfly optimization to solve the emergency vehicle routing problem with relief materials in sudden disasters. Open Geosciences, 11(1): 391–413.

Zhan, S. L., Liu, S., Ignatius, J., Chen, D., and Chan, F. T. 2021. Disaster relief logistics under demand-supply incongruence environment: A sequential approach. Applied Mathematical Modelling, 89: 592–609.

21 Disaster Management with Fuzzy Sets and Engineering Economic Analysis

Eda Boltürk
Istanbul, Türkiye

CONTENTS

21.1 Introduction....................313
21.2 Fuzzy Sets and Extensions....................314
21.3 Literature Review....................315
21.4 Interval-Valued Spherical Fuzzy Present Worth Analysis....................316
21.5 Engineering Application in Disaster Management....................318
21.6 Conclusion....................319
References....................321

21.1 INTRODUCTION

Due to climate changed, disaster management is a big concern. Countries need to take precautions to prevent big economic losses and their effects on human life. For this reason, the fields of forecasting and investment management gain great importance. Countries are required to manage their disaster-management plans and measures, including financial analysis according to their geopolitical situation. Engineering economics has been used with economic principles in engineering decisions. Present worth analysis (PWA), cost–benefit ratio analysis, future worth analysis, rate of return analysis, and annual worth analysis are significant investment analysis methods in engineering economics. These techniques provide direction to investors as to whether invest in the short-, medium-, or long-term. Ordinary fuzzy logic is composed of membership degrees and non-membership degrees. After the introduction of fuzzy sets, various extensions have been proposed and new parameters introduced to deal with the complexity of human thought. One of these extensions has been proposed as "spherical fuzzy sets" (Kutlu Gündogdu and Kahraman 2019a). Spherical fuzzy sets (SFSs) provide a wider area to experts with spherical fuzzy numbers that consist of three parameters: membership, non-membership, and indeterminacy. Later,

DOI: 10.1201/9781003212904-21

Kutlu Gündogdu and Kahraman (2019b) introduced interval-valued spherical fuzzy sets (IVSFSs). An interval-valued spherical fuzzy number (IVSFN) of the universe of discourse U is given as A˜ in Eq. (21.1).

$$\tilde{A} = \{<u,([\mu_A^-(u), \mu_A^+(u)], [v_A^-(u), v_A^+(u)], [\iota_A^-(u), \iota_A^+(u)], | u \in U>\} \tag{21.1}$$

where $0 \le \mu_A^-(u) \le \mu_A^+(u) \le 1$, $0 \le v_A^-(u) \le v_A^+(u) \le 1$, and $0 \le (\mu_A^+(u))^2 + \le (v_A^+(u))^2 + \le (\iota_A^+(u))^2 \le 1$.

Fuzzy sets are used in order to eliminate the vagueness of investors/experts. As a result, engineering economics techniques and fuzzy sets are used to achieve better results based on human thought, and improve accuracy. The aim and importance of this study is to introduce the interval-valued spherical fuzzy PWA method as a new extension of fuzzy engineering economics, and to use the extension in the field of disaster management. The main contribution of this study to the literature is to extend the PWA method with interval-valued SFSs and to use this fuzzy method in economic analysis of disaster management. The main purpose of the proposed method is to indicate that interval-valued SFSs can be extended in the PWA method in disaster management. The paper is organized as follows. Fuzzy sets and extensions are given in Section 21.2. In Section 21.3, a literature review based on disaster management and engineering economics is summarized. Proposed interval-valued spherical fuzzy PWA (IVSPWA) is presented in Section 21.4. Section 21.5 provides an engineering application in disaster management, and Section 21.6 provides conclusions with future suggestions.

21.2 FUZZY SETS AND EXTENSIONS

Fuzzy sets were introduced by Lötfi Zadeh in 1965. Fuzzy sets theory (Zadeh 1965) is a significant mechanism that captures the indefiniteness in human thought and determines them as linguistic assessments. The most significant principle of the fuzzy set theory is the extension principle, which allows the use of traditional mathematical operations in a fuzzy environment. In fuzzy sets, the addition, multiplication, subtraction, division, power operation, integration, and derivation are performed as these extended mathematical operations. Since then, fuzzy sets have been extended to its various extensions. These extensions are type 2 fuzzy sets (Zadeh 1975), interval-valued fuzzy sets (Grattan-Guiness 1975; Zadeh 1975; Sambuc 1975; Jahn 1975), intuitionistic fuzzy sets (Atanassov 1986), fuzzy multisets (Yager 1986), intuitionistic fuzzy sets of second type (Atanassov 1989), neutrosophic fuzzy sets (Smarandache 1998), nonstationary fuzzy sets (Garibaldi and Ozen 2007), hesitant fuzzy sets (Torra 2010), Pythagorean fuzzy sets (Yager and Abbasov 2013), picture fuzzy sets (Cuong 2014), q-rung orthopair fuzzy sets (Yager 2016), fermatean fuzzy sets (Senapati and Yager 2020), SFSs (Kutlu Gundogdu and Kahraman 2019a), and circular intuitionistic fuzzy sets (Atanassov 2020). IVSFS is an interval-valued version of SFSs and illustrated by Kutlu Gundogdu and Kahraman (2019b).

21.3 LITERATURE REVIEW

Investment analysis in disaster management is generally examined under the following headings: risk analysis, infrastructure resilience, disaster-sensitive urbanization, information and computer technologies, knowledge management and numerical analysis, economic analysis, environmental analysis and sociological analysis, etc.

In this section, a literature review based on disaster management and engineering economics is given. Marshall et al. (2004) published optimization protection that resulted in devastating damage against natural hazards. They assessed the risk of hazards and possible damage; identified alternative risk mitigation strategies, management practices, and financial mechanisms; and evaluated the lifecycle economic impact. Ganderton (2005) used benefit–cost analysis for disaster mitigation costs to provide evidence and justify the use of public funds and agencies. Lechtenbörger (2006) worked on estimating and modeling monetary losses from disasters and their consequences on infrastructure, nature, and human beings. Geldermann et al. (2008) illustrated a reference installation approach, a methodology that grants a transparent and consistent evaluation of individual industrial asset values in natural disasters. Tian et al. (2010) presented a cost–benefit analysis of two categories, macro analysis and definite analysis, for engineering projects for disaster reduction decision-making. Tsai (2011) presented an economic impact analysis framework for tourism industry natural disaster risk management and proposed an economic benefit-constrained maximization model. Jing and Zhang (2011) suggested reducing the impact of natural disasters on economic growth by decreasing the replacement investment in disaster losses and increasing the investment in fixed assets for China. Yu et al. (2011) introduced the capital asset pricing model for pricing analysis of engineering earthquake insurance. An engineering earthquake insurance pricing model was handled from capital market perspective. Kenny (2012) investigated disaster risk reduction from the perspective of costs, benefits, and institutions for developing countries because of a cost–benefit ratio that is adverse as compared to other attempts. Novoa (2013) studied economic analysis for investments in flood protection infrastructure for mitigating increased disaster risk due to climate change and urbanization, based on socio-economic, hydrologic, and engineering considerations in Ho Chi Minh City. Probabilistic cost–benefit analysis was studied for the economic viability of infrastructure design for flood protection. The study showed economic methods in investment assessment for climate change adaptation and disaster risk reduction. Hallegatte (2014) estimated costs for potential disasters with indirect economic consequences risk-management strategies. Xiao et al. (2015) illustrated an integrated economic model for disaster prevention port investments for coastal and marine disasters climate change. Rong et al. (2016) evaluated the earthquake disaster loss evaluation of a railway tunnel with a program. Chen et al. (2016) assessed the economic value of the debris flow control engineering in tourist towns by integrating both disaster and welfare economics with a cost–benefit method. Karasev and Solozhentsev (2016) presented a new hybrid logical and probabilistic model for risk management in socio-economic systems. They proposed the concept of invalidity by analogy with reliability and safety in engineering and used logical and probabilistic risk models. Azimi et al. (2019) presented a flood disaster scenario and an assessment based on identifying disaster risk factors in construction industry. Heo and Heo (2019) studied an economic analysis for short- and long-term effects of investment in a disaster-safety related project, and the effects of damage reduction were analyzed before and after project implementation to assess the short-term effects and perform a cost–benefit analysis. Analysis was performed to assess the long-term effects. The results showed that disaster prevention projects reduce damage both in the short and long term. Therefore, investing in preventive projects to deal with disasters was significant to maximize return on investment. Wu et al. (2020) conducted a comparative performance analysis based on sustainable development goals for five countries: China, Saudi Arabia, the USA, Russia, and Canada. These countries are major oil-dependent economies. The oil industry is the backbone of the economic profile of any country, which is also directly connected to power generation, industrialization, and the environment. The performance of these countries was examined based on the environment and economics during the COVID-19 pandemic. In addition, a management policy was advised to maintain economic management and oil pricing for sustainable development and engineering economics. The literature review is summarized in Table 21.1. It is seen that there has not been any study with IVSPWA for disaster management.

21.4 INTERVAL-VALUED SPHERICAL FUZZY PRESENT WORTH ANALYSIS

In the literature, PWA has been extended with fuzzy sets such as a spherical fuzzy PWA method (Boltürk and Seker 2021; Onar et al. 2020), hesitant fuzzy PWA (Kahraman et al. 2015), Pythagorean fuzzy PWA (Kahraman et al. 2017), fermatean fuzzy PWA (Sari and Sergi 2021), intuitionistic fuzzy PWA (Kahraman et al. 2018), type-2 fuzzy PWA (Kahraman et al. 2018), and picture fuzzy PWA (Bolturk 2021). We extend the PWA method with IVSFSs in this study. PWA method briefly gives the steps of proposed interval-valued spherical fuzzy PWA (IVSPWA) method.

- **Step 1.** Collect parameter values from experts. The parameters of first cost (FC), salvage value (SV) annual benefit (AB), annual cost (AC), interest rate (i), and life (n) are given with interval-valued spherical fuzzy membership numbers (IVSFN) in Eqs. (21.2–21.6).
- **Step 2.** Aggregation of expert opinions with interval-valued spherical geometric mean (IVSWGM), was proposed by Kutlu Gündoğdu and Kahraman (2019b) as in Eq. (21.7).
- **Step 3.** Defuzzification of aggregated membership functions. The defuzzification formula and score function of interval-valued spherical fuzzy numbers (Kutlu Gündoğdu and Kahraman 2019b) are given in Eq. (21.8).
- **Step 4.** Normalization of defuzzified values.
- **Step 5**. Obtaining weighted parameter values with normalized values.
- **Step 6.** Calculation of PWA value based on values obtained in Step 5 as in Eq. (21.9).

$$\tilde{FC} = \{\langle fc_1, IVSFN_1, \ldots, IVSFN_m \rangle, \langle fc_2, IVSFN_1, \ldots, IVSFN_m \rangle, \ldots, \langle fc_k, IVSFN_1, \ldots, IVSFN_m \rangle\} \tag{21.2}$$

TABLE 21.1
Literature Review on Disaster Management and Engineering Economics

Author	Year	Subject
Marshall et al.	2004	Optimization protection in devastating damages against natural hazards
Ganderton	2005	Benefit–cost analysis for disaster mitigation costs
Lechtenbörger	2006	Estimating and modeling monetary losses from disasters
Geldermann	2008	Reference installation approach for natural disasters
Tian et al.	2010	Cost–benefit analysis for disaster reduction decision-making
Tsai	2011	Economic impact analysis framework for tourism industry natural disaster risk management
Jing and Zhang	2011	Reducing the impact of natural disasters on economic growth by decreasing the replacement investment and increasing the investment in fixed assets
Yu et al.	2011	Capital asset pricing model for pricing analysis of engineering earthquake insurance
Kenny	2012	Disaster risk reduction investigation from the perspective of costs, benefits, and institutions
Hallegatte	2014	Estimation costs for potential disasters with indirect economic consequences
Xiao et al.	2015	Integrated economic model for disaster-prevention port investments
Rong et al.	2016	Earthquake disaster loss evaluation of railway tunnel with the VB program
Chen et al.	2016	Economic value of the debris flow control engineering by integrating both disaster and welfare economics with cost–benefit method
Karasev and Solozhentsev	2016	Hybrid probabilistic and logical models for system risk management
Azimi et al.	2019	Flood disaster scenario and evaluation based on identifying disaster risk factors in construction industry
Heo and Heo	2019	Economic analysis for short- and long-term effects of investment in a disaster-safety related project, and the effects of damage reduction
Wu et al.	2020	Disaster-management policy development and engineering economics analysis with the game-changing impact of COVID 19

$$\tilde{AB}=\{\langle ab_1, IVSFN_1,\ldots, IVSFN_m \rangle, \langle ab_2, IVSFN_1,\ldots, IVSFN_m \rangle,\ldots, \langle ab_k, IVSFN_1, \ldots, IVSFN_m \rangle\} \tag{21.3}$$

$$\tilde{AC}=\{\langle ac_1, IVSFN_1,\ldots, IVSFN_m \rangle, \langle ac_2, IVSFN_1,\ldots, IVSFN_m \rangle,\ldots, \langle ac_k, IVSFN_1, \ldots, IVSFN_m \rangle\} \tag{21.4}$$

$$\tilde{i}=\{\langle i_1, IVSFN_1,\ldots, IVSFN_m \rangle, \langle i_2, IVSFN_1,\ldots, IVSFN_m \rangle,\ldots, \langle i_k, IVSFN_1, \ldots, IVSFN_m \rangle\} \tag{21.5}$$

$$\tilde{n}=\{\langle n_1, IVSFN_1,\ldots, IVSFN_m \rangle, \langle n_2, IVSFN_1,\ldots, IVSFN_m \rangle,\ldots, \langle n_k, IVSFN_1, \ldots, IVSFN_m \rangle\} \tag{21.6}$$

$$IVSWGM_w(IVSFN\tilde{}_1, IVSFN\tilde{}_2,\ldots,IVSFN\tilde{}_n) = \{[\Pi_{j=1}^{n}(\mu_j^-)^{wj}, \Pi_{j=1}^{n}(\mu_j^+)^{wj}], [1\text{-}\Pi_{j=1}^{n}(1\text{-}(\upsilon_j^-)^2)^{wj})^{0.5}, (1\text{-}\Pi_{j=1}^{n}(1\text{-}(\upsilon_j^+)^2)^{wj})^{0.5}], [(\Pi_{j=1}^{n}(1\text{-}(\upsilon_j^-)^2)^{wj} - \Pi_{j=1}^{n}(1\text{-}(\upsilon_j^-)^2\text{-}(I_j^-)^2)^{wj})^{0.5}, (\Pi_{j=1}^{n}(1\text{-}(\upsilon_j^+)^2)^{wj} - \Pi_{j=1}^{n}(1\text{-}(\upsilon_j^+)^2\text{-}(\iota_j^+)^2)^{wj})^{0.5}] \quad (21.7)$$

$$\text{Defuzzification(IVSFN)} = ((\mu^-)^2 + (\mu^+)^2 - (\upsilon^-)^2 - (\upsilon^+)^2 - (\iota^-/2)^2 - (\iota^+/2)^2)/2 \quad (21.8)$$

$$PW_{IVSFN}\tilde{} = -FC_{IVSFN}\tilde{}-AC_{IVSFN}\tilde{}[(((1+i_{IVSFN}\tilde{})\text{^}n_{IVSFN}\tilde{})-1)/(i_{IVSFN}\tilde{}\times((1+i_{IVSFN}\tilde{})\text{^}n_{IVSFN}\tilde{})] + AB_{IVSFN}\tilde{}[(((1+i_{IVSFN}\tilde{})\text{^}n_{IVSFN}\tilde{})-1)/(i_{IVSFN}\tilde{}\times((1+i_{IVSFN}\tilde{})\text{^}n_{IVSFN}\tilde{})] + SV_{IVSFN}\tilde{}\times((1+i_{IVSFN}\tilde{})\text{^}(-\,n_{IVSFN}\tilde{}) \quad (21.9)$$

21.5 ENGINEERING APPLICATION IN DISASTER MANAGEMENT

In this section, an exemplificative engineering application example for preventing disaster is given. Two experts experienced in disaster prevention give their opinions with interval-valued spherical fuzzy membership functions for possible values for each investment analysis (see Table 21.2).

The weights for Expert 1 and Expert 2 are 0.6 and 0.4, respectively. The first cost is the preventive construction cost ,and the annual cost consists of maintenance cost, disaster insurance, and disaster planning. The aggregated membership functions,

TABLE 21.2
Values and Thoughts of Two Experts for Disaster Prevention Investment

			Expert 1	Expert 2
			0.6	0.4
Parameter	W	Possible Values ($)	Membership Functions ([μ⁻, μ⁺], [υ⁻, υ⁺], [ι⁻, ι⁺])	Membership Functions ([μ⁻, μ⁺], [υ⁻, υ⁺], [ι⁻, ι⁺])
FC	0.4	1,000,000	([0.1, 0.3], [0.3, 0.5], [0.2, 0.3])	([0.2, 0.3], [0.1, 0.3], [0.1, 0.2])
	0.3	1,200,000	([0.4, 0.6], [0.2, 0.3], [0.1, 0.3])	([0.4, 0.5], [0.1, 0.2], [0.2, 0.3])
	0.2	1,300,000	([0.5, 0.6], [0.1, 0.3], [0.1, 0.2])	([0.6, 0.7], [0.3, 0.4], [0.2, 0.3])
	0.1	1,400,000	([0.4, 0.6], [0.2, 0.3], [0.2, 0.3])	([0.3, 0.5], [0.2, 0.4], [0.1, 0.2])
AB	0.4	1,000	([0.4, 0.5], [0.3, 0.6], [0.2, 0.3])	([0.2, 0.3], [0.1, 0.5], [0.1, 0.2])
	0.3	2,000	([0.5, 0.6], [0.1, 0.5],[0.1,0.3])	([0.5, 0.7], [0.2, 0.3],[0.2,0.3])
	0.2	3,000	([0.3, 0.5], [0.2, 0.6],[0.1,0.2])	([0.4, 0.5], [0.4, 0.5],[0.1,0.2])
	0.1	1,500	([0.5, 0.6], [0.4, 0.5],[0.2,0.3])	([0.4, 0.5], [0.1, 0.5],[0.2,0.4])
AC	0.4	500,000	([0.4, 0.5], [0.1, 0.2], [0.2,0.5])	([0.5, 0.6], [0.3, 0.4], [0.2,0.4])
	0.3	600,000	([0.5, 0.6], [0.3,0.4],[0.2,0.4])	([0.4, 0.5], [0.2,0.4],[0.1,0.3])
	0.2	450,000	([0.4, 0.6], [0.2,0.5],[0.1,0.3])	([0.5, 0.6], [0.3,0.4],[0.2,0.4])
	0.1	550,000	([0.5, 0.6], [0.1,0.2], [0.1,0.2])	([0.4, 0.5], [0.2,0.3], [0.2,0.3])
SV	0.4	10,000	([0.4, 0.5], [0.2, 0.5],[0.2,0.4])	([0.4, 0.6], [0.3, 0.5],[0.1,0.3])
	0.3	20,000	([0.5, 0.6], [0.3, 0.4],[0.2, 0.4])	([0.4, 0.6], [0.3, 0.4],[0.1, 0.2])
	0.2	15,000	([0.4, 0.5], [0.4, 0.5],[0.2,0.3])	([0.4, 0.5], [0.3, 0.5],[0.2,0.3])
	0.1	20,000	([0.5, 0.6], [0.1, 0.2], [0.2, 0.5])	([0.4, 0.6], [0.2, 0.3], [0.1, 0.4])

			Expert 1	Expert 2
			0.6	0.4
Parameter	W	Possible Values ($)	Membership Functions $([\mu^-, \mu^+], [\upsilon^-, \upsilon^+], [\pi^-, \pi^+])$	Membership Functions $([\mu^-, \mu^+], [\upsilon^-, \upsilon^+], [\pi^-, \pi^+])$
i	0.4	1.1	([0.3, 0.6], [0.1, 0.2], [0.1, 0.4])	([0.5, 0.6], [0.1, 0.2], [0.1, 0.4])
	0.3	1.3	([0.3, 0.4], [0.4, 0.6], [0.1, 0.4])	([0.4, 0.5], [0.3, 0.4], [0.2, 0.4])
	0.2	1.4	([0.4, 0.5], [0.1, 0.2], [0.3, 0.4])	([0.5, 0.6], [0.1, 0.2], [0.3, 0.4])
	0.1	1.6	([0.2, 0.4], [0.1, 0.3], [0.4,0.5])	([0.3, 0.5], [0.2, 0.3], [0.3,0.4])
n	0.4	20	([0.1, 0.7], [0.3, 0.4], [0.1, 0.2])	([0.4, 0.6], [0.3, 0.5], [0.1, 0.2])
	0.3	22	([0.5, 0.6], [0.3, 0.5], [0.2, 0.3])	([0.6, 0.8], [0.2, 0.3], [0.1, 0.2])
	0.2	25	([0.4, 0.5], [0.1, 0.4], [0.4, 0.5])	([0.4, 0.5], [0.3, 0.4], [0.4, 0.5])
	0.1	24	([0.1, 0.3], [0.1, 0.2], [,0.2, 0.3])	([0.4, 0.6], [0.2, 0.3], [,0.2, 0.3])

defuzzified and normalized values for expert membership functions, and weighted parameter values for investment parameter are given in Table 21.3.

The final PWA value is calculated based on the values in Table 21.4 and the PWA obtained is $–1,656,862.31.

21.6 CONCLUSION

Climate change and geopolitical situations push countries to take precautions in order to manage disasters effectively. Unforeseen natural disasters such as storms, fires, and floods have great impacts on human life in many areas. For this reason, investment analyses for mitigating disasters attract great attention and it becomes important to bring together the ideas of experts working in different fields, such as economics, meteorology, civil engineering, water management, risk management geophysics, etc., in investment analysis. IVSFSs are powerful extensions of fuzzy sets extensions and allow experts to evaluate their opinions in a broader field. In this study, a new fuzzy engineering economics extension called IVSPWA is proposed. The IVSPWA method has first been proposed and used in the field of disaster management. The proposed method has been provided with a sample illustrative application for the analysis of investments made to reduce disaster losses based on the opinions of experts from the disaster-management sector. The preliminary information gained from IVSFS is used in calculations with the IVSPWA method. According to the proposed fuzzy extension IVSPWA, the final present worth value is $–1,656,862.31. This value is negative as there are very small benefits for each year and a large economic advantage without any loss in the end. The finding of this study is that the interval-valued spherical expert opinions and interval-valued SFSs can be used with the PWA method in disaster-management investment analysis. For further research, fermatean fuzzy PWA, picture fuzzy PWA, and q-rung orthopair fuzzy PWA methods can be compared and the most suitable method for this field can be seen in disaster management. In addition, sensitivity analysis can be performed for each method and the applicability of the methods for variable parameters can be found based on real data.

TABLE 21.3
Aggregated, Defuzzified, and Normalized Values for Membership Functions and Weighted Parameter Values for Investment Parameters

Parameter	Aggregated Membership Functions ([μ⁻, μ⁺], [υ⁻, υ⁺], [ι⁻, ι⁺])	Defuzzified Values	Normalized Values	Weighted Values
FC	([0.514, 0.568], [0.243, 0.436], [0.170,0.271])	0.157	0.162	161,832.422
	([0.565, 0.631], [0.168, 0.265], [0.148,0.300])	0.296	0.306	366,729.918
	([0.538, 0.570],[0.207, 0.344], [0.151,0.248])	0.216	0.223	289,947.760
	([0.619, 0.631], [0.200, 0.344], [0.168,0.263])	0.300	0.310	433,331.604
AB	([0.679, 0.662], [0.243,0.564], [0.170,0.270])	0.248	0.339	338.605
	([0.576, 0.570], [0.149,0.436], [0.149,0.301])	0.209	0.284	568.874
	([0.504, 0.576],[0.300,0.564], [0.100,0.200])	0.083	0.113	338.127
	([0.618, 0.631],[0.321,0.500], [0.200,0.345])	0.194	0.264	396.373
AC	([0.516, 0.538], [0.207,0.300], [0.200,0.461])	0.180	0.193	96,496.104
	([0.618, 0.631], [0.265, 0.400], [0.169,0.364])	0.255	0.274	164,296.329
	([0.516, 0.600],[0.246, 0.464], [0.150,0.342])	0.158	0.169	76,249.955
	([0.618, 0.631],[0.149, 0.246], [0.149,0.247])	0.339	0.364	200,054.929
SV	([0.565, 0.538], [0.246, 0.500],[0.167,0.365])	0.129	0.167	1,665.603
	([0.618, 0.600],[0.300, 0.400],[0.168,0.338])	0.228	0.294	5,887.493
	([0.565, 0.576], [0.364, 0.500],[0.200,0.300])	0.118	0.153	2,288.524
	([0.618, 0.600], [0.149, 0.246],[0.167,0.463])	0.299	0.386	7,729.936
i	([0.446, 0.600], [0.100, 0.200],[0.100,0.400])	0.233	0.384	0.422
	([0.504, 0.516], [0.364, 0.535],[0.147,0.402])	0.027	0.045	0.058
	([0.516, 0.538], [0.100, 0.200], [0.300,0.400])	0.221	0.364	0.509
	([0.503, 0.516], [0.149, 0.397], [0.363,0.460])	0.127	0.208	0.333
n	([0.325, 0.658], [0.300, 0.444], [0.100,0.200])	0.119	0.232	4.639
	([0.538, 0.542], [0.265, 0.436], [0.169,0.271])	0.149	0.289	6.354
	([0.565, 0.576], [0.207, 0.400], [0.400,0.500])	0.173	0.336	8.409
	([0.325, 0.396], [0.149, 0.246], [0.200,0.300])	0.074	0.143	3.429

TABLE 21.4
Scored Parameters for Present Worth Analysis Value

Parameters	Values
FC	1,251,841.7
AB	1,642.0
AC	537,097.3
SV	17,571.6
I	1.3
N	22.8

REFERENCES

Atanassov, K. 1986. Intuitionistic fuzzy sets. *Fuzzy sets and Systems*, 20(1):87–96.

Atanassov, K. 1989. More on intuitionistic fuzzy sets. *Fuzzy Sets and Systems*, 33(1):37–45.

Atanassov, K.T. 2020. Circular intuitionistic fuzzy sets. *Journal of Intelligent & Fuzzy Systems*, 39:5981–5986.

Azimi, M. A., Syed Zakaria, S. A., and T. A. Majid. 2019. Disaster risks from economic perspective: Malaysian scenario. Paper presented at the IOP Conference Series: Earth and Environmental Science, 244(1). doi: 10.1088/1755-1315/244/1/012009.

Boltürk, E. 2021. Container ship investment analysis using picture fuzzy present worth analysis. *Journal of ETA Maritime Science*, 9(4):233–242. doi: 10.4274/jems.2021.89156.

Boltürk, E. and S. Seker. 2021. Present worth analysis using spherical fuzzy sets. In International Conference on Intelligent and Fuzzy Systems. 777–788. Springer, Cham.

Chen, M., Hu, G., Chen, N., Zhao, C., Zhao, S., and D. Han. 2016. Valuation of debris flow mitigation measures in tourist towns: A case study on Hongchun gully in southwest China. *Journal of Mountain Science*, 13(10):1867–1879.

Cuong, B.C. 2014. Picture fuzzy sets. *Journal of Computer Science and Cybernetics*, 30(4):409.

Ganderton, P. T. 2005. 'Benefit-cost analysis' of disaster mitigation: Application as a policy and decision-making tool. *Mitigation and Adaptation Strategies for Global Change*, 10(3):445–465.

Garibaldi, J.M. and T. Ozen. 2007. Uncertain fuzzy reasoning: A case study in modelling expert decision making. *IEEE Transactions on Fuzzy Systems*, 15(1):16–30.

Geldermann, J., Merz, M., Bertsch, V., et al. 2008. The reference installation approach for the estimation of industrial assets at risk. *European Journal of Industrial Engineering*, 2(1):73–93.

Grattan-Guiness, I. 1975. Fuzzy membership mapped onto interval and many-valued quantities. *Z. Math. Logik. Grundladen Math.*, 22:149–160.

Hallegatte, S. 2014. Modeling the role of inventories and heterogeneity in the assessment of the economic costs of natural disasters. *Risk Analysis*, 34(1):152–167.

Heo, B-Y. and W.-H. Heo. 2019. Economic analysis of disaster management investment effectiveness in Korea. *Sustainability*, 11(11):3011. doi: 10.3390/su11113011.

Jahn, K. U. 1975. Intervall-wertige Mengen. *Math. Nach*, 68:115–132.

Jing, P. and G. Zhang. 2011. Short-term impact of natural disasters on china's economy – empirical analysis based on Harrod-Domar model. 2011 International Conference on Information Management, Innovation Management and Industrial Engineering, 197–200.

Kahraman, C., Onar, S. C., and B. Oztaysi. 2017. Present worth analysis using Pythagorean fuzzy sets. In Advances in Fuzzy Logic and Technology 2017, 336–342. Springer, Cham.

Kahraman, C., Onar, S. Ç., and B. Öztaysi. 2015. Present worth analysis using hesitant fuzzy sets. https://www.atlantis-press.com/proceedings/ifsa-eusflat-15/23549

Kahraman, C., Onar, S. Ç., Öztayşi, B., Sarı, İ. U., and E. İlbahar. 2018. Wind energy investment analyses based on fuzzy sets. In Energy Management—Collective and Computational Intelligence with Theory and Applications, 141–166. Springer, Cham.

Karasev, V. V. and E. D. Solozhentsev. 2016. Hybrid logical and probabilistic models for risk management of systems. *SPIIRAS Proceedings*, 5(48):125–150. doi: 10.15622/sp.48.7

Kenny, C. 2012. Disaster risk reduction in developing countries: Costs, benefits and institutions. *Disasters*, 36(4):559–588.

Kutlu Gündoğdu F. and C. Kahraman. 2019a. Spherical fuzzy sets and spherical fuzzy TOPSIS method. *Journal of Intelligent & Fuzzy Systems*, 36(1):337–352.

Kutlu Gündoğdu, F. and C. Kahraman. 2019b. A novel fuzzy TOPSIS method using emerging interval-valued spherical fuzzy sets. *Engineering Applications and Artificial Intelligence*, 85:307–323.

Lechtenbörger, C. 2006. CEDIM-Risk Map Germany: First results. In RISK21-Coping with Risks due to Natural Hazards in the 21st Century, 149–156. CRC Press, Boca Raton, FL.

Marshall, H. E., Chapman, R. E., and C. J. Leng. 2004. Risk mitigation plan for optimizing protection of constructed facilities. *Cost Engineering (Morgantown, West Virginia)*, 46(8):26–33.

Novoa, D. C. 2013. The economics of climate change, urbanisation, and long-term flood protection. In: Leal Filho W. (ed.) Climate Change and Disaster Risk Management, 483–496. Springer, Berlin, Heidelberg.

Onar, S. C., Öztayşi, B., Kahraman, C., and I. Otay. 2020. Engineering economic analysis of solar energy investments using spherical fuzzy sets. In Developments of Artificial Intelligence Technologies in Computation and Robotics: Proceedings of the 14th International FLINS Conference (FLINS 2020), 182–189. World Scientific.

Rong, M., Wang, H., and H. Rong. 2016. VB program design application in railway tunnel earthquake disaster loss evaluation. *Electronic Journal of Geotechnical Engineering*, 21(10):3937–3949.

Sambuc R. 1975. Fonctions φ-floues. Application l'aide au diagnostic en pathologie thyroidienne. PhD Thesis, University of Marseille, France.

Senapati, T. and R.R. Yager. 2020. Fermatean fuzzy sets. *Journal of Ambient Intelligence and Humanized Computing*, 11:663–674.

Sergi, D. and I.U. Sari. 2021. Fuzzy capital budgeting using fermatean fuzzy sets. In: Kahraman C., Cevik Onar S., Oztaysi B., Sari I., Cebi S., Tolga A. (eds) Intelligent and Fuzzy Techniques: Smart and Innovative Solutions. INFUS 2020. Advances in Intelligent Systems and Computing, 448–456. Springer, Cham.

Smarandache, F. 1998. Neutrosophy, Neutrosophic Probability, Set, and Logic. American Research Press, Rehoboth.

Tian, H., Nan, H., and Z. Yang. 2010. Benefit-cost analysis of geological disaster mitigation and example. *Disaster Advances*, 3(4):163–165.

Torra, V. 2010. Hesitant fuzzy sets. *International Journal of Intelligent Systems*, 25(6):529–539.

Tsai, C. 2011. Optimum arrangement between tourism industry natural disasters insurance and risk control. 2011 International Conference on Information Management, Innovation Management and Industrial Engineering, 3:295–298.

Wu, H., Shah, S. M. A., Nawaz, A., et al. 2020. Disaster management policy development and engineering economics: an analysis of game-changing impact of COVID 19 on oil-power industry, environment, and economy. *Revista Argentina de Clínica Psicológica*, 29(5):550.

Xiao, Y., Fu, X., Ng, A. K. Y., and A. Zhang. 2015. Port investments on coastal and marine disasters prevention: Economic modeling and implications. *Transportation Research Part B: Methodological*, 78:202–221.

Yager, R.R. 1986. On the theory of bags. *International Journal of General System*, 13(1):23–37.

Yager, R.R. 2016. Generalized orthopair fuzzy sets. *IEEE Transactions on Fuzzy Systems*, 25(5):1222–1230.

Yager, R.R., and A.M. Abbasov. 2013. Pythagorean membership grades, complex numbers, and decision making. *International Journal of Intelligent Systems*, 28(5):436–452.

Yu, T., Liu, H., Wang, D., and G. Yang. 2011. Applied research on capital asset pricing model in premium rate making of engineering earthquake insurance. 2011 International Conference on Information Management, Innovation Management and Industrial Engineering, 3: 82–86.

Zadeh, L. 1965. Fuzzy sets. *Information and Control*, 8(3):338–353.

Zadeh, L. 1975. The concept of a linguistic variable and its application to approximate reasoning-I. *Information Sciences*, 8:199–249.

Index

Note: Figures are indicated by *italics*. Tables are indicated by **bold**.

A

accommodation 192, 200, 223, 290, **292**, **295**, 300
actions 44, 48, 51, 72–4, 76–7, 84, **87–8**, 90, 108, 110, **297–9**, 305
active fault 259, 284
AFAD 31
age 94–5, **99**, 102–3, 124, 283
AHP 4, 6–8, 20, 22–5, **23–6**, 43–51, 64, 74, 107–11, 113, **115**, 117, 157–9, 173–4, 194–5, 236, 251–3, 255–7, 259–64, 267–8, 270–6
AHP-TOPSIS 7, 25, 174, 195
aid 3–4, 6–8, 33, 38, 76, 83, 180, 195, 232, 252, 304–5, **305**, 307
aim 24–5, 33, 73, 89, 94, 113, 125, 164, 198, 216, 225, 252, 264, 314
airport 71–6, 83–4, 88–90
algorithm **35–6**, 38–9, 64, 193–4, 245, 303, 304
allocation 32–4, **36**, 38, 40, 236, 284, 290, 303–4, **305**
alternative 2–4, 7, 19, 39, 43–51, 61, 66, 79, 95, 113, **117**, 126, 128, 130, 133, 136, 138–50, 157–61, 172, 174–5, 178–9, 191, 193–4, 196–9, 202–5, 213–14, 216–18, 224–7, 229, 231–2, 239–41, 246–7, 251, 253, 258–9, 261–2, *263*, **263**, 264, 274–5, 277, **277**, 315
alternative evaluation 258, 262, **263**, 274, 276
analytic hierarchy process *see* AHP
analytic network process *see* ANP
annual cost 316, 318
ANP 4, 6, 7, 20, 23–4, **24**, 25, **26**, 47, 49–50, 64, 95
ant colony 303
arcgis 241, *242*, 290
assembly area 23, 252, 261
assessment 4–5, 45–6, 49–51, 64, 71–3, 75, 77, 79, 81, 83, 85, 87, 89–90, 94–5, **99**, 103, 107–13, **116**, 117, 125, 138, 169, 172–4, 177, *180*, 181–2, 191–3, 195–8, 205–6, 214–16, 218, 224–5, 232, 236, 239, 252, 291, 314–15
atom 288
avalanche 251
aviation 71–2, 78, 83–4, 89
awareness 75–7, 88–9, 195, 205

B

baltic dry index 60, 63, 65
basic needs 252
Bayesian theory 171
benefit 32, 34, **36**, 39, 47, 64, 73–4, 126–8, 138–40, 145, **149**, 156–7, 160–1, 165, 216–17, 219, 225, 227, 232, 237, 239–40, 313, 315–17, 319
benefit-oriented criteria 156
BWM 2, 4, 6, 8, **47**, 64, 71, 73–4, 77, 79, 80–4, 88, 89, 169–71, 174–5, *176*, 179–85, 193–5, 295, 300
best-worst method *see* BWM
biological 1, 95, 98, 155, 224
blank sailing 61, 63, 65–6
branch and bound 305, 307, 309
brick 224, 228, **229**, **231–2**, 233, 287–8
budget 38, 40, 76, 156, 223
buffer zones 238, 246

C

cancellations 72, **75**
capacity 1, 7, 17–18, 31, 60–2, 64, 66, 73, 75, **137**, 157, 170, 172, **173**, 212–15, 218, 220–1, 225, 228, 236, 247, **254**, **262**, **269**, 270, **276**, **305**, 307–9
capacity of oxygen 220–1
cargo capacity 269–70, 276
catchball 78
central anatolia region 113, **116**
characteristic 44, 93, 103, 156, 174, 251, 269, 282
circular intuitionistic fuzzy sets 314
classification 2, 4, 11, **291**
climate 2, 18, **26**, 44–5, 51, 62, 71–7, 83–4, 88–90, **115**, 251, 315, 319
climate change **26**, 44–5, 51, 62, 71–7, 83, 88–9, 108, 115, 315, 319
climate-resilient 74–5, 88–9
climatological 1, 109, 224
cluster **36**, 38, 232, **245**, 246
CoCoSo **47**, 191–8, 202, **203**, 205–6, 223–4, 226, **231**, 235, 237, 239, 241, 247
combined compromise solution *see* CoCoSo
coefficient 159, 161, 226
combination 7, 21–5, 34, 45–6, 48–9, 55, 175, 193, 224, 239, 307, 309
communications 155
comparison-based methods 3, 4
complex 5, 20, **36**, 38, 46, 48, 56, 95–6, 98, 136, 138, 156, 172, 213, 235, 273, 282, **292**, 294, **295**, **297**, **299**
compromise **47**, 48, 49, 156, 157, 159, 161, 192, 193, 198, 224, 237–9, 247, 274, 275

compromise solution **47**, 157, 159, 161, 192–3, 224, 237–9, 247, 274–5
concrete 193, 228–9, **231**, **232**, 233, 287
confirmed cases 124
congestion 32, **37**, 39, 57, 60, 63–4, 66, **254**, **262**
consensus 78, 96–7, **99**, 145, 156, 161, 274
container shipping 57, 60–1, 63–5
contaminations **75**
contribution to the economy 224, 230–2
control distance 156–9, 161, 163
controlling 73, 75
coordination **36**, 39, 62, 75, 150, 215, 235, 246, 270
COPRAS 5, 7
complex proportional assessment *see* COPRAS
coronavirus 95, 102, 124
cost 18, **35**, 39, 55, 58, 65–6, 74, 127–8, 139, 156–7, 160–1, 205, 216–17, 223, 225, 227, 230, 232–3, 239, 252, 261, **263**, 304–9, 313, 315–8
COVID-19 2, 5, 7–9, 18, 55, 59–66, 72, 93–5, 98, 123–5, 170–4, 194, 211–16, 218, 220–1, 283, 300, 316
cred 17, 18
critic method 123, 125, 128, 132, 193
crisis management 8, 150, 192, 212, 282–5, 290, 295–6, 300
criteria weight 84, 113, **115**, 117, *118*, 127, 132, 142, **144**, 164, 171–2, 175, 177, 181–3, 191, 193–4, 196, 198, 206, 218, **220**, 224, 226, 232–3, 251–3, 256, 257–62, 274–6, **296**
current expenditure on health per capita *see* GDP
custom-made drones 156

D

damage 1, 3, 17, 44–6, 49, 64, **75**, 107–8, **137**, 192, 225, 228–30, 232, 252, 268, 284, *285*, 287, *287*, 288, 300, 304, 315, 316, **317**
damaged buildings 3, 198–200
dams 45, 46
data 5–6, 18, 20, 49, 59–60, 74, 96, 98, 113, 123–5, 142, 157, 159, 161, **162–3**, 174, 177–8, 183, **184**, 192, 194–5, 214, 216, 226, 228, 231, 233, 236–8, 264, 283, 285, 288, 300, 321
data envelopment analysis *see* DEA
DEA 5, 124–5, 194
debris 3, 223–5, 228–33, 283, 317
decision analysis framework 235, 237, *238*, 240, 247
decision making 3–8, 11, 19, 20, **21**, *22*, **23**, **24**, **26**, *27*, 38, 43–5, 48–51, 64–5, 74, 77, 79, 94–6, 108, 124, 136, *138*, 140, 156–8, 171–2, 174, 175, *176*, 180–1, 185, 192–4, 212, 216, 224, 228, 252, 256, 262, 264, 268, 275, 283, 290, 315, **317**
decision-making trial and evaluation laboratory *see* DEMATEL
decision matrix 126–7, 138–9, 142, 145–7, 159–61, **162**, 192, 196–8, 200, **203**, 205, 216, 218, **219**, 226–7, 229, 238–9, 241, **242**, 246, 274
DEMATEL 5–7, 20, 23, **24**, 64, 93–6, 98, 103, 193
defuzzification 98, 113, 316, 318
delays 72, **75**, 83, 88–9
delivery 56, 60, 78–9, 150, 158, **254**, 261, **262**, 268–70, 276–7, 304–5
Delphi 7, 93–6, 103, 290, 300
demand 20, **35–6**, 38–40, 56, 59–61, 63–6, **75**, 107, 170, **181**, 195, 213, 221, 223, 225, 246–7, 285, 288, 303–5, **305**, 307
demographic 93–4, 103
demonstration 237
desertification 89
digitalization 59, 61, 66
digitizing 237–8
disaster area 39–40, 136, 268, 270, 277, 305–6
disaster crisis management 150
disaster management 2–5, 8, 10, 11, 17–25, 27, 32, **35**, 40, 55, 64, 94–5, 155, 157–8, 165, 192, 205, 215, 223–5, 229, 236–8, 241, 247, 252–3, 264, 313–19, 321
disaster management cycle 2–5
disaster management operations 4, 165
disaster risk 5, 8, 20, 95, 150, 235, 246–7, 315, **317**
disruption 56, 58–9, 65, **75**, 124, 225, 270
distance-based methods 3, 4
distance to medical facilities **137**, 140, **144**, **146–7**
distribution 5, 8, *10*, 22, *22*, 23, **23**, *24*, **24**, 25, **26**, *27*, 31, 38, 60, 73, 157–8, 175, 193–4, 252–3, **254**, **262**, 267, 276, 283–4, 303–5
distribution center 193, 252–3
district **141–2**, 145, 149, 191–2, 195, 198–200, 204–6, 240, 261, 264, 284, 290
division 113, 307, 314
doctor 33, 124, **173**, **181**, 214, 218, 220–1
drainage systems 45, 73, **75**
drone capabilities 156
drone evaluation criteria set 156
drones 156–9, 161, 164–5
drone technology 156
drought 7, **18**, 107–10, 113–15, 158, 224
dynamic 34, **36**, 56, 90, 288, 304–5

E

eaps 235–8, 240–1, 244–7
early warning system 3, 4
earthquake 6, 7, **26**, 32, 58, 109, **137**, 191–2, 195, 198–200, 204–6, 223–4, 228–9, 231–3, 241,

259, 261, 264, 270, 282–5, *286*, 287, 288, *289*, 290, 300, 303, 305, 309, 315, **317**
earthquake effects 282, 284, 300
earthquake zone 259
economic analysis 314–16, **317**
economic factors 94, **99**
economical 156, 251
edmond folk festival 34, **35**
education 72, 98, **99**, 103, 172, 180, 232, 282
effective 1, 2, 8, 10, 18–19, 32, 38, 40, 49, 66, 73, 75–6, 79, 89, 95, 109, 115, 130, 150, 156–9, 165, 170–1, 174–5, 213, 215–16, 221, 235, 241, 252, 261, 264, 268, 277, 283–4, 307
efficient distribution 252
ELECTRE 4, 6, 8, 20, **24**, 25, 47, 49, 159
ELECTRE III 7, 49
elevation 109, **115**
elimination et choice translating reality *see* ELECTRE
emergency aid 3, 5, 252
emergency assembly points 235, 236
emergency department 7, 169–70, 174, 180, 215
emergency management 5, 7, 19–21, **21**, **23**, 229
emergency response 7, 171, 252
emergency supply depot 251–3, 256, 259, 261–2, 264
engineering 45–6, 48, 72, 74, 79, 98, 169, 206, 281, 284, 313–19
engineering economics 313–16, **317**, 319
enterprise dynamic 288
entropy 5–7, 23, **24**, 49, 110, 125, 191–2, 194–8, 200, **202**, 203, 205–6, 211–13, 217, 219, **220**, 291, **294**, 300
environment 2, 3, 17–18, 24, 38, 44–5, 48–51, 56, 63, 94, 98, 107, 109, 124, 137, 156–9, 165, 171–2, 174, 193, 216, 223, 229, 252, 254, **260**, 261, 314–16
erosion 49
ethnic diversity 98, **99**
euclidean distance 179
evacuation 5, 6, 48–9, 157, 236–7, 239, 241, 244, 247, 252, 285, 290, 304
evaluate 5–8, 19–20, 24, 33, 38, 40, 44, 48–9, 61, 64–5, 72, 74, 76, 89, 94–5, 99, 103, 108–10, 113, 115, 125, **126**, 129–30, 132, 156–8, 171–4, 184–5, 193–5, 212, 214–16, 218, 221, 224, 225, 228, 230, 232–3, 235–8, 241, 244, 246, 252–3, 257, 259–62, 264, 268, 274–5, 283–4, 288, **289**, **291**, 295–6, 300, 307, 315
evaluating goods 282
entropy weighting method *see* EWM
EWM 211–13, 216–18, 221
experience 19, **37**, 39, 50, 59, 64–6, 158, 172, **173**, 180, 206, 213, 223, 226, 229, 264, 283, 290, 303, 318
expert 4, 19, 50–1, 77, 84, 96, 136, 158, 172–5, 177, 179–80, 182, 185, 193, 195, 224–7, 231, 233, 252, 253, **254**, 259–64, 268–9, 272, 275–6, 283–4, 290–1, 295–6, 300, 313–14, 316, **318**, 319
expert opinion 179–80, 185, 231, 233, 253, 259, 269, 272, 276, 296, 316, 319
extreme 71–3, 76–7, 89, 198, 200, 224

F

fermatean fuzzy sets 314
field hospital 135–8, 140, **141–3**, 145, 149, 150
field of view 156, 159, 161, **162**, **164**, 165
fire **26**, 33, **36**, 38, 57, **75**, 76, 108, 157, 283, 285, 290, **291**, **292**, **294**, 319
fire station **36**, 38, 290, **291**, **292**, **294**
first cost 316, 318
fixed-wing 155–6
flexibility 76, 157, 165, 237, **269**, 270
flight 72, **75**, 83, 89, 156, 158–9, 161, **162**, **164**, 165
flight time 156, 158–9, 161, **162**, **164**, 165
floating model 284
flood 1, 4–8, **18**, 24, 25, **26**, 43–5, 48–9, 51, 73, **75**, 95, 108, 136–7, 157, 224, 251, 290, 315, **317**, 319
flood risk management 45–6, 48
flow: deluge 44, 48, 61, 315, 317; money 58, 299; information 55, 78; product 65, 76, 180; method 38; activity 251
food 137, 172, 252, 283, 285, 288
food pack 285, 288
Fothergill, A. 33
framework 19, 25, 32, 34, 39, 40, 46, 48–51, 74, 78, *81*, 94–5, 179–81, 194, 215, 224, 233–5, 237, 240, 247, 283, 305, 315, 317
freight 59, 62–3, 65–6, 72
fresh water 252
fuzzy 5, 7, 20, 24–5, 51, 64–5, 93–9, 100–3, 107–8, 110–13, 117, 157–8, 174–6, 193–5, 253, 267, 269, 271–4, 277, 290, 300, 313–16, 318–19
fuzzy AHP 7, 24–5, 64, 107–8, 111, 115, 267
fuzzy Delphi method 290
fuzzy logic 51, 109, 253, 313
fuzzy multisets 314
fuzzy sets 94, 96–7, 107–8, 110, 117, 174, 194, 253, 267, 271, 277, 313–14, 316, 319
fuzzy SWARA 5, 194
fuzzy TOPSIS 7, 64, 268
fuzzy WASPAS 5, 107–8

G

Game Theory 305
Garcia, C. 34

GDP **126**, 129, 132
gender 94, 99, 103
General Assignment Problem 37, 39
geographic information systems 282
geography 156–7
geological 251
Germany 36, 95, 126, 129, 130, 132
GIS *see* geographic information systems
glass 229, 231–2
global 2, 35, 39, 56–66, 71–2, 108, 114, 124, 170, 172, 182, 194, 214
global issues 72
global warming 71
goal programming 35–8
government 94, 99, 103, 173, 211, 214, 224, 261
greedy 303
Grey Relational Analysis 4–5, 47, 125
groundwater level 109

H

H1N1 94, 123, 125
hazard 1, 2, 5, 7, 8, 17, 24–5, 31, 45–6, 48–9, 51, 55, 73, 95, 109, 192, 195, 225, 228, 239, 282, 315, **317**
HCWs (health care workers) 212, 214–15
health care 72, 94, 136, 238
health indicators 125
heatwave 18
heliport **141–9**, 150
hesitancy 96, 258, 271
hesitant fuzzy sets 314
HFNC (High Flow Nasal Cannula) 214, 218, 221
hierarchical criteria structure 252
hierarchy 2, 4, 20, 46, 64, 74, 108, 157, 173, 194, 236, 252–3, 267–9, 272, 276, 283
Hoshin Kanri 71, 73–4, 77–80, *82*, 83, *88*, 89
Hoshin Planning 77–8
hospital treatment 199, 205
human activities 45, 51, 251
human resources 149, 218, 252
human thought 274, 313–14
humanitarian 5–7, 31–40, 252, 260, 303
humanitarian relief **36**, 252, 260
humidity 74, **75**, 90, 254, **262**
hurricane 1, 18, **26**, 58, 223–5
hydrological 1, 45, 49, 51, 107, 224, 236

I

Iceland **126**, 129, 130, 132
income 59, 93–4, **99**, 108, 110
industrial accidents 23
integer programming 34, **35**, **36**, 38
interaction 3, 4, 17, 59, 95–6, 216, 309
interaction-based methods 3, 4
International Journal of Disaster Risk Reduction 8
interval numbers 112, 157–61
Ireland **126**
Istanbul 24, 191–2, 195, 198, **199**, 204–5, 223–4, 228–9, 231–3, 251–2, 259, 261, 264
IVIF AHP 107, 110–11, 113, **115**
IVIF WASPAS 107, 110, 112–13, 115
IVN-AHP 251–3, 256, 259, 261–2, 264
IVN-WASPAS 251–3, 258, 264
improvement 61, 76, 78, 204
indeterminacy 111, 253–4, 273, 313
infant mortality 124, **126**, 129, 130, 132
information layers 283, 290–1
infrastructure 3, 5, 7, 45, 57, **75**, **76**, 94, 192, 223, 228, 236, 238, **254**, **260**, 261, **261**, 262, 282–5, *285*, *286*, 288, 300, 315
initial decision matrix 192, 196–8, 200, **201**, **202**, **203**, 205, 216, 218–19, 226–**7**, **229**
injured people 199, 200, 205, 227, 285, 288
intensive care unit (ICU) 170, 172, **173**, 214
interest rate 316
interval-valued fuzzy sets 314
interval valued intuitionistic fuzzy numbers 111
interval valued neutrosophic numbers 253, 255, 258–9, 262
interval-valued Pythagorean fuzzy sets 267
interval-valued spherical fuzzy sets 174, 313–14
intuitionistic fuzzy sets 107–8, 253, 271, 314
intuitionistic fuzzy sets of second type 314
investment analysis 313, 315, 318–19

J

JICA 284–5, 287–8, 300

K

Key Performance Indicators 77, 83
key role 156, 264
K-Means 36, 38, 245
K-Medoid 36, 38
Korea 126, 173

L

laboratory 5, 20, 64, 94, 172, 180, 193, 218
land use 109, 110, **115**
landslide 2, 8, 24, 25, **26**, 109, 136, 158, 224, 251, 283
levees 45–6
lifestyle 95–8, 102–3, 232
Likert scale 81–3, 85–7
limitation 20–1, 27, 32–4, **35–6**, 38–9, 61, 192, 206, 225, 264, 300

liner shipping 61–3, 66
linguistic 65, 96, 97, 99, **99**, 111, 112, 157, 158, 183, 193, 194, **257**, 258, 260, 262, 272, **273**, 274, **274**, 275, 276, 314
linguistic term 183, 194, 256–8, 260–2, **273**, 274, 276
literature review 3, 5–8, 11, 17, 19, 31, 33–4, 40, 51, 93–4, 107–8, 136, 140, 169, 171, 175, 179, 180, 191–2, 211, 213, 215, 218, 223–5, 246, 252–3, 300, 315–16
Littman, M.L. 38
loading **269**, 270, 305
location accuracy **269**, 270
location selection 8, 64, 135, 194, 251–3, 259, 260
lockdown 59, 60
logistics 4, 7, 25, 38, 58, 60–5, 155, 157, 193–4, 235, 300, 303–5
Lowe, S. 33
lower limit 160

M

MADM 45, 49–51, 169, 171–4, 179
maintenance 44, 59, **75–6**, 172, **269**, 318
managerial insight 157, 184, 237, 240, 244
man-made 2, 3, 8, 17, 27, 31, 40, 136, 251
man-made disaster 2, 3, 8, 17, 27, 40, 136
map 6–7, 20, 24–5, 94, 109, 192, 236–7, 239, 241
maritime 55–66
Markov-chain Monte Carlo approach 177
Markov Decision Process **36**, 38–9
MATLAB 177, 182
maximization 34, **36**, 38, 315
max speed **162**, **164**
Mayorga, M. 38
MCDM *see* multi-criteria decision making
measure 2, 7, 9, 38, 46, 48–9, 51, 59–61, 65, 73, 94, 96, 109, 113, 115, 124, 145, 157, 170–2, **173**, 174, 175, *176*, 180–5, 192–4, 204, 211, 216, 227, 245, 252, 284–8, 294, 296, 313
medical supply 267, 270, 275–6
medical supply warehouse 270
medicine 64, 72, 74, 138, 174, **181**, 252, 283
Meijs, M. 39
membership 108, 110, 253–4, 271–2, 313–14, 316, 318–19, **320**
membership function 253–4, 272, 316, 318–20
MERS-CoV 123–5
meteorological 1, 72–3, 107, 224
methodology 2, 3, 5, 17, 20, 31–5, **36**, 40, 48–9, 95, 107, 110, 135–8, 141–2, 145, 150, 155, 157–8, 161, 165, 169, 171, 175, 185, 251–3, 256, 258–60, 262, 264, 267, 270, 275, 305, 315
minimization 34, **35–6**, 38–9
mitigation 2–4, 8, 19, **23**, **24**, 25, 40, 44–5, 48, 94, 108–9, 117, 315, **317**
mixed 22–5, **23**, **24**, **26**, 39, 225, 240, 304
mixed integer 304
mobile phone 39, 61
monitoring 59, 73, 75, 172, 215
Mosha fault model 284
multi-criteria decision analysis 2, 5, 17, 19, 214, **216**, 235–7, 247
multi-criteria decision making 5, 19, 44, 45, 64, 95, 108, 124, **138**, 157, 171, 192, 212, **214**, 216–17, 224, 252, 268
multi-criteria decision method 125
Multi-Method Research 37, 39
Multi-Objective Optimization on the basis of Ratio Analysis (MOORA) 5
multi-rotor 155–6

N

natural disaster 2, 18, 25, **26**, 31, 44–5, 56, 77, 83, 94, 107–9, 115, 224, 251–2, 259, 282–4, 290, 304, 317, 319
natural gas pipeline 199
natural hazards 1, 45, 48, 51, 315, 317
negative ideal solution 4, 160, 274
Netherlands 18, 126
network structure 47, *286*, 300
neutrosophic fuzzy sets 314
neutrosophic logic 252–3, 256, 264
new global search algorithm 35, 39
nonstationary fuzzy sets 314
normalization 128, 196–7, 202–4, 316
North Tehran fault model 284
Norway 126
number 8, 18, 27, 31, 33–40, 44, 49, 55, 57–8, 60–1, 72, 89, 95, 111–12, 124, 126, 129, 130, 132, 135, 137, 145, 150, 157–61, 174–6, 195–6, 198–200, 205, 212, 214, 218–21, 235, 241, 245–6, 253–9, 262, 264, 271, **273**, **274**, 277, 283, 285, 288, *289*, 290, 300, 303–5, 307–9, 313–14, 316
number of damaged points in: drinking water pipeline 199, 200
number of physicians per 1000 population 126, 129, 130, 132
nurse 180–1, 218

O

objective evaluation method 127, 216
OECD countries 123–6, 128–32
open areas 236–41
operations 2–5, 7, 55–6, 62, 72–6, 83, 111, 113, 155–6, 167, 172, **173**, 185, 224–5, 235–6, 241,

247, 260, 268, 270–1, 276, 285, 290, 300, 304, 314
operations research 2, 3, 165
Opricovic 4, 46, 137, 273, 275
optimal Solution 309
optimization 5, 31, 35–6, 38, 40, 46, 79, 138, 193, 281–3, 295, 304–5, 315, 317
Ordered Weighted Averaging (OWA) 5
organization 2–4, 7, 19, 20, 32–3, 35–40, 47, 55–7, 59, 62, 72, 74–8, 124, 135–6, 171, 193–4, 212–14, 224, 254, 262, 264, 284, 290, 296, 300
other methods 3, 4, 5, 23, 25, 51, 233
outbreak 60, 62, 124, 169, 171–2, 175, 179–81, 185, 214
outranking-based methods 3, 4
Özel, B. 39

P

pairwise comparison 3, 4, 46–7, 79, 84, 96, 111, 113–14, 174–5, 180, 256, 260–1, 272–3, 291, 296
pandemic 2, 3, 8, 10, 18, **26**, 55, 57, 59, 63, 65–6, 72, 93, 95, **99**, 102–3, 123, 125, 128–9, 170, 174, 185, 194, 211, 216, 218, 221, 316
PAPRIKA **47**
passengers 72, 76
patient 124, 170, 172, 180–1, 185, 212–13, 215, 218, 221
pavements 45, 75–6
PDCA 78
penalty **35–6**, 38
people 3, 17–18, 31, 44–5, 59, 61–2, 93–5, 99, 102–3, 123–4, 145, 149–50, 157, 172, 174, 195, 199–200, 205, 211–12, 221, 232, 235–6, 241, 246–7, 252, 261, 270, 277, 282, 284–5, 288, 294, 296, 304–5
per diem **34–6**, 38
performance 4, 7, 37, 46, 51, 55–6, 64, 73, 77–8, 83, 98, 123–5, 128–33, 156–61, 172, 174, 178, 180–1, 184–5, 193–5, 212, 216, 221, 237, 241, 245, 268–70, 277, 281–2, 284, 288, 316
picture fuzzy sets 314
planning 3, 5, 19, 32, 46, 48, 50–1, 59, 62, 72, 77–8, 95, 170, **173**, 204, 215, 225, 233, 252, 283, 300, 304, 318
plastic 229
policy: deployment 77–8; makers 3
political factors 94
population: density 49, 98–9, 137, 140, 144–7, 149, 195, 235, 240, 261, 296–8; size 137, 140, 144–7
port 57–64, 66, 317
Portugal **126**, 129, 132
post-disaster 4, 8, 19, 136, 150, 192, 223–5, 229, 233, 236, 247, 252, 261, 264, 267–8, 276, 303
positive ideal solution 160
practicality 269–70
precautions 76, 225, 232, 235, 244, 313, 319
precipitation 74–5, 89–90, 108–9
pre-disaster 23–4, 191–2, 225, 252
Preference Ranking Organization Method for Enrichment Evaluations (PROMETHEE) 4, 6–7, 20, **23–6**, **47–8**, 296, **299–300**
preferences 2, 32, 34–6, 38–9, 195, 233
preliminary evacuation areas 252
preparedness: phase 155, 157; assessment 169, 172, 180–1, 215–16; performance 125, 128–9, 174
present worth analysis 313, 316, **320**
PRISMA 5, *6*, 10, 20
price 50, 60, 156–7, 159, 161–2, 164–5, 308
prioritized 44, 49, 65–6, 194–5, 213, 244, 291, 296, 300
priority 19, 79, 112, 115, 172, 174, 231, 237, 246, 284, 290, 294, 296–300
probability 20, 55, 78, 175, 177–8, 214, 283
procedures 3, 5, 17–18, 76, 89, 96, 171–2, 175, 183, 195, 206, 214, 282
profiles 157, 159, 161, 164–5
profitability 62–3, 66
PROMETHEE I **47**
PROMETHEE II 7, **47**
propeller 269–70
proposed methodology 171, 252–3, 259, 262, 267, 270, 275
public: policies 27; transport 56, 60; resources 94; health 103, 225; services 150; hospital 212, 218; emergencies 303
pythagorean fuzzy sets 117, 174, 194, 267, 271, 314

Q

q-rung orthopair fuzzy sets 314, 319
QUALItative FLEXible (QUALIFLEX) 5–6, 8
questionnaire 60, 180, 213–15, 262
queue **36**, 39

R

rainfall 44, 46, 73, **107–10**, **115**, **254**, **262**, 283
Ray fault model 284
recharge 269–70
reconnaissance 155
recovery 2, 8, 19, 23–4, 31, 40, 48–9, 59, 61–2, 65, 73, 192, 225, 268, 303
region 5, 11, 18, 44, 48, 77, 83, 89–90, 93–5, 98, 108–9, 113, 115, **116**, 117, 124, 191, 195, 236–8, 240–1, 245–6, 252, 259, 261, 268, 270, 284, 291–2, 295, 300, 305

relief: centers 282–5, 290–1, 294–6, 299–300; teams 238, 241
relocation 38
remanufacture 230, 232
resilience 3–5, 7, 27, 55–9, 61–3, 65–6, 71–7, 79, 83, 89–90, 94–5, 239, 315
resource: allocation 34, 36; scarcity 7; reallocation 34; optimization 36, 38; planning 48, 50; capacity 75
response 2, 4, 7–8, 17, 19, **23–4**, 31–2, 38, 40, 61, 136, 149–50, 156–7, 170–2, **173**, **181**, 214–15, 225, 237, 241, 244–5, 247, 252, 268, 284
review 3, 6, 8, 11, 17, 19–21, 31–6, 40, 45, 45–6, 48, 50–1, 73, 76–8, 93–4, 107–8, 136, 140, 169, 171, 175, 179–80, 191–2, 211, 213, 215, 218, 223–5, 246, 252–3, 269, 300, 314–16
RIM approach 171, 175
risk: assessment 46, 72, 109, 191–2, 196, 198, 205–6; description 75, 86–9; reduction 5, 8, 25, 27, 300, 315, 317
risky area 252
road **137**, **140–2**, **144**, **146–7**, 150, 195, 206, 230, 237–8, **241–2**, 245, 285, 290, 294, 304
robust 51, 76, 283, 304
robustness 76, 89, 232
rockfall 251
route 61, 66, **75**, **173**, 230, **254**, **262**, 270, 285, 303–4
runways **75–6**

S

Saaty 4, **46–7**, 256
safety 58–9, 64, 76, 171, 181, 185, 225, 269–70, 277, 281–3, 315–17
salvage value 316
SAW 8, 46, 49, 224
scenarios 35, 39–40, 158, 164–5, 228, 268, 282–3, 300, 307
schedule **35–6**, 38, 304
sea level rise 71, 73, 89
security 7, 55, 57–8, 84, 172, 194, 214, **254**, **261–2**, **269–70**, 283, **296–8**
sedimentation 49
selection 5, 7–8, 47, 64, 135–9, 145, 155–9, 164–5, 193–5, 206, 213, 217, 236, 244, 247, 252–3, 259–60, 264, 267–8, 277, 282
sensitivity analysis 51, 115, 117, 157, 159, 161, **232**, *238*, 321
separability 229–32
service bed 218, 220–1
Shaskolsky, L. 33
shelter 7–8, 95, 157, 228, 236, 239, 247, 252, 283
shortage 34–6, 38, 60, 64–5, 89, 214, 285, 304
SIR 47
silhouette score 245, 248
simulation 46, 51, 177, 281–4, *288–90*, 300
size 137, 140–1, 144–7, 170, 177, 223, 232, 236–8, 241–2, 261, 300, 310
skill 35–6, 38, 40
slope 108–9, 113, 115, 236, 238, 283, 290–2, 294
social 17–18, 46, 49, 51, 55–6, 60, 93–6, 98–9, 102–3, 109, 124, 170, 181, 195, 211, 214, 251, 254, 260–1
social vulnerabilities 95
soil structure 108, 113
solution process 309
Spain 126
specification sheet 158–9
speed limit 269–70
spherical fuzzy sets 174, 313–14
spontaneous 32–9
stages of disaster management 2, 23
stakeholders 20, 44, 50–1, 55–6, 62, 75–6, 238, 241
steel 228–9, 231–3, 287
Stepwise Weight Assessment Ratio Analysis 193, 224
Stochastic Multi Criteria Acceptability Analysis (SMAA) 5, 193
storms 2, 8, 89, 319
strategy deployment 74, 77–8, 83
stressors 74–5, 84, 89–90
sub-problems 307
suitability scores 236–8, 241, 244–6
suitable 3, 7, 20, 36, 38, 40, 45, 49–50, 150, 156–9, 164–5, 228, 230, 236–7, 246–7, 261, 268, 270, 275, 277, 284, 290, 300, 321
supply chain: management 5, 7; risks 55–7, 65
surplus 34–5
sustain 73
SWARA 5–6, 193–4, 224–5, **230**
Sweden **126**
swine flu 123
symptoms 123–4, 171
systematic classification 2, 10

T

task 35–6, 38–40, 172, 238, 268, 270, 275–6
technological: disasters 2, 225; developments 59, 268; integration 62; events 251
Tehran earthquake283–5
teleworking 61, 63, 66
temperature **74–5**, 89–90, **108–9**, 113, **115**, 224, **254**, **262**
temporary accommodation needs 192
TOPSIS 2, 4, 6–8, 20, *22–6*, **47–50**, 64, 117, 124–5, 158–9, 174, 178, 195, 214, 268, 291, **294–5**, 300

total hospital beds per 1000 population **126**, 129, 132
total loss of life 124
training 3, 36, 40, 76, 155–6, 165, 170, 214–15, 245, 282–3
transportation 7, **36**, 38, 57–8, 60, 63–4, 72, 149–50, 169, 195, 223, 230, 232, 252, **254**, **261–2**, 268, 270, 276–7, 285, 303, **305**
tsunami 2, 7, 8, **26**, 44, 58, 109
Turkey 31, 49, 107–9, 113, 135, 171, 180, 212, 214, 224, 236, 240, 251, 259, 264, 304
type 2 fuzzy sets 117, 314

U

uncertainty 20, **36**, 38, 51, 96, 108, 110, 158–9, 171, 212–13, 215–17, 221, 253, 274, 277
unemployment 99, 103
United Kingdom 126
unloading **269–70**
unmanned aerial vehicles 268
unpredicted event 55
unsuitable 236–7, 246
upper limit 160
urbanization 229, 235, 246–7, 315
utility-based methods 3, 5

V

vagueness 253, 314
valuation matrix 178–9, 183, 258, 275, **277**
variable neighborhood 305
variance 193, 241, **243–5**
vehicle 7, 57, 155, 193–4, 224, 268, **303–9**
ventilation **75–6**
VIKOR 2, 4–6, 20, **23–5**, 46, **47**, 49–50, 117, 135–9, 141–2, 145, 150, 155, 157, 159–60, 165, 174, 178, 195, 267–8, 270, **273–6**
virtual EAP 238, 246
volatility 156–8
volcanic eruption 26
volcanoes 2, 8
volunteerism 32, 40
vulnerability assessment 4, 94–5, 103, 107, 109, 113

W

war 59, 62, 65, 136
warehouse management 252
WASPAS 5, 107–8, 110, 112–13, 115, 117, 123, 125, 127–30, 132, 195, 197, 251–3, **258–9**, 262, 264
wastewater pipeline 200
watershed 49
weather condition 57, 72, 89
weight 5, 19, 73–4, 79–81, 83–5, 88–9, 95, 108–10, 112–13, **115**, *117–18*, 125, **127–9**, 132–3, 139, 142, **144**, 156–7, 159–62, **164–5**, 171–5, 177–85, 191–8, 200, **202–4**, 212–14, 216, 218, 220–1, 224, 226–7, 230–2, 238–9, 241, 243, 245, 251–3, 256–62, 268–70, **273–6**, 283, 291, **294–7**, 300, 304, 318
Weighted product method 108, 197
Weighted sum method 46, 108, 197
World Health Organization (WHO) 214, 224
wind **74–6**, 89–90, 270
wood 62, 228–9, **231–2**, 288
WPM 112, 127–8, 197–8
WSM 5, 7, 20, **47**, 49, 112, 127–8, 197–8

X

X matrix 73, 78, 81, *82*, 83, *88*

Y

Yager 271, 314